Managing the Oil Wealth

To the memory of

Eleanor

My wife, moral support, and mentor

Managing the Oil Wealth

OPEC's Windfalls and Pitfalls

Jahangir Amuzegar

I.B.Tauris *Publishers*
LONDON • NEW YORK

Paperback edition published in 2001 by I.B.Tauris & Co Ltd
6 Salem Road, London W2 4BU
175 Fifth Avenue, New York NY 10010
www.ibtauris.com

In the United States and Canada distributed by St. Martin's Press
175 Fifth Avenue, New York NY 10010

ISBN 1 86064 648 4

A full CIP record for this book is available from the British Library
A full CIP record for this book is available from the Library of Congress

Library of Congress catalog card: available

Typeset in Caslon by Dexter Haven, London
Printed and bound in Great Britain by MacKays of Chatham

Contents

Preface to the Paperback Edition

With crude oil prices reaching ten-year highs in mid-2000 – resulting in widespread consumer outrage in the United States and paralysing fuel protests in Europe – the role of the Organization of the Petroleum Exporting Countries once again attracted the media spotlight. The potential impact of the high price run-off on the global economy – slower growth, higher inflation, currency turmoil, external imbalances and lower corporate profits – once again raised the spectre of a new oil crisis. The prospects of the new oil windfalls being squandered on further white elephant projects or extravagant military outlays raised new issues regarding world income redistribution. The position and posture of OPEC members in the new monetary power equation thus became once again the focus of attention in world political and financial circles.

The Summary of Findings at the end of the 1999 edition of this book, covering the crucial 1974–94 period, detailed what the oil bounty had done *to* the OPEC members in addition to what it had done *for* them. Those pointers reflected the strategies and policies which the oil-blessed countries could (and should) have followed during the two decades to escape avoidable pitfalls.

The macroeconomic challenges facing the members on the anniversary of their 20-year experience with the oil's boom-and-bust cycles were implied in that summary. To make up for the lost opportunities and reverse the setbacks in the previous 20 years, they had to: reduce state dominance over the economy and lower dependence on hydrocarbon exports; pay attention to domestic absorptive capacity in allocating the oil windfalls; contain the industrialisation drive within the confines of the available skilled labour, technological capability and management know-how; strive towards a fairer distribution of wealth and income; coordinate domestic and foreign economic policies; strengthen the economy's institutional base; increase bureaucratic transparency and accountability; and cut military expenditures in favour of human capital formation and preservation of the environment.

Specific policy measures designed to meet these challenges were largely known and familiar. Leading the roster were effective and corruption-free privatisation of inefficient public enterprises; expansion of non-oil exports; a thorough revision of the domestic tax base and other sources of state revenues; price decontrol and overall economic deregulation; the overhaul of the comprehensive subsidy system in favour of an equitable social safety net; reform of the judiciary as the backbone and guarantor of structural reforms; and establishment of a reserve fund to cope with oil price fluctuations as well as inter-generational equity considerations.

The supplement presented in this edition attempts to show how the member countries coped with these challenges in the rest of the decade. It covers ten current

OPEC countries, as Ecuador and Gabon have already left the association, and data for Iraq are too fragmentary and dated to allow comparative treatment. A short table at the end of the discussion shows selected quantitative figures. The reference sources consulted for this update have been the same as those for the main study. Easier access and ampler use of the country and consultation reports prepared and disseminated by the World Bank and the International Monetary Fund (IMF) under their new transparency policy have made the task more productive. Additional data have been sourced from the Economist Intelligence Unit's latest country profiles.

Preface

No economic event in the second half of the twentieth century perhaps attracted so much worldwide attention as the four-fold increase in the price of crude oil during 1973–74. The Organization of Petroleum Exporting Countries (OPEC) that was widely castigated for creating the first global energy crisis became the focus of universal curiosity and concern in the market, the media and international financial circles. Wild predictions were made regarding the fortune and clout of the OPEC members, the industrial world's reversal of growth and prosperity, and the instability of the international monetary system.

There were dire forecasts about OPEC's future power in setting oil prices, dictating the terms of transactions with the major oil companies, and even denying Western access to much of the world's oil. OPEC members were depicted as capable of determining future world inflation and recession; becoming the world's major bankers bent on establishing a new economic order in their own favour; setting an example for other commodity-producing Third World countries to follow; and exerting significant influence over foreign policies of the industrial powers.[1] Future OPEC trade surpluses, and the rest of the world's deficits and debts, were seen skyrocketing. 'With the possible exception of Croesus,' said a respected oil expert, 'the world will never have seen anything quite like the wealth which is flowing, and will continue to flow, into the Persian Gulf'.[2] No less venerable an institution than the World Bank projected the accumulated OPEC gains to exceed $635 billion by 1980 and over $1,200 billion by 1985. There was speculation in the press as to how much of Western financial assets could be acquired by OPEC wealth each year, month, week or day.

Gigantic trade deficits were expected to bankrupt the industrial West and the oil-less developing economies. In the absence of urgent remedial actions, it was feared the West would resort to the old-fashion 'beggar-my-neighbour' policy, and trigger the collapse of the already (post-1972) fragile monetary system. A prominent oil analyst later went on to declare, "the world as we know it now will probably not be able to maintain its cohesion, nor able to provide for the economic progress of its people against the onslaught of future oil shocks – with all that this might imply for the political stability of the West, its free institutions, and its internal and external stability'.[3]

With the transfer of immense global resources, the dawn of a Midas-type affluence was predicted for the group's own economies whereby OPEC members could not only finance their infrastructural and industrial projects at home and buy into Western industrial and financial assets abroad, but also escape the common plight of the fast developing countries. Their oil export proceeds were assumed to reduce domestic inflation through the supply of oil-financed imports. Rising investments in productive ventures were expected to provide gainful employment for the rising population and labour force. Vast sums of steady income in hard currencies were believed capable of financing all domestic development and defence needs of members without the need for foreign borrowing or aid. Rapid, comprehensive prosperity and progress was thought to be attainable without forced savings, belt-tightening, or escalating budget deficits. A steady stream of oil income was projected to fill the gaps in national saving, foreign exchange earnings, and public revenues that constituted traditional constraints on rapid and sustained economic growth. National welfare and rising living standards, in turn, were assumed to guarantee eventually greater individual freedoms, enhanced national political stability, more social cohesion, improved civil society and, hopefully, both participatory democracy and political maturity. These far-reaching forecasts were based on the assumptions that: the total foreign exchange reserves of the oil-importing countries were not sufficient to pay for their oil bills beyond one or two years; the money paid for imported oil might not come back to the oil-importers' coffers for a long time because OPEC had limited capacity for imports; and oil importers might never be able to pay back their debts because their deficits would be accumulating year after year.[4]

The hollowness of these assumptions, and the strength of industrial economies to withstand the temporary impact of the oil price rise, however, became evident shortly after the events, and even as early as 1976–77[5] when these projections turned out to be no more than mere fairy tales. Most of the best and brightest petroleum analysts proved wrong in almost all their prognostications: their estimates of global need for oil; OPEC's oil-producing capacity; oil-importers' misfortunes; and oil exporters' amassed wealth and prosperity. Alarmist and exaggerated predictions about the industrial world's vulnerability to the energy crunch were even more fully refuted with respect to the rosy projections about OPEC's power and fortune. The oil windfalls resulted in real exchange-rate appreciation; caused immense and enduring changes in domestic economic structure, institutions and expectations; and proved disastrous when oil fortunes was reversed. Within less than a decade, most OPEC members were plagued by economic chaos, social unrest and political turmoil. Planned targets were unfulfilled due to production bottlenecks, mismatched investments, enterprise inefficiencies, resources waste and corruption. Prices, exchange rates, deficits and external debts increased. Dependence on oil was not effectively diminished. Terms of trade turned against oil exports. The real inflation-adjusted price of oil gradually declined. Economic growth became distorted and unbalanced. And, even the principle of oil nationalisation and independence from foreign oil majors was called into question in the end.

This study attempts to examine and evaluate the experience of 13 OPEC members during the crucial 1974–1994 period. Within the framework of some major hypotheses regarding the dynamics of economic development in an oil-based economy, several fundamental questions are raised: how did the members allocate their oil windfalls among competing needs? What strategies and policies did they pursue in optimising returns on their fortunes during the oil boom, and controlling the damage during the oil bust? And, to what extent did those allocations and strategies help them achieve their stated or implied national socioeconomic agenda?

To this end, a brief introductory chapter discusses OPEC members' underlying national characteristics and structural differences in order to show how their astonishingly similar experiences defied such divergences. As a prelude to the subsequent examination of their approaches to development, Chapter II sets out a theoretical framework for the treatment of 'oil rent', and the process of adjustment to oil revenue fluctuations. Chapter III traces the development of OPEC as a commodity organisation, and its dealing with booms and busts during the 1974–1994 period in the light of fundamental policy differences among members with respect to oil prices and supplies. Chapter IV provides in some detail the allocation strategies and adjustment policy adopted by each OPEC member during the period under review. Chapter V is devoted to the investigation of economic performance in each country in terms both of its own stated national socio-economic agenda and certain universal criteria. Chapter VI offers some cross-country comparisons among the 13 members. Chapter VII summarises the study's findings. Chapter VIII reviews the study's conclusions in the light of certain conventional hypotheses.

In the preparation of this study, in addition to reliance on the OPEC Secretariat's monthly and annual reports, as well as major published works on each country (listed in the bibliography), use has been made of other sources, including primarily official national statistics, as well as annual and periodic statistical data published by international organisations e.g., the United Nations, the World Bank, the International Monetary Fund, the UN Development Program, UN Conference on Trade and Development and other UN affiliated agencies. To fill data gaps for certain categories or for some years, resort has been made to the information provided by major non-governmental publications (e.g., the Economist Intelligence Unit, the *Middle East Economic Digest*, the *Middle East Economic Survey* and other oil-focused journals). Furthermore, whenever necessary, the standard global data sources e.g., the Library of Congress's area handbooks, Central Intelligence Agency's *Fact Book*, Europa Publications, and various world statistical abstracts have also been consulted. For most of Chapter IV, the study draws heavily from periodic 'consultation mission' reports of the International Monetary Fund and the World Bank on individual countries. These reports have restricted internal distribution, and their contents are officially not for attribution except when they are condensely published.[6]

The rich variety and diversity of these sources, while enhancing the quantity and reliability of collected data, have presented a nearly insurmountable problem, namely the irreconcilability of facts and figures, even when they are presumably based on national official sources.[7] Such discrepancies are particularly (but not exclusively) pronounced when national currency data are converted into US dollars for easier and more meaningful comparisons. Due to the prevalence of multiple exchange rates, however, these conversions often produce strange results and render any sensible conclusions impossible.[8]

Moreover, statistical data have not been uniformly available or reliable for all 13 countries in all comparable categories in all years. Lack of statistical data has been especially unfortunate in such essential areas as the performance of public enterprises, the magnitude of direct and indirect state subsidies, domestic income distribution, and size of the 'informal economy' that escapes public scrutiny. As a result, country-specific accounts in Chapter V have often varied in length, variety and quality of treatment, making an overall comparison of performance based on quantitative data subject to significant conceptual and methodological shortfalls.

Finally, drawing relevant and definitive conclusions from OPEC members' 20-year experiences is made tangentally difficult because at least seven members – Algeria, Iran, Iraq, Kuwait, Libya, Nigeria and (to some degree) Venezuela – were in varying degrees affected during the period by events or situations largely extraneous to oil revenue management and internal economic development. Iran was plagued with the 1979 revolution, a long and costly war with Iraq, and Western (particularly US) trade and credit sanctions. Iraq had a debilitating war with Iran, a devastating conflict with Kuwait and the allied forces, ending up with ruinous UN sanctions and domestic deprivation. Nigeria went through repeated military coups d'etat. Libya was under severe UN sanctions for alleged complicity in international terrorism. Algeria faced a violent internal opposition in the early 1990s, and Venezuela went through the impeachment and conviction of its president. How these countries would have managed their oil windfalls in the absence of these exogenous elements is not easy to ascertain. No such counterfactual evaluations ever are.

Due to the shortcomings in the availability, reliability and comparability of collected data, extreme caution should thus be exercised in drawing definitive conclusions from quantitative values. Although figures on some major indicators of economic and social development are given in detail (and in some cases in decimal digits), they should be interpreted as indicating basic trends rather than precise quantifications of those differences. And, despite considerable efforts made to harmonise the data obtained from different sources for inter-country comparisons, certain ambiguities and even inconsistencies between official statistics and the study's figures might have remained unavoidable.

NOTES ON PREFACE

1   See 'OPEC's Imperium' in Daniel Yergin, *The Prize*, (New York: Simon and Schuster, 1991) pp 633–634.
2   James E. Akins, 'The Oil Crisis', *Foreign Affairs*, April 1973, p 480.
3   Walter J. Levy, 'Oil and the Decline of the West', *Foreign Affairs*, Summer 1980, p 1015.
4   Cf Jahangir Amuzegar, *The Impact of Recent Oil Developments on the World Monetary System*, Princeton Near East Paper no. 19, 1975.
5   See, for example, Jahangir Amuzegar, 'Oil Wealth: A Very Mixed Blessing', *Foreign Affairs*, Spring 1982.
6   Due essentially to space limitations, the common sources of international comparative data are cited only as sources in the tables and not repeated for each country in the discussions of Chapters IV–VI. Specific reports dealing with each country are cited only in Chapter IV, and omitted in the subsequent chapters. A selected bibliography at the end of the study offers further references that are consulted on each country, but not repeated in each chapter.

7   For example, two publications of the same institution (e.g., the World Bank's *Atlas* and *World Development Report*) sometimes give different figures for the same category of the same country in the same year.

8   For example, Iran's reported national output for the year 1992/93 ranges from $48 billion (*The Economist*) to $57 billion (IMF); $113 billion (OPEC Secretariat); $131 billion (World Bank); and $338 billion (UNDP)!

I

A Heterogeneous Group with Shared Objectives

Of the world's major commodity clubs or cartels, none has proved as active and enduring as the Organisation of Petroleum Exporting Countries, in spite of highly improbable odds. Initially, OPEC faced a strong resistance from the major oil concessionaires who controlled global oil supplies and trade; it encountered open hostility from powerful oil-consuming governments which still stubbornly defended their vanishing colonial power; and it was only half-heartedly supported by other developing countries which envied its wealth and clout. Yet in spite of the rather unique composition of its membership, the organisation prospered and survived. This introductory chapter attempts to describe some of the significant differences in the physical, demographic, economic, technological and socio-political characteristics of the thirteen nations – Algeria, Ecuador, Gabon, Indonesia, Iran, Iraq, Kuwait, Libya, Nigeria, Qatar, Saudia Arabia, the United Arab Emirates and Venezuela – that kept OPEC alive, while the organisation's obituaries were being written time after time.[1]

OPEC members were at no time a homogenous group except for two principal features: reliance on petroleum export revenues as a mainstay of their economies; and primacy of the state in the ownership, acquisition and disposition of oil revenues. The centrality of state power, in addition to the authoritarian nature of government in nearly all members during most of the 1974–94 period, was underpinned by a number of other factors. The legal and constitutional ownership of mineral deposits by the state empowered governments to make crucial decisions regarding the magnitude of oil exploitation and the distribution of oil income. Oil revenues belonging to the nation as a whole (and not in private hands) could be optimally used by the state to serve national objectives both over time and across alternative uses.[2]

1

Considerations based on the concept of oil reserves as exhaustible assets rather than permanent sources of income mandated the need for a central authority to create viable and productive alternatives to oil. And, the search for such alternatives required, initially at least, a substantial degree of government initiative, guidance, direction and support in the choice of investment outlets.

Except for these two common characteristics, OPEC members differed widely in geographic area, size of population and labour force, degree of urbanisation, dependence on oil revenues, availability of non-oil resources, stage of economic development, the size of the domestic market, and such other factors as ethnic and religious origins, type of paternalistic government, intergovernmental associations, socio-cultural proclivities, and entrepreneurial tradition and experience. The combination of these factors tended to determine each country's domestic absorptive capacity. Internal output capabilities, in turn, influenced the choice of windfall allocations and the selection of the development strategy (see Chapter IV).

Physical diversities included the scope of geographic area and endowment with national resources, including oil. As Table 1 shows, on the eve of the oil price surge, OPEC members differed in land area from Qatar and Kuwait with less than 20,000 square kilometres to Saudi Arabia and Algeria with more than 2 million. Among the middle-size countries with more than a million square kilometres – Indonesia, Iran and Libya – the first two had considerable agricultural potential while Libya had precious little.[3]

Demographic differences were reflected in the size and density of population, the size of the labour force, rate of population growth, composition of population and extent of urbanisation. In 1974 countries like Gabon, Kuwait, Qatar and the UAE had populations of less than one million, compared to Nigeria with 73 million and Indonesia with 132 million (Table 2). In terms of population density, the range spread from 1.1 persons per square kilometre in Libya to 50 in Kuwait, 69 in Indonesia and 79 in Nigeria. Annual population growth and the sectoral composition of employment also differed widely among group members. Of the total inhabitants in each country, about 20 percent or less was classified as urban in Indonesia and Nigeria relative to 56 percent in Iraq, 72 percent in Venezuela and 82 percent in Kuwait.

Table 3 shows the magnitude of proven petroleum deposits and peak daily crude extraction within OPEC.[4] As seen from this table, Saudi Arabia in 1974 possessed more than twice as much proven oil deposits as either Iran or Kuwait; five times as much as the United Arab Emirates; nearly ten times as much as Venezuela; more than twenty times as much as Algeria, Ecuador or Qatar; and 265 times as much as Gabon. At the 1974 rate of extraction, existing petroleum deposits were expected to last well into the 21st century in Iraq, Kuwait, Saudi Arabia and the UAE, but only a few decades in Algeria, Ecuador, Gabon, Indonesia and Qatar. Furthermore, the life expectancy of reserves in some countries (e.g. Iran, Iraq, Qatar, Saudi Arabia, UAE and Venezuela) was supposed to increase substantially through discoveries of new oil fields, while such prospects in Algeria, Ecuador, Indonesia, Libya and Nigeria were not considered equally bright under existing technology.

In terms of dependence on oil, for current consumption, development and defence needs, the differences were again considerable. On the basis of the three major criteria – the ratio of petroleum exports to total exports, the contribution of petroleum industry to GDP, and the share of oil-export revenues in total government revenues – the dependence varied from moderate to nearly total.[5] Members' oil exports as percentage of total exports in 1973 ranged from as little as half in the case of Indonesia and Ecuador to 90 percent or more for Iran, Iraq, Kuwait, Libya, Qatar and Venezuela – reaching as high as 99 percent for Saudi Arabia (Table 5). In its contribution to GDP, the oil industry, even after the oil price rise in 1974, accounted for less than 15 percent in Ecuador and Indonesia, and less than 30 percent in Algeria and Nigeria; the oil share was upward of 70 percent for Kuwait, Qatar, Saudi Arabia, and UAE (Table 4). And, as a source of government revenues, oil exports income accounted for less than 30 percent in the case of Indonesia, but upward of 90 percent in Kuwait, Qatar, Saudi Arabia and the UAE – with the rest ranging from 40 percent to 80 percent.

Non-energy factor endowments included cultivatable land, water, mineral deposits, marine resources and, not least, manpower. With regard to non-energy resources, Algeria, Indonesia, Iran, Iraq, Nigeria and Venezuela were endowed with a fairly large supply of agricultural land and water. By contrast, the barren, flat and salty soil, plus scant rainfall and harsh arid climate in much of Saudi Arabia, Kuwait, Qatar and the UAE precluded normal agricultural and farming activities. Among the countries with notable agricultural potential, Algeria had the capacity for increased output of cereals, citrus fruits, vegetables, grapes and dates. Ecuador had bananas, cocoa, coffee and shrimp. Indonesia produced sizeable quantities of rice, rubber, palm oil, copra, coffee, cocoa, tea, tobacco, sugar and various spices. And a wide range of climate and soil conditions allowed for the production of many different crops (e.g. cereals, cotton, tobacco, groundnuts, citrus fruits, cocoa, coffee, tea, rubber and palm products) in other group members.

Some of the countries in the group, both large and small, were endowed with considerable commercial quantities of other mineral resources. Algeria, for example, in addition to substantial reserves of gas, had high-grade iron ore, phosphates, lead, zinc, coal, copper, salt, uranium and mercury; Gabon was rich in uranium and manganese; Indonesia had extensive deposits of tin, bauxite, copper and nickel; Iran had iron ore, chromium, copper, lead, manganese and zinc; Iraq had phosphates and sulphur; Nigeria possessed coal, tin, uranium, phosphate, limestone and iron ore deposits; and Venezuela had very large and rich mineral deposits of iron ore, coal bauxite, gold and diamonds.

With respect to the sectoral structure of production, Table 4 shows that on the eve of the oil price hike, agriculture provided a substantial portion of GDP in Ecuador, Indonesia and Nigeria, while this share was less than 5 percent in Kuwait, Libya, Qatar, Saudi Arabia, the UAE, and Venezuela. More than 45 percent of the labour force was engaged in farming and livestock activities in Algeria, Ecuador, Gabon, Indonesia, Iran, Iraq, Nigeria and Saudi Arabia (Table 2). The large labour contingent was present even in Algeria, Gabon, Iran, Iraq and Venezuela where agriculture constituted no more than 12 percent of GDP, and in Saudi Arabia, where farming made up only one percent of total output (Table 4). As the same table shows, industry had a relatively notable share in GDP of about 15 percent or more in

3

Algeria, Ecuador, Indonesia, Iraq, Nigeria and Venezuela, but 6 percent or less in Kuwait, Libya, Saudi Arabia and the UAE – with a negligible magnitude in Qatar.

In terms of the stage of economic development as measured by the availability of infrastructural facilities, differences among the group were again noticeable. Measured by the extent of paved road and rails in the country relative to its size, Algeria, Indonesia, Nigeria and Venezuela were comparatively more developed than Ecuador, Gabon and Iraq among countries for which data were available. Compared by the magnitude of electric power generating capacity relative to the size of the population, Venezuela was by far ahead of the pack followed by Iran, while Indonesia and Nigeria were decidedly at the bottom. (Table 1)

As to the quality of human development, as measured by the United Nations criteria of longevity (life expectancy), knowledge (adult literacy) and standard of living (GDP per capita)[6], OPEC members displayed distinct differences as shown in Tables 6, 7 and 8. Life expectancy at birth was lowest in Gabon, Indonesia, Nigeria, and Saudi Arabia (below 50 years), and highest in Kuwait and Venezuela (65 years and over). Adult literacy was fairly low (i.e. about or below 35 percent of the population) in Algeria, Gabon, Iraq, Nigeria and Saudi Arabia, but about 50 percent or more in Indonesia, Iran and Kuwait, and highest in Ecuador and Venezuela. In other indicators of social development, the number of population per physicians was highest in Indonesia and Nigeria, and lowest in Kuwait, Libya, the UAE and Venezuela. Algeria and Saudi Arabia also ranked relatively badly in the group, while Ecuador and Iran occupied a better position. Infant mortality (i.e. of under-fives) was less than 50 per thousand in Venezuela; but it was as high as nearly 140 in Algeria and Gabon, and more than 160 in Nigeria. In education, more than 90 percent of the age group attended primary schools in Ecuador, the UAE, and Venezuela, while the figure was as low as 37 percent in Nigeria and 45 percent in Saudi Arabia. Attendance in institutions of higher learning was also almost seven times the average for the rest of the group in Ecuador, and nearly four times the group average in Venezuela – with the lowest figure for Nigeria.

In standard of living, disparities were most conspicuous. A generation before the oil price rise, Kuwait, Qatar, Saudi Arabia and UAE were among the poorest countries in the world, with little prospects for economic progress due to their small and largely illiterate populations, poor health facilities and sanitary conditions, and harsh environments (i.e. vast deserts, high temperatures, strong winds and limited arable land or water supplies). Many other members were also among the World Bank's low-income groups. As indicated in Table 8, in 1974, per capita income was still as low as $170 in Indonesia and $280 in Nigeria. By contrast, the corresponding figures were $4,440 in Libya, $7,240 in Qatar, more than $10,000 in Kuwait and more than $11,000 in the UAE.[7] Energy use per capita as a measure of high living standards ranged from less than 40 kg of oil equivalent a year in Nigeria to more than 2,000 kg in Venezuela, 4,000 kg in the UAE and 7,000 kg in Kuwait (Table 8). Access to safe drinking water, sanitation, and health care by percentage of the population showed similarly wide ranges (Table 6).

In other areas not easily quantifiable, the disparities were equally noteworthy. For example, the number of skilled workers, competent technicians, and managerial talent

– even adjusted for the size of population and the domestic market – differed considerably among members. In terms of credit-worthiness and access to world capital markets – although itself partly a function of domestic resources and per capita income – some of the Persian Gulf members with virtually convertible currencies were in a very different position from, say, Nigeria, to tap Eurocurrency funds or to service their external debt.

OPEC members also presented wide differences in their political systems, ideologies, alliances and special relationships with the rest of the world. For some, their share of petroleum output in the global energy picture was more important than oil prices; for many the reverse was the case. Still others put more stock in non-monetary considerations than purely pecuniary gain.[8] Some members' sovereignty and independence dated only from the post-World War II period (i.e. Algeria, Gabon, Nigeria, Kuwait, Qatar and the UAE). Other members, namely Indonesia, Iraq, Libya and Saudi Arabia, traced their statehood to the end of World War I. Still others (Ecuador and Venezuela) had been sovereign nations long before. And Iran's history went back several thousand years.

Political systems ranged from a strict theocracy with Islam as the guiding light (Saudi Arabia and post-1979 Iran) to an intermittently functioning Western-type democracy (Ecuador, Gabon and Venezuela). In between, there were 'hereditary emirates' in Kuwait, Qatar and UAE; military rule (at frequent intervals in Iraq, Libya, Nigeria and, in a modified form, in Indonesia); and a one-party system with military overtones in Algeria. The countries had embraced at least two major religions, and spoke several different languages. They also belonged to various political and military alliances. Algeria, Iraq and Libya initially belonged to the Arab socialist camp, having close relations with the Soviet Union, and a 'rejectionist' position toward Israel – and, by implication, political aloofness toward the United States. By contrast, Kuwait and Saudi Arabia along with small countries of the Persian Gulf, were conservative Arabs, friendly to Washington and to US business interests. Iran, in the 1970s, forged a special relationship with the United States, while maintaining correct and business-like dealings with the Soviet Union. Kuwait, Qatar, Saudi Arabia and the UAE were the most consistently 'accommodationist' members of OPEC *vis-à-vis* the North, while Algeria and Libya routinely behaved as confrontationist hawks, and Iran joined one or the other group as its national interests dictated at the time. Indonesia, Nigeria and Venezuela, by contrast, frequently adopted a moderate position in-between.

DOMESTIC ABSORPTIVE CAPACITY

Due to the foregoing differences in size, population and resources, OPEC members were naturally different in their absorptive capacity for domestic development. Some, with a relatively small population, scant water, poor farmlands, few complementary resources, a shortage of skilled labour and a small domestic market were 'low absorbers'. They were often referred to as 'watch-makers', 'hot-house growers' and 'rentier states'. Others, with large populations, good agricultural potentials, non-oil mineral resources, a large and better trained workforce, and a generally more diversified economy were characterised as 'high absorbers'. Differences in absorptive capacity frequently affected the members' rate of oil extraction, choice of domestic allocation and selection of development strategy.

For the purposes of this study, four separate groups within OPEC might be identified. The first and the most highlighted group, consisting of Kuwait, Saudi Arabia and the UAE, had enormous petroleum reserves which, despite large daily extraction, could last eighty years or more; they had few other matching resources to absorb their sizeable export earnings in domestic investments. These countries possessed certain favourable features for measured economic development such as: a sizeable foreign exchange; a resilient (albeit small) workforce; other unused (albeit scant) resources; relative political stability; an eager, confident and active private sector; and a pragmatic and largely non-ideological approach to development. Contrasting with these positive factors, however, were such growth-inhibiting elements as a small economic base, shortages of talent and skills, absence of indigenous technology and, above all, water shortage, a harsh climate and poor soil conditions. As already indicated, the initial absorptive capacity in all of these countries was limited. But, for some (especially larger ones such as Saudi Arabia), sufficient public investments in infrastructure were likely to open up certain possibilities for industrial diversification over time. These possibilities existed particularly in oil-related industries, high-energy-based projects, trade and services. The maximisation of added value in the energy sector (oil refining and petrochemicals) made better sense than agricultural pursuits. Geography, climate and oil riches, however, subjected countries in this group to an inescapable long-term imperative: they could hardly become self-sufficient in most of their non-energy needs. Dependence on the outside world for imports, investment outlays and expatriate labour had to continue for some time.

The second group – Algeria, Ecuador, Gabon, Indonesia and Nigeria – were, by contrast, endowed with much smaller oil reserves, but owned substantial non-oil resources. Barring unexpected new oil discoveries, petroleum reserves in these countries were not expected to last more than a few decades, despite relatively low daily production. For these countries, a strategy emphasising agricultural development and the expansion of labour-intensive industries was more appropriate.

The third category – Iran, Iraq and Venezuela – were relatively more promising members who had both significant oil (and gas) reserves, lasting well beyond the middle of the next century, as well as other resources and the absorptive capacity to use the oil proceeds productively. For these countries, a combination of import-substitution industries and other resource-based export industries (including agriculture) seemed the right course to follow. But the ultimate profitability of the latter still depended on the extent of comparative advantage in a number of other complementary variables (e.g. low transport costs to potential markets and low cost of 'borrowed' technology) as well as cost advantages in labour, management and financial services.

Finally, Qatar and Libya – with relatively less oil than the first group, and much less of anything else (except gas for Qatar) compared to the second group – constituted the fourth. For these countries, trade and services were the primary investment venues at home, and assets accumulation was the overall choice outlet abroad.

Altogether, only a very few countries among OPEC members fit into the 'oil-only' category. Even there, possible technological breakthroughs in water desalination, hydroponic cultivation and industries based on raw material imports offered profitable investment outlets. Furthermore, these countries could develop service industries (e.g. banking, insurance, tourism, transhipment etc.). For the majority of OPEC members, where non-oil complementary resources were in varying degrees available (and, in some countries, abundant), the development of a secure, non-oil production base was not only a possibility but a necessity.

BASIC NATIONAL OBJECTIVES

Despite the absence of homogeneity among OPEC members in terms of history, geography, culture, size of population, oil reserves, non-oil resources, stages of development and political structure, these countries shared certain objectives and aspirations which reflected their common status as energy-based developing economies. These national goals were either clearly stated in their development plans and annual budgets or, otherwise, implicitly observable in practice.

The foremost overall objective of all OPEC members was the establishment of a sustainable base for a viable post-oil economy. The right choice of a long-term development strategy was crucial because their initial, resource-based, comparative advantage was short-lived at best: industries and activities that were dependent on cheap local energy were bound to lose their privileged position once oil reserves were diminished or exhausted. While this consideration was absolutely critical for such high-absorbing countries as Algeria, Indonesia and Nigeria (and even, perhaps, Iran and Venezuela), the super oil-rich countries (e.g. Saudi Arabia and other Persian Gulf members) had much greater latitude. Nevertheless, all members were cognisant of the fact that, sooner or later, either they would run out of oil or the world might no longer need as much of theirs.

This concern over the eventual exhaustibility of oil reserves prompted all members to follow a policy of wealth conservation, namely the exchange of their natural wealth (i.e. oil and gas) for man-made capital (i.e. modern infrastructure, productive plant and equipment, technical skills or financial assets). In following this policy, as a universally common objective – and an article of faith – all OPEC members adopted the diversification of their domestic production base as their prime goal. Reinforcing the compelling need for adopting such a guiding principle were the hazards of exposing the country's development efforts to fluctuations in the global demand for oil and hence its price. This vulnerability further enhanced the need for greater financial security through diversified exports. Still further, the incentive for diversification reflected petroleum's inability, at least in early phases of development, to generate production and consumption linkages with the other economic sectors or to create extensive employment opportunities. Finally, lurking behind the emphasis on product variety, were certain national security considerations such as food self-sufficiency, defence preparedness and external political commitments. An essential concomitant of diversification was the enhancement of national productive capacity and expansion of domestic output.

7

The realisation of these objectives required substantial modernisation and development of infrastructural facilities, and improvements in labour productivity. Implied in the latter goal were both a guarantee of high employment and the development of skilled human power to meet high-productivity requirements. Hence, in countries with large and fast growing populations (e.g. Algeria and Iran), emphasis had to be placed on the need to expand and improve the quality of education and vocational training. In others, with small populations and scant skills (Kuwait, Qatar, Saudi Arabia and the UAE), the development of indigenous talent, and reduced dependence on foreign (imported) manpower, was a primary objective. Increased access to modern and advanced technology was thus a standard objective among all, in order to gain greater national self-sufficiency in basic needs, expand home-grown research, and enhance national standing in the world community.

Implicit in the objectives of steady growth and high employment was a universal desire for a sustainable rise in real per capita income and palpable improvements in living standards. Starting generally from a very low per capita income base at the time of their commercial oil discoveries, almost all member governments were compelled to place the objective of greater social welfare and improved social amenities at the top of their national agenda. Social justice, equity and fairness in income distribution thus became other important common goals for OPEC members. For almost all, a major task to be tackled was a fuller provision of basic human needs (i.e. better nutrition and health care, more decent housing, more and better education, and satisfactory employment).

Explicitly stated or not, strengthening national security and defence against real or perceived enemies and hostile neighbours was, again, a common concern. So was the maintenance of balance-of-payment equilibrium as a measure of financial security and avoidance of external debt. Protection of the environment was another emphasised goal. Elevation of the spiritual, social, cultural and intellectual needs of the citizenry was emphasised or alluded to in the periodic economic and social development plans of most members. Some of the corollaries of these basic objectives also included: a narrowing of the technological gap with the industrial countries; the maintenance of solidarity with political, geographic or military allies; and co-operation with other developing countries in the North/South context.

In short, all members, with varying degrees of firmness and with greater or less emphasis, adopted as their national economic agenda such objectives as: improving living conditions (nutrition, health, housing, education etc.); reducing economic dependence on the outside world; catching up with the West; reducing income inequities; increasing efficiency of productive factors; and maintaining established cultural values and social stability. Some members with deep-rooted Islamic traditions (Islamic Iran, Saudi Arabia, and other small Persian Gulf countries) were emphatic about wanting to avoid 'unwanted social change' and preserve traditional cultural and spiritual values.

A successful achievement of these primary objectives necessitated the pursuit of other goals as intermediary means. These intermediate objectives included domestic price stability, a sustainable balance in foreign obligations, as well as fiscal and monetary discipline. The attainment of these secondary goals was crucial for orderly and sustained private capital formation, preservation of national currency value and avoidance of burdensome external debt.

NOTES ON CHAPTER I

1   For the story of OPEC's endurance record, see Jahangir Amuzegar, 'OPEC's Seventh Life', *Middle East Policy*, September 1997.
2   See Alan Gelb, 'From Boom to Bust', *IDS Bulletin*, Vol, 17, No. 4, 1986.
3   The intention here is not to present a detailed picture, but to highlight the magnitude of differences.
4   The Society of Petroleum Engineers defines oil reserves as estimated volumes of crude oil, condensates, associated natural gas and gas liquids, and other substances that are commercially recoverable from known accumulations under existing economic conditions by established operating practices and under existing government regulations. Accordingly, reserve estimates change as existing reserves are depleted, new reservoirs are discovered, additional geological and/or engineering data become available, economic conditions (costs and prices) change, or new operating practices (technology) become available. See *OPEC Bulletin*, March 1993, p 8.
5   Cf Jahangir Amuzegar (1983).
6   See *Human Development Report 1994* (New York: UNDP, 1994), p 91.
7   The measurement and comparison of OPEC members' GDPs, GNPs and per capita incomes, by conventional accounting methods, involve sizeable distortions since the so-called 'reproducible' component of national output differs considerably from country to country depending on the share of (exhaustible) oil in total product. In countries that depend heavily on petroleum for their livelihoods (e.g. Kuwait, Qatar, Saudi Arabia and the UAE), this component is estimated to be fairly low (e.g. as little as a quarter to a half of reported GDP in Saudi Arabia).
8   It is believed that some OPEC members' decision to move downstream and invest in foreign oil facilities has been based more on strategic considerations than profitability criteria. Kuwait, for example, is reported to have made 'no real profit from their foreign downstream operations'. See *Middle East Economic Digest*, 1 April 1994, p 9.

II
The Framework of Analysis

As the two eventful decades of 1974–94 unfolded, history played a clever joke on all soothsayers regarding OPEC's fortunes (and everyone else's misfortunes). Contrary to all the predictions (outlined in the Preface) the major oil-importing industrial countries managed to reverse their adverse terms-of-trade with OPEC after a brief pause. In contrast, OPEC members ended the 20-year period with the real (inflation-adjusted) price of crude oil lower than the 1973 level, their treasuries in hock to their developed trade partners, and their economies facing more formidable challenges than ever before.

What happened? What went awry? Why did everyone err in contributing to what seemed to be the consensus at the time regarding OPEC's future? This chapter intends to argue that the reversal of fortunes was due to three crucial and somewhat unrelated elements. First, petroleum was a unique strategic raw material, difficult to substitute in the short-run, and thus highly vulnerable to short-term shocks from changes in supply and demand. Second, major petroleum deposits were initially concentrated in a few small geographic areas, making cartelisation easier for oil than other strategic commodities, but rising oil prices gradually opened up new fields elsewhere and turned the oil cartel into a paper tiger. And third, state ownership of oil reserves in all OPEC members made misallocation, inefficient use, waste and misappropriation much easier than if the reserves had been in private hands.

The task of this chapter is to provide a brief theoretical background to the later analyses of OPEC's booms and busts during the 1974–94 period. The chapter attempts to tie together (1) the uniqueness of petroleum as a strategic commodity responsible for the rise of a very large 'economic rent' (2) the nature and mechanics of allocating the oil 'rent' among alternative claimants (3) the 'process of adjustment' involved in such allocations and (4) the policy options open to OPEC members in such an adjustment process.[1]

An extensive and sophisticated literature has emerged since the early 1970s on the optimum use of oil windfalls[2] and the effect of oil booms and busts on the economies of the major petroleum exporting countries. The intention here is neither to review this literature nor to add to its size, but to provide an introductory back-drop to the chapters that follow. According to this conventional literature, optimal resource management in a depletable resource-based economy requires the judicious conversion of a dormant stock of exhaustible deposits into a production base for generating permanent flows of income in the future. Optimality, in turn, would be attained when the marginal rate of return in each of the 'uses' of the exhaustible resources (i.e. extraction or conservation) is equalised. This conventional principle of the economics of exhaustible resources, first enunciated by Hotteling and Robinson[3] implies that if an oil exporting country expects the future rate of increase in the world price of oil, net of extraction costs, to be higher than the expected rate of long-term interest, it has good reason, other things remaining equal, to keep oil underground. Changes in the current price of oil would then depend largely on changes in expectations about future market conditions, both for the global oil balance and for prospective interest rates. This relationship will, of course, hold only if interest rates themselves are not directly or indirectly influenced by oil price changes – a consideration which cannot be assumed away in developed economies that are heavily dependent on oil imports, and more particularly in a world where global economic interdependence tends to link national interest rates to events beyond national borders or trade flows. In the real world, there are also other exogenous factors – such as the dependence of extraction costs on the rate of production, uncertainties regarding future supply and demand for oil and oil substitutes, various market imperfections and, above all, non-economic and non-financial considerations that render the decision on optimum extraction rate extremely complicated – and in OPEC's case, unique.

THE GEOECONOMICS OF OIL

Petroleum or 'rock oil' embodies a number of specific characteristics that are absent in non-mineral raw materials, as well as other strategic minerals. Mineral exporters, in general, are different from other primary exporting LDCs in three significant respects.[4] First, export revenues from minerals are considered subject to wider fluctuations than some other primary commodities. Second, mineral deposits are exhaustible over time, necessitating not only different national accounting criteria but also a replacement provision for future income. And third, forward and backward linkages between the mineral sector and other sectors are relatively small. The main linkage between mineral extractions and other economic activities would be fiscal, whereby 'economic rent' accruing to the state is distributed among other sectors through public investments, government services, subsidies or other transfer payments.

11

Oil is viewed to be further different from other minerals. A number of specific features exclusive to petroleum play a significant role in creating the so-called oil 'imperium', compared to history's other commodity arrangements, associations and cartels. In the words of an engaging storyteller, '[The] 20th century has been completely transformed by the advent of petroleum'.[5] The first distinctive feature of oil is that no other internationally-traded strategic mineral – aluminium, bauxite, chromium, copper, nickel, phosphate, tin or zinc – matches petroleum's singular importance to the world economy in war or peace. On no other mineral (with the arguable exception of chromium) are industrial countries so strategically dependent on developing economies for the maintenance of their military strength, economic competitiveness or high standard of living. Victories and defeats in World War II were frequently influenced by the availability of fuel supply. Oil still remains critical to military prowess and super-power status. Oil availability and cost are an integral part of international comparative advantages in many product lines. And Western lifestyle, if not indeed Western lifeblood – from fertilisers for food, gasoline for transport, and plastics and chemicals as the building blocks of modern life – is vitally dependent on oil.

Second, in the volume of trade, oil is the world's biggest multinational business, and the largest traded commodity.[6] Over one third of the largest 20 companies in the Fortune 500 are in the oil business. Along with food items, oil prices constitute one of the sharpest (and most volatile) determinants of consumer price indices in the developed countries. The latter's rate of economic growth is profoundly affected by major oil price movements. Third, there are as yet few cost-effective, short-term substitutes for the flexible and versatile petroleum, particularly in transportation, and more specifically in aviation. Natural gas, coal, wood, nuclear power and exotic sources of energy (sun and wind) may serve as substitutes for some uses of oil, but not all – and not always as efficiently. Fourth, oil lends itself more easily to supply control and trans-shipment on short notice. Supertankers on high seas not only carry oil the cheapest way but can change course at a moment's notice, and may also serve as floating reservoirs. Fifth, the capital-intensive nature of oil production, refining and shipping makes the petroleum industry immune to labour disputes. The need for large worker layoffs in depressed economic periods is thus free from union pressures for maintaining production levels and glutting the market. Sixth, in no other minerals business do risk and rewards so doggedly defy corporate strategy, technological imperatives and cost calculations; or so often fall prey to chance and adventure.

Finally, for no other mineral is the difference between average unit cost of extraction in the rich-reserves areas and the selling price in the world markets – the 'economic rent' – as large or as enduring.[7] Oil is the prime example of nature's bounty creating 'economic rent' because it is non-renewable (albeit both discoverable and capable of enhanced recovery). Any price over and above the cost of its extraction at the well is therefore considered 'pure economic rent'. Among major minerals, petroleum extraction is even more specifically associated with the Ricardian notion of rent because of its many extraordinary and unique features and the concentration of its deposits in a few geographic enclaves. Due also to the oligopolistic structure of the oil industry, and the difficulties of entry and exit, oil rents are sometimes divided into two parts: differential rent and monopoly rent. Differential rents are further broken down into four separate sources: quality, position, mining and technology.[8] In short, oil provides the best overall energy source for consumers in terms of accessibility, transportability, flexibility and cost. For these reasons, oil has become the dominant source of primary energy in the latter part of the 20th century. In 1925, petroleum accounted for a little over 13 percent of total global energy use. In 1950, oil constituted nearly 30 percent. By 1975, the share reached nearly 48 percent. Thus, within five decades, petroleum consumption's share rose more than two-and-a-half times.[9]

At the same time, by a quirk of nature, the bulk of the earth's known oil reserves is located in certain specific areas while the rest of the planet goes almost totally without. No other strategic material (with the possible exception of chromium in Zaire and South Africa) is so heavily concentrated in a relatively small geographic area of the globe as is oil. The world's known major oil reserves are located in a few Middle East countries where domestic demand for oil is dwarfed by the abundant supply. By contrast, with the exception of a few (United Kingdom, Norway and Russia), nearly all industrial countries and some of the newly industrialised nations in Southeast Asia and Latin America are partially or almost totally dependent on imported oil. By the end of 1994, of the world's total proven recoverable crude oil reserves of 1,020 billion barrels, 76.3 percent belong to OPEC.[10] Yet, while OPEC produced 41 percent of global oil production, they consumed only 7 percent.[11]

Because of these uncommon characteristics, the oil market also presents certain features of its own. To begin with – at least since the 1928 Achnacarry Agreement (see Chapter III) whereby world demand for oil was to be met by the three leading Western oil companies – the oil market has always been imperfect, and basically managed, albeit not always very efficiently. Due to heavy initial capital requirements, complex and evolving technology, and marginally profitable retail outlets, world oil was, up to the early 1970s, concentrated in the hands of a few vertically integrated giant companies which owned and/or controlled crude reserves, refineries, tanker fleets and retail service stations. Furthermore, the petroleum market was 'inherently' imperfect and volatile due to the 'inability of economic agents to take the long view'.[12] Oil's strong convulsions were, in turn, due to an asymmetry between the time pattern of reserve discoveries (i.e. discrete jumps at irregular intervals) and the time pattern of extraction (i.e. regular, continuous depletion). Large discoveries tend to cause a temporary glut and low prices, encouraging consumption and choking off needed long-term investments. When the glut is reabsorbed by continuous and larger extraction, a shortage will ultimately re-emerge, prices rise, new investment overshoots the target, and a glut reappears.

The oil market in the 1950s through 1980s clearly reflected such cycles.[13] During the 1950s and 1960s, the oil majors kept oil prices competitively low under their short-term, myopic perspective, failing to anticipate the impact of low prices on the future oil balance. Their fortuitous oil discoveries led them to seek short-term profits from large sales at small margins above cost. Although it was increasingly clear that the rapidly rising consumption rate was not sustainable and required a price rise, the oil concerns refrained from raising prices because (a) they feared the ire of consumers in the industrial countries who were instinctively suspicious of their power and (b) they did not wish to whet the producers' appetites for higher and higher shares of the profit. In other words, the oil majors simply failed to put their long-term interests ahead of short-term considerations.

THE OWNERSHIP FACTOR AND CHOICES FOR ALLOCATION

The allocation of oil revenues,[14] and their impact on the oil-producing economy depend on who owns the oil deposits: the government or private entities. Where oil wells are privately owned (as in the United States), there is no room for consideration of the national interest or social welfare in oil income distribution. In countries where, by law, tradition or ideology, mineral resources belong exclusively to the state, the national government is expected to act as a trustee for the welfare of current and future generations. In all OPEC economies, where the ownership and exploitation rights of oil resources belong to the state, decisions regarding the extraction of oil and disposition of its sales proceeds have been made by government authorities, presumably in the direction of optimum national welfare. This institutional feature has had a significant bearing on both the scope and extent of state intervention in the economy, and on the economy's response to the government's initiatives as well as public incentives. Oil's exhaustibility also had a primary influence on the disposition of windfall gains. Political leaders and economic planners had primary responsibility for compensating a permanent loss in the national stock of wealth by other forms of social utility (e.g. fixed capital formation in other productive sectors). The choice of windfall allocation thus became a central issue for the government.

In theory, the oil rent received by the government in a typically oil-based economy must be allocated among competing uses (and over time) in an optimal fashion. The state, assumed to be both rational and free from vested interests of its own, is expected to allocate oil windfalls in accordance with certain consensual public welfare criteria in such a way as to optimise popular satisfaction and to best serve the public's long-term interests. In practice, however, none of these assumptions quite holds. Economic optimality is commonly constrained by traditional, ideological and social considerations. A real-world government is also neither always rational nor frequently without interests of its own. Moreover, the contemporary state is hardly ever a monolithic entity, and is often composed of competing groups with vested interests of their own – particularly under a Western-type democracy. And, finally, no government's 'long-term' horizon usually extends beyond its own expected tenure – which, again, is much shorter under participatory democracies than in so-called Oriental autocracies.

The alternatives faced by OPEC members for the disposition of their oil windfalls thus fell within a wide spectrum. The primary decision involved the optimum rate of daily crude extraction, i.e. to produce or to conserve. Factors influencing this decision were the size of oil reserves, maximum daily output capacity, need for revenues, individual country quotas assigned by OPEC, influence of domestic pressure groups, and certain foreign political considerations (i.e. political use of the 'oil weapon').

Once the level of extraction was determined, the next decision involved increased domestic consumption *vs* additions to fixed domestic capital formation. This decision was affected by such factors as the prevailing standard of living and per capita income, the regime's own felt security and time horizon (i.e. the need to appease the populace through welfare handouts), and ultimately a deep-rooted cultural propensity toward conspicuous consumption *vs* productive capacity formation. Increased current consumption was also often favoured both for raising the national standard of living, as well as strengthening internal security and external defence – depending on the ideological and political make-up of the regime in power.

The portion of the oil windfalls set aside for capital formation was subject to a new series of decisions: to invest in domestic development projects or to 'save abroad' (i.e. go into foreign direct or portfolio investments). In the domestic arena, the choices were influenced by both national comparative advantage as well as non-economic considerations (including the so-called 'spoils system'). Public investments were divided between infrastructural facilities, the physical stock of social capital, human capital formation and economic diversification. In the latter's case, ideological and political factors played a role in the choice between agriculture, industry and services. Domestic investment was undertaken directly by the state or through participation of the private sector with various types of assistance from the government (i.e. grants, low-cost credit, tariff protection etc.).

When there were limitations in short-term domestic absorptive capacity, the choice of windfalls allocation turned in favour of 'sterilising' foreign exchange receipts by insulating them from domestic money supply, namely recycling the oil revenues in foreign ventures. These ventures ranged from simple interest-bearing foreign certificates of deposit to highly sophisticated monetary and real foreign assets. Finally, part of the windfall gains was set aside for assistance to poorer non-oil developing countries for humanitarian, political or commercial reasons (see Chapter IV).

Underlying country-specific choices within the foregoing range were a host of technical, economic and political factors. On the whole, however, national allocation decisions were based on domestic absorptive capacity and the magnitude of annual oil revenues. Yet choices were not always (a) ideally free from historical or socio-political biases (b) strictly logical from a national-interest point-of-view (c) consistent across current alternatives or over time and (d) in the best long-term interest of the public beyond the incumbent administration's life.[15] For the group members who enjoyed not only sizeable oil deposits but also substantial arable land, water, other natural resources and trained labour (see Chapter I), increased domestic investment was a preferred option. For the 'only oil' members with no realistic opportunities for a non-oil output base, the rational choice was to accumulate sufficient financial assets abroad that could sustain the economy once the oil wealth was exhausted, in the so-called pure 'rentier-state' status, living off the returns on earlier foreign investments. Richer countries were obviously more generous with their poorer neighbours and friends by dispensing foreign aid; they could also afford a more lavish lifestyle, more sophisticated arms build-up and fancier investment projects (see Chapter IV).

The combination of policy alternatives adopted by each country, while intrinsically based on national resource endowments and technological capabilities, did, at times, override considerations of strict economic rationality or long-term comparative advantage. Normally, a broad mixture of available alternatives was adopted and followed. Each alternative was also pursued through different public policies. And each policy presented its own strengths and limitations.

THE ADJUSTMENT PROCESS

The need for adjustment arises whenever the external current account is not in balance, i.e. showing either a deficit or a surplus. Current account imbalances are normally offset by movements in the capital account, i.e. by borrowing or lending abroad (including foreign aid and grants). The main instrument of adjustment is increased purchases of foreign goods, services or financial assets when there is a surplus, and reduced imports, sale of assets and borrowing abroad when deficits develop.

The process of adjustment is commonly analyzed in terms of (1) the impact of sudden wealth (or abrupt shortfall) on aggregate domestic expenditures (2) the effects of increased (or decreased) domestic expenditures on various sectors of the economy and (3) the results of shifts in the composition of consumption and investment for the balance-of-payments. OPEC economies fit this analytical framework ideally because the state ownership of oil reserves gives the government the sole initiative in increasing or decreasing the volume of expenditures out of the oil money.

Adjustment in a typical oil-based economy is described by three mainstream theories – two structural and one neo-classical.[16] The so-called structural bottleneck theory holds that economic growth in a typically developing nation is constrained by the scarcity of basic productive resources as well as certain technical and institutional rigidities which can not be remedied by the market (price) mechanism. Scarcities or 'gaps' are most critical in the supply of key factors, e.g. plant and equipment, raw materials, processed inputs and modern technology. Some of these gaps and rigidities could be overcome via imports. But the paucity of domestic savings, as well as limited foreign exchange earnings, preclude this solution. Growth-impeding shortages of domestic savings and foreign exchange form the core of the 'two-gap' model. The addition of fiscal shortage (i.e. inadequate taxes) introduces the 'three-gap' model wherein economic development is held back by the government's inability to increase its revenues. Implicit in these models are the postulates that the economy is in its long-term full-employment equilibrium, and greater or more efficient government intervention to close the savings gap is not a possibility.

Within this structuralist approach, oil income may thus have both immediate and potential contributions to growth. In the near-term, oil earnings can help alleviate some of the bottlenecks that constrain the economy, increase structural flexibility, and create new demand for surplus domestic resources (e.g. labour). Oil's potential contribution lies in the dynamic effects of increased domestic expenditures on the patterns of long-term structural change via linkages with other sectors of the economy and through a strong stimulus to new entrepreneurship.

In the case of OPEC members, oil export earnings were expected to remove or greatly alleviate all three bottlenecks by providing increased (painless) savings and investable funds; ample foreign exchange in the form of hard currencies; and additional fiscal revenues without imposing back-breaking higher taxes on the (poor) nation. Furthermore, since oil receipts accrued directly to the state, the oil impact was likely to be quicker and more goal-oriented than if market forces alone were relied upon to facilitate the absorption. As a result, the government's role in the process of adjustment to oil shocks – usually assumed to be neutral or marginal in the gap models – becomes immensely enhanced.

The second structural theory dealing with windfall effects is the linkage theory. This approach maintains that growth, particularly in an underdeveloped environment, is likely to result not so much from an optimal use of existing and known resources as from the discovery and employment of hidden, scattered or badly utilised resources and abilities.[17] And such a stimulus can be provided by a leading industry or activity through production linkages with other sectors of the economy, both upstream (backward) or downstream (forward). Backward linkage refers to a profit opportunity created for upstream, supplying industries through the expansion of demand for raw materials and intermediary goods and services generated in the leading industry or activity. Forward linkage is a profit opportunity created for downstream processing industries by lower-cost inputs produced by the advancing industry or sector. The magnitude of these linkages depends on the nature of the leading industry itself, availability of other resources in the country, and technical synergies between them.

When production linkages with other sectors of the economy are weak because (1) the leading sector is rather small in relation to gross domestic product (2) its demand for labour and other domestic inputs is rather limited or (3) the demand for its product comes from abroad, and not from other domestic industries, growth cannot be stimulated through direct contact. This weakness is particularly noticeable in the case of the oil sector in OPEC economies not only because of the 'enclave' nature of the industry itself but more because the oil revenues overwhelmingly accrue to the government, and the state's economic involvement and intervention in the economy is anything but marginal or inconsequential.

Under such circumstances, production linkages are replaced by fiscal, or indirect, linkage. Here the state contributes to the growth process by influencing the patterns of aggregate demand or the magnitude of total supply. Aggregate demand is impacted by the distribution of the oil revenues through reduced taxes or increased subsidies and welfare outlays. Aggregate supply can be increased by removing structural rigidities (i.e. establishing or expanding infrastructural facilities); placing funds in the hands of private entrepreneurs through development banks; undertaking public agricultural or industrial projects; or otherwise improving supply-side externalities on the economy. The ultimate benefits (or blessings) from the oil windfalls would depend on the efficiency and efficacy of state intervention and involvement.

The fiscal linkage, in the final analysis, will be weak or strong depending on the manner in which the oil revenues are absorbed. The reason for the weak linkage is that governments – and particularly governments in the developing countries – are not better equipped to pick 'winners' among industries or activities than are private entrepreneurs who risk their money or their reputation. Public infrastructural projects also may not always follow optimal allocation and may be undertaken for political, prestige, patronage and other non-economic considerations. Furthermore,

17

the quality of public undertakings in both infrastructure and productive sectors may suffer from a number of different factors: inadequacies of planning; concentration of allocative decisions in the hands of a few inexperienced, biased or ambitious civil servants; the preponderance of grandiose, capital-intensive projects in the overall plan; and the choice of public investment decisions based on friendship, partisan politics, kickbacks, log-rolling and similar considerations. Last, but not least, given the volatility of the global oil market, a sharp decline in the price of oil and a drastic decline in the state revenues would wreak havoc with long-gestation (and half-finished) public projects started during the boom period.

On the other hand, fiscal linkage may be strong when it enhances the quality of overall domestic resource allocation by compensating for market deficiencies and consumers' profligacy. If the oil windfalls were to be widely appropriated across the population through individual vouchers, for example, there might be an immediate increase in welfare and subsequent production shifts in accordance with consumers' wishes. But, given the relative poverty, and the extremely high consumption and import propensity in a majority of OPEC economies, the reliance on consumers' preferences would, in all likelihood, tend to inhibit productive investments. Without the state paying due attention to intergenerational utility functions, the likelihood of domestic production being undermined by high imports, 'rent-seeking' activities and leakages from the system would be enhanced. The free market mechanism, by giving short-term signals, would often fail to serve the nation as a whole, from a long-term development perspective.[18]

The third adjustment theory describing the windfalls' effect is the 'booming-sector' theory which, within the neo-classical framework, assumes rationality on the part of economic agents, substitutability of productive factors and flexibility of the price mechanism. Simply put, this model describes how a sudden surge in an export activity or a substantial rise in export prices (e.g. of oil and gas) can be expected to cause a slowdown in the export-oriented or import-competing sector (e.g. manufacturing). This theory deals with sectoral reallocation of resources and structural transformation rather than the dynamic growth process itself. The export boom is expected to affect the domestic economy in two ways: the 'spending effect' and the 'resource-movement effect'. In the first stage, larger domestic incomes arising from the boom lead to additional domestic expenditures. Increased spending, however, affects different goods (and services) differently. Since the prices of traded goods are determined by international market forces, they remain unaffected despite the extra domestic expenditure. But the prices of non-traded goods are determined by internal market forces and consequently tend to go up. As a result, traded goods become relatively less profitable and non-traded goods relatively more lucrative to produce. The second stage involves the movement of productive factors from traded to non-traded sectors, lured by higher prices and bigger profits. Reallocation of capital and labour between the economy's two main sectors thus takes place as a result of changes in relative prices and profits in those sectors.

The process is commonly explained either through a rise in the real exchange rate or a change in real wages. In a monetarist approach, focusing on external balance, the shift occurs as the export boom produces a trade surplus and increases the supply of foreign currencies. With the nominal exchange rate unaltered, the value of foreign exchange in terms of local currency declines. Relatively cheaper foreign currencies would raise the demand for imports, gradually absorbing the surplus. However, cheaper imports means a relative fall in price of traded goods (such things as industrial and agricultural imports) and a corresponding rise in the relative price of non-traded goods (such as land, housing and services) at home.[19] Under neo-classical assumptions, and with the economy in full-employment equilibrium, labour and other domestic resources would shift away from the production of traded goods toward the non-traded sector – reducing the volume of locally produced traded commodities where profit is declining. The shortfall would be gradually substituted by imports. The contraction of the traded sector is termed the 'Dutch Disease' because of the Netherlands's experience with the large inflow of North Sea gas export revenues in the 1970s and a subsequent recession in the Dutch manufacturing industries.[20]

An alternative theoretical explanation follows the domestic spending route. In this version, the change in the real exchange rate is portrayed through a change in the price index of non-traded goods compared to the price index of the tradables. Prices of traded goods are normally determined in the world market and are subject to the law of 'one price' for everybody (except, of course, for tariffs and transport costs). The supply of traded goods for an importing country is almost perfectly elastic, i.e. unaffected by the country's increased imports. Prices of non-tradables, however, could differ substantially across borders, and their supply elasticities are often less than perfect. Thus, as the proceeds from exports are spent at home, the prices of non-traded relative to traded goods will increase. With the nominal exchange rate remaining fixed, a rise in the ratio of the non-traded price index to the traded price index would be the same thing as an up-valuation of the real exchange rate (i.e. cheapening of imports). As a result, demand shifts toward the tradables, and production to the non-tradables. Real GDP is increased by the larger output of non-tradables and higher volume of tradable imports. The trade balance will be restored when the transfer of resources from abroad – the windfall gains – is reversed by increased purchases abroad.

The 'Dutch Disease' model is based on the neo-classical assumptions of initial general equilibrium (i.e. full employment of resources), external balance, wage/price flexibility, ease of entry and exit (i.e. absence of market distortions), mobility of labour internally, and immobility of labour and capital across national frontiers.[21] It also basically describes the situation in a small, open economy with state intervention in non-conventional areas at a minimum. The availability of unemployed or under-employed workers – a prominent feature of the Third World economies, including OPEC members – is thus ruled out. Yet the existence of such large pools of idle or semi-idle labour can obviously offset the necessity of resource transfers lured by expenditure and the price effects of export booms. Under such circumstances, non-tradable sectors (e.g. construction, services and others) can experience significant growth without reducing output and employment in the tradable sectors (e.g. agriculture and manufacturing).

In the OPEC economies where oil revenues (and windfall gains) accrue to the state, and other neo-classical assumptions actually do not hold, the impact of

19

government expenditure on the sectoral supply response (and particularly on the demand for imports) becomes the key variable. This impact would be compounded if public spending is concentrated in domestic construction and local services whose supply elasticities are relatively low. The relative price (and profit) changes in favour of the non-traded sector may result in stagnation of agriculture and manufacturing, while public services, construction and real estate actually boom. Rising real wages in the latter activities presumably contribute to the exodus of labour from farming and unprotected small industries toward construction jobs in the cities. Higher real wages in combination with an overvalued real exchange rate would reduce international competitiveness in export-oriented or import-substitution industries, and there would be political pressure for export subsidies and import protection. However, under any of the three approaches – gaps, linkages, or the booming sector – it would be specific government spending that alters the relationships between domestic economic sectors and influences the size of GDP at the macroeconomic level. For this reason, the role of government policies in the OPEC economies assumes the utmost importance.

As shall be seen later, none of the three adjustment models – particularly the Dutch Disease – fully applied to OPEC economies mostly because state subsidy and protection policies often softened the impact of the windfall gains on domestic economic sectors. Furthermore, these models' emphasis on the equilibrium of the real exchange rate at the outset, real-wage flexibility and the immobility of capital between sectors made them particularly weak in the case of OPEC economies where exchange rates were not initially in balance, the labour market was seldom allowed by the government to be cleared by market forces alone, and structural adjustment policies in both booms and slumps involved some intersectoral reallocation of scarce capital.[22]

PUBLIC POLICY OPTIONS

As the foregoing discussions indicated, oil booms and busts required adjustment. In a typical OPEC economy, the adjustment process did not take place automatically, i.e. through the market mechanism. The process was normally initiated, guided, helped or influenced by the government. State intervention was in the form of direct allocation of the (oil) surplus among competing alternatives. Or, it involved indirect intervention through various policy measures designed to influence the decisions of private firms and households.[23] Fiscal, monetary, exchange, trade and country-specific policies were used in varying measures to implement the adjustment process. These policies, of course, were not mutually exclusive, and, in fact, were often used jointly.

The design of these strategies and policies and the quality of their implementation ultimately defined the adequacy of managing oil booms and busts. The difficulties associated with such management – loosely code-worded Dutch Disease – involved: (i) *ad hoc* and rapid disposition of large foreign exchange inflows (ii) the pursuit of inward-looking development projects and a neglect of the non-oil traded goods sector (iii) the 'waste' of oil wealth through public sector 'white elephants' and (iv) the promotion of rent-seeking and other unsavoury behaviours.

NOTES ON CHAPTER II

1 General readers who are not particularly keen or interested in abstract discussions may skip this chapter and go directly to the next.

2 The concept and magnitude of oil windfalls have been subjected to analysis using some elaborate and sophisticated econometric models. Since the conclusions obtained from these complex counterfactual models are hardly different from the plain, common-sense, nature of observable results (although with arguably greater precision of detail), this study treats windfalls and shortfalls as simply increases or declines in the government's annual receipts from oil and gas exports.

3 Harold Hotteling, 'The Economics of Exhaustible Resources', *Journal of Political Economy*, April 1931, pp 137–75; and T.J.C. Robinson, *Economic Theories of Exhaustible Resources* (London: Routledge, 1989).

4 See Gobind Nankani, 'Development Problems of Mineral Exporting Countries', *World Bank Staff Papers*, No. 354, August 1979.

5 Daniel Yergin (1991), p 13.

6 In 1974, the industrial countries' total imports of all non-oil raw materials amounted to $40.5 billion while their petroleum imports alone reached $100 billion.

7 Economic rent is commonly defined as a return to a factor of production in excess of the remuneration needed to secure its participation in a given activity: it is the difference between the market price and the opportunity cost of an input.

8 For fuller discussion of these technical characteristics, see J.M. Chevalier, 'Theoretical Elements for an Introduction to Petroleum Economics', in A.P. Jacquemin and H.W. Dejong (eds), *Market, Corporate Behavior and the State* (The Hague: Nijhoff, 1976). Cf B. Fitch 'OPEC: The Big Cartel that Couldn't...' *Against the Current*, vol. 1 1982; A. Krueger, 'The Political Economy of Rent-Seeking Society', *American Economic Review* vol. 64(3), 1974; and T. Wilson, 'The Price of Oil: A Case of Negative Marginal Revenue', *Journal of Industrial Economics*, vol. 27(4), 1979.

9 In 1994, oil constituted more than 40 percent of the world's primary energy consumption compared to less than 29 percent for coal, 21 percent for gas, 7 percent for nuclear power, and about 3 percent for hydro.

10 Proven and recoverable reserves are estimates of the amount of petroleum that can be extracted through known technology at prevailing world prices. A second category, called probable reserves, may be recovered by more advanced techniques. Still a third variant, called potential deposits 'in place', are retrievable only with newer technology and at higher prices.

11 All data in this section are drawn from British Petroleum, *Statistical Review of World Energy*; the United Nations, *Statistical Yearbook*; the OECD *Energy Statistics and Balances of Non-OECD Countries*; and OPEC *Facts and Figures*, various issues.

12 See Robert Mabro, 'The Long and the Short of the Oil Market', *OPEC Bulletin*, September 1985, p 15ff for the elaboration of this thesis.

13 See the discussion in Chapter IV.

14 What is generally referred to as oil 'revenues' are not really income in the sense of a permanent cash flow. Petroleum revenues are in reality 'cashed' worth of a given stock of capital, and the transformation of one asset into another – or capital liquidation. For an elaboration of this concept, see Saleh El-Serafy, 'Absorptive Capacity, the Demand for Revenue, and the Supply of Petroleum', *Journal of Energy and Development*, Autumn 1981, pp 73–88; and T.R. Stauffer, *Accounting for Wasting Assets*, (Vienna: OPEC Fund for International Development, November 1984).

15 Cf Alan Gelb (1988) Chapter 3 for further theoretical discussions.
16 These theories are briefly examined with respect to some OPEC members by Alan Gelb (1988); Ahmad Jazayeri (1988); and Massoud Karshenas (1990).
17 See A.O. Hirschman (1958); and *Essays in Trespassing* (New York: Cambridge University Press, 1981), Chapter 4.
18 For specific reasons calling for government intervention in the developing economies, see H.A. Myint (1964), Chapters 7–8.
19 The division between traded and non-traded goods is neither precise nor static. With changes in relative prices (via tariffs, transport costs etc.), some hitherto non-tradables may move in international trade, and the movement of some tradables may become prohibitively costly. Furthermore, within agriculture and manufacturing, which are commonly classified as traded sectors, there are certain items that are not traded (e.g. local food staples or heavy local construction materials). Finally, certain highly differential traded goods may not involve a two-way trade: Persian carpets, caviar and pistachio nuts are exported from Iran but hardly ever imported by that country. See Jazayeri (1988), p 157.
20 For a quick review of the vast literature on the Dutch Disease syndrome, see W.M. Corden, 'Booming Sector and Dutch Disease Economies: Survey and Consolidation', *Oxford Economic Papers* 36 (1984); J.P. Neary and S.V. Wijnbergen, 'Can Higher Oil Reserve Lead to a Recession?' *Economic Journal* (June 1984) and *Natural Resources and the Macroeconomy* (Oxford: Blackwell, 1986).
21 See Philip Daniel, 'Mineral Exporters in Boom and Slump,' *IDS Bulletin*, October 1986 and J.J. Struthers, 'Nigerian Oil and Exchange Rates,' *Development and Change*, April 1990.
22 See, David Evans, 'Reverse Dutch Disease and Mineral Exporting Developing Economies,' *IDS Bulletin*, October 1986.
23 For a more detailed discussion of these strategies, including analytical references, see Jahangir Amuzegar (1983).

III

OPEC's Birth, Growing Pains and Maturity

OPEC's history, in a nutshell, is the saga of the oil giants' greed and petulance overwhelmed by the oil producers' frustration and rage. The Organisation of the Petroleum Exporting Countries owes its emergence to a combination of circumstances: the over-weening arrogance, if not clear miscalculation, by some Western oil executives after World War II; the producers' ingenuity, daring and solidarity in challenging the entrenched and powerful oil oligopoly; and, finally, certain historic developments beyond the reach or control of both oil countries and oil companies. This chapter traces the origin and development of OPEC as a commodity association or 'cartel'.[1]

OPEC IN THE MAKING

Shortly after the end of World War II, a group of oil-rich but economically poor developing countries – many of them the old colonies of Western superpowers – began to see their economic salvation and financial prosperity in following Iran's footsteps toward energy autonomy. Up until the Iranian oil nationalisation in 1951, the bulk of the world's proven and recoverable petroleum reserves were either owned by a few major Western oil companies or were in hock to them under long-term proprietary concessions. The so-called Seven Sisters – Exxon (the old Standard Oil of New Jersey), Texaco, Royal Dutch/Shell, Mobil (the old Socony Vacuum), Gulf, Standard Oil of California (Chevron) and British Petroleum (the old Anglo-Iranian Oil Co.) – plus a French state company, Campagnie Française des Petroles (CFP) –

had an almost total control over the exploration, extraction, shipping, refining and marketing of world oil in the non-Communist world. But burgeoning independence movements and decolonisation demands in Asia and Africa presaged looming troubles for the Western-dominated world oil industry.

Since its very inception in the latter half of the nineteenth century, the 'rock oil' industry was never a tranquil enterprise. Repeated cycles of boom and bust in its early development during the 1860s in the US were followed by interminable acts of violence – entrapment, arson, sabotage and armed clashes – throughout the 1870s. Rival attempts for control of world oil production and prices led to global conflicts over the strategic mastery of petroleum as a source of security and prosperity. In no other business in modern times, perhaps, were so many fortunes made and lost so quickly and so dramatically as in oil. The oil business also exemplified market capitalism in its vilest, most brazenly selfish and least socially-concerned caricature. And nowhere was greed as evident as in the oil companies' relations with the oil-rich countries over the shares of the oil bounty accruing to company coffers as profits and to the oil exporting countries as royalties and other levies.[2]

The successive discovery of giant oil fields in the Middle East and elsewhere after World War II, and the resulting global oil glut, created price competition for markets, and shifted attention from production to marketing. Petroleum being a fairly undifferentiated commodity, rivalry among the oil companies focused on price cutting as a means of capturing markets. And the companies, being the members of an informal oligopoly under the 1928 Achnacarry Agreement,[3] could easily make up their lost revenues from downstream operations by reducing royalties paid to the oil countries. As concessionaires, the oil majors had power and authority over production, transportation and downstream operation.

During the 1950s, oil production capacity in the Persian Gulf region was growing faster than rising demand for oil. Thus, the only way cash-starved oil exporting countries could raise their oil revenues was to sell more oil at falling prices. At the time, the foreign oil companies operating in the Middle East had an official price for oil – called the 'posted price' – at which the country's 'take', i.e. taxes and royalties, was calculated. Actual market prices at which oil was sold globally usually showed various discounts from the posted price. Thus, while the countries had their takes on the contractual basis, their realised share was often more.

With the imposition in the United States of oil import quotas in 1958, the largest world market for crude was partly shut to the international oil companies, and the pressure on price discounts intensified. The re-entry of the Soviet Union – by then the world's second largest oil producer after the United States – into the global oil exports market became a new destabilising factor. The USSR's aggressive oil barter deals in Europe (particularly with Italy) at prices (delivered at Black Sea ports) of about half the posted price for Mideast oil were believed at the time to be part of the Cold War manoeuvres by the Soviets to derail the West's oil relations with the Middle East governments. Unwilling to bear the brunt of new price cuts to meet the Soviet challenge in Europe (which was the main market for Mideast crude), the oil majors decided to share the fall in their profits with the oil producers by lowering the posted price. British Petroleum – the successor to the old Anglo-Iranian Oil Company nationalised by the Mossadeq government in 1951 – took the lead. In February 1959, the posted price for Venezuelan oil was cut by as much as 25¢/b. The Persian Gulf price was lowered by 18¢ a barrel (or 10 percent) from $2.04/b to $1.86/b. The new official price in the Persian Gulf was thus 16 percent below the $2.22/b price posted as late as 1948. On 9 August 1960, in a decision that later proved to be a short-sighted blunder, the Standard Oil Company of New Jersey unilaterally cut between 10¢ and 14¢ a barrel (up to about 7 percent reduction) in the posted prices of Mideast crudes, down to $1.78/b. Of this barrel price, only 80¢ was the oil exporters' share.[4] The other companies followed suit albeit with some hesitation.

The competition among the oil majors to capture a bigger share of the market by producing more oil from their concession areas and selling it at a discount was clearly against the national interest of the producing countries. Internal political pressure on the incumbent governments to counter the oil majors' ruinous race was becoming irresistible. The kernel of the idea of an oil producers' association, to combine national economic power and to act in concert to exploit their petroleum resources had emerged as far back as 1949 when Venezuela approached the four major Persian Gulf producers – Iran, Iraq, Kuwait and Saudi Arabia – and suggested an 'exchange of views' and 'closer communications' on oil matters. But in the highly charged and fluid atmosphere of the Iranian oil nationalisation, the suggestion was overtaken by events. During the protracted, complicated and bitter negotiations between Iran and the former Anglo-Iranian Oil Company (AIOC) in 1949–51, however, Venezuela played a crucial part. By giving Tehran significant information on its own 50-50 profit sharing agreement with the oil majors, Caracas showed the Iranians how many more concessions could be gained from the AIOC; how foolish it was for the producers to cut each other's prices through discounts in order to gain larger market shares; and how the oil exporters could profitably improve their lot by banding together. The suggestions were not lost on Iranian negotiators.

Ten years later, in a calmer political atmosphere in the Middle East and elsewhere, when domestic economic development was the top priority among struggling oil exporters, the companies' evident miscalculations played into the producers' hands. Since most oil major exporters depended predominantly on oil revenues for their foreign exchange needs, development finance and even current budgetary outlays, the two successive price cuts for the Persian Gulf crude provided the last straw. Under a Saudi-Venezuelan initiative, the major oil exporters were suddenly jolted into action. And, within a short time, OPEC was born.

In retrospect, the birth of an OPEC-like organisation could easily be envisaged for several reasons. First, in addition to the untenable relationships between the oil companies and host countries, the global trend in energy and oil balance could not be sustained. Oil consumption was rising at 7 percent a year during the 1950s, and each additional percentage growth in gross world product required an extra percentage point in the use of energy. At the same time, the oil producers, bent on raising their national output and standard of living, were pressuring the oil companies for larger offtake, and were ending up receiving the same or a lower income as rising oil supplies depressed oil prices. The obvious rational course of action was to reverse the process.

Furthermore, the Iranian oil nationalisation in 1951 – and the Arab oil embargo of 1956 – had shown the producers that the oil majors could always successfully off-set smaller offtake in one country by larger extraction from another, thus punishing the non-cooperative or the recalcitrant. A clear counter measure was for the countries to close ranks in order to preempt such company stratagems. Still further, a strong trade association like OPEC could ward off the threat of ruinous competition from new or potentially hostile low-cost producers, thus keeping the oil price high enough not only to help members but also the major companies.

Finally, the postwar emergence of smaller independent oil companies (e.g. AMOCO, CONOCO, ARCO and Occidental in the United States), the establishment of state oil companies in oil-importing industrial countries (e.g. ENI in Italy, Petrofina in Belgium and Gelsenburg in Germany), the formation of national oil companies among oil exporters (e.g. NIOC in Iran) and the re-entry of the Soviet Union into the world oil market were all instrumental in challenging the oil majors' hegemony. The independents were ready to offer large royalties and taxes to the producers. Some governments in the oil-less countries bypassed the oil majors through direct purchase from oil exporters. And, the oil majors' strategy of reducing posted prices to ward off the independents, encourage oil consumption, and improve their own profits was diametrically opposed to the producers' interests. The latter could not sit still forever.

THE DAWN OF A NEW CLUB

The ground was thus conveniently prepared for a decisive move on the part of the oil countries. There was both theoretical and practical justification for this move. Theoretically, many oil analysts, in and out of the producing countries, argued that the cycles of good and bad times in the global oil industry were 'natural' and inevitable. This basic instability was likened to a similar up and down process in animal husbandry which agricultural economists called the 'hog cycle'. The essence of this unstable process was that for some commodities it took years for the markets to respond to price signals, and by the time producers, unaware of their competitor's similar endeavours, simultaneously brought forth their supplies, it was always too late: the market would be flooded beyond the original expectations. And a reverse process toward an eventual shortage would get underway almost immediately.

In the case of oil, it took years for new reserves to be found and brought into production in response to a sudden price rise. It also took years for conservation and efficiency measures to reduce consumers' demand for oil. But by the time the new wells came on-stream, and the lifestyle was conveniently altered, supply would almost inevitably overshoot demand. Prices fall, new explorations are discouraged, marginal producers go broke, and the stage is set for prices to rise again.

Up until 1972, the oil majors dealt with these market swings through their own integrated corporate structures and their interlocking connections. By 'owning' oil deposits directly or through long-term concessions, they controlled the rate of exploration and development. Through their gentlemen's agreements, they reduced inter-company competition and administered oil prices. And, thanks to their large size, marketing power and ample finances, they were able to hold prices up during gluts and down in shortages. Many in the oil industry thus believed that a free oil market, left to its own devices, would inevitably swing from boom to bust. OPEC founders were persuaded that the organisation would act as a 'benevolent manager' of oil markets – maintaining the industry's stability on the lines of the old oil company consortia.[5]

This idea was behind the Venezuelan initiative, as far back as in 1949, to have a new producers' club replacing the oil majors. Later, Juan Pablo Pérez Alfonzo, Venezuela's knowledgeable and shrewd Minister of Mines and Hydrocarbons, in coalition with Sheikh Abdullah Tariki, the ardent nationalist Saudi Oil Minister, took the lead in persuading Persian Gulf producers to demand higher royalties or taxes from the companies. As a self-taught oil analyst, and an experienced foe of the international oil majors, Pérez Alfonzo knew that the only way the oil countries could get a better return on their wasting assets was through a system of inter-country pro-rationing and market allocation, modelled after the Texas Railroad Commission in the United States.

Each of these two iconoclast ministers had his own additional reasons for the initiative. Venezuela, with heavy oil and higher per-barrel cost of extraction (about 80¢ /b), had every reason to fear an all-out competition with Middle East producers (with 20¢ /b cost). Output quotas were the only way Venezuela could protect its oil industry against cheap foreign oils, and an easy way for cheap oil suppliers to exact larger shares of their concessionaires' profits. A pro-rationing scheme, sanctioned by oil-producing governments, would deny the companies the luxury of pitting one country against another by deciding where to produce more oil. Saudi Arabia, anxious to increase its principal source of income – and having toyed with the idea of oil nationalisation and the expansion of downstream operations – came to the conclusion that the only way to stand up to the companies' arbitrary decisions was control over prices and production. This decision was particularly prompted by the fact that companies were producing more oil in Kuwait with a very small population (and need) than in Saudi Arabia, plagued with more poverty and deprivation.

In an Arab Petroleum Congress, which had already been scheduled to open in Cairo in April 1959, Pérez Alfonzo and Tariki succeeded in convincing the delegates from Iran, Kuwait and Iraq (in a secret meeting unrelated to the conference's agenda) to 'recommend' to their respective governments certain measures designed to improve their bargaining position *vis-à-vis* the oil majors. The recommendation in this secret Gentlemen's Agreement included: the establishment of an Oil Consultation Commission; a common stand in the defence of the oil price structure; a change in the 50-50 profit sharing principle in the countries' favour; and a concerted move toward integration of production, refining, marketing and downstream operation in order to secure reliable markets and stable prices. While no concrete result followed the Cairo Congress, the seeds for a collective action by the oil exporting countries against the dominant oil giants were sown.

In volumes of briefs and thousands of speeches in defence of their economic independence, political sovereignty and nationalistic aspirations, OPEC leaders accused 'foreign companies of both bad faith and foul play'. The concessionaires were charged with exploiting the host country's irreplaceable heritage, stifling their development efforts, acting in a self-centred, arrogant, and condescending manner, interfering in local politics, corrupting public officials, and, altogether, putting the interests of colonial power before those of host governments. Right or wrong, they felt politically maltreated, financially cheated and even contractually short-changed.

The dormant resentment against the Western oil companies, awakened by Nasser's Arab nationalism and Mossadeq's heroic nationalisation of the Anglo-Iranian Oil Company, thus came to a boiling point after the second oil price cut in 1960. Pérez Alfonzo and Tariki moved to bring the signatories of the Cairo agreement together. Iraq took the initiative, and invited all to Baghdad. Between September 10 and 14, 1960, the delegates from Iran, Iraq, Kuwait, Saudi Arabia and Venezuela met in the Iraqi capital to lay the foundation of the Organisation of Petroleum Exporting Countries. The primary objective of the organisation was to defend the price of oil, and restore it to pre-crisis levels. The action was presented to the world as a *mouvement de resistance*, or a defence league, against the oil majors' arbitrary decisions. OPEC signatories pledged their solidarity in case the companies resorted to sanctions against any one of them; they demanded bilateral consultation on oil price decisions; and they called for a system of production regulation to ensure oil price stability.

According to an OPEC publication, OPEC was born of protest against the high-handedness of the major international oil companies – the Seven Sisters – which had created 'states within states' in the oil producing countries, controlling the amount of oil extracted, how much and to whom it was sold and at what price. On these vitally significant matters, the countries were never consulted. While the companies built huge financial empires at the expense of producers, the latter received a meagre royalty.

OPEC's principal objectives in the document registered at the UN Secretariat in 1962 were declared to be (1) to coordinate and unify the oil policies of member countries (2) to determine the best means of safeguarding their individual and collective interests (3) to seek ways and means of ensuring the stabilisation of prices in international oil markets with a view to eliminating harmful and unnecessary fluctuations (4) to provide an efficient, economic and regular supply of petroleum to consuming nations and (5) to obtain a fair return on capital to those investing in the oil industry.[6] In a word, OPEC was founded essentially 'to defend the interests of its single-product and economically weak member countries in a world dominated by the established, economically advanced Western consumer societies.[7]

Managing world oil trade thus became a high-wire act for the Seven Sisters. At the global level, they had to contend with the Soviet Union and the independent oil concessionaires. In their relations with host countries, they had to deal with national politico-military ambitions, intra-regional rivalries for leadership and power, political pressures from their own governments siding with one or another contestant, and their own long-term financial interests as commercial enterprises. Among themselves, they had their own competing and divergent interests: as partners in different countries – Exxon, Mobil, Chevron and Texaco in Saudi Arabia; the four majors, plus Gulf, Shell and British Petroleum in Kuwait; the Seven, plus the French CFP, in Iraq; and the Eight, plus some independents, in post-1954 in Iran – they had to negotiate daily offtake for each company among themselves and at all times. The task became particularly complex as OPEC membership grew eventually to include all major developing country exporters. Qatar joined in 1961, Indonesia and Libya in 1962, the United Arab Emirates in 1967, Algeria in 1969, Nigeria in 1971, Ecuador in 1973, and Gabon later (see Table 3).

OPEC'S PRICE TAKEOVER

In the first ten years of its life, OPEC was successful only in keeping the posted price from being further cut, while the oil majors shifted their strategy of global competition to other and more subtle forms. Between 1960 and 1971, the concessionaires refused to deal with OPEC as an intergovernmental entity, and OPEC members also continued to deal with the oil companies indirectly and separately, refusing to cede their 'sovereignty' to the organisation and let it bargain on their behalf. Meanwhile, competition among oil producers to obtain larger revenues (at prices over which they had no control) kept the market oversupplied.[8] Within OPEC itself, attempts to regulate output faced stiff resistance by the Saudis who had a dominant position in the market.

During this period, the oil majors' main concern was with the Russian challenge, local US oil politics and the oil glut inundating the globe from newcomers and new finds. Particularly troublesome was Libya with its sudden and spectacular surge as a major producer of light crude, close to the European markets, and partly in the hands of independent companies which had no stake or obligation in the Middle East.

Once again, world oil developments turned in OPEC's favour. Soon after the conversion of the American economy from a war footing to a civilian course, the demand for gasoline surged beyond all expectations, rising by over 40 percent in the five years between 1945 and 1950. Within this brief period, the United States was transformed from a net exporter to a net importer of oil. War-ravaged Europe, being helped by the Marshall Plan, also could not be reconstructed without new oil from the Middle East. Thus, the demand for petroleum rose by nearly 7 percent a year during the last half of the 1960s, and by more than 9 percent in the first six months of 1970. A series of extraneous circumstances – the closure of the Suez Canal in 1967, damage to the oil pipeline in Syria in 1970 and Libya's production cutbacks shortly after – helped tighten the supply-demand balance. The rising new demands tipped the scale in the producers' favour, and created a sellers' market never before experienced or expected.

The impetus for the decisive action began with OPEC's Declaratory Statement of Petroleum Policy issued at Caracas in 1968.[9] The statement's preamble emphasised the inalienable right of oil producers to exercise permanent sovereignty over their natural resources. It also called on government participation in the oil industry to secure greater control over foreign concessionaires and to exact larger income from them (e.g. by using the 'posted price' as the basis for assessing the oil companies' income). The next step was taken again at Caracas in 1970 by adopting certain specific declaratory objectives, including the 55 percent formula as the minimum tax rate on the oil companies' income. Also, in accordance with this resolution a new five-year price-fixing agreement was signed in Tehran on 14 February 1971 between 23 multi-national oil companies and 6 Persian Gulf oil producers (Abu Dhabi, Iran, Iraq, Kuwait, Qatar, and Saudi Arabia). The 1971 Tehran agreement abrogated the 20-years-old 50-50 profit sharing formula. The new accord included: a flat across-the-board increase of 33¢/b in the posted prices; a flat increase of 5¢ annually in lieu of claims relating to adjustment in product prices; an increase of 2.5 percent annually for inflation; full expensing of royalties; elimination of the marketing allowances; and, above all, an increase in the tax rate to 55 percent. All posted prices in the Persian Gulf, based on quality differentials, were also to be realigned, using the newly adjusted price of Arabian Light – $2.18/b – as a reference.[10]

With a significant triumph in hand under the 1971 Tehran Agreement – abolishing the tradition of unilateral price-fixing by the oil companies, and establishing the principle of price determination through bilateral negotiation – OPEC took the upper hand in designing a new international oil order. By mid-1973, OPEC's membership had extended to virtually all major developing petroleum exporters. The group was thus in a position to turn the table against the oil majors. *The Economist*, in a nostalgic lament, later recalled that while 'a few years before, the companies largely produced what quantities of oil they wanted, from where they wanted, at a price largely set themselves', the process has been turned upside down.[11] By their own time table, the oil producers were moving steadily to assert national sovereignty over their resources.

A number of successive 'victories' followed the 1971 Tehran Agreement. In January 1972, a new concession was obtained from the oil companies for the adjustment of the producers' royalties and taxes (received in US dollars) against dollar depreciation *vis-à-vis* nine major currencies. In March of the same year, the principle of 'participation' by producer governments in the operation of the companies – long resisted by the latter – was accepted for the Persian Gulf countries. The posted price was thus raised to $2.90/b in mid-1973. Following the outbreak of the Yom Kippur war on 6 October 1973, and the imposition of an oil embargo against the West by OPEC's Arab members shortly after, the absence of spare capacity anywhere else produced an extremely tight oil balance and caused an energy crisis. On 16 October 1973, buoyed by the impact of the 'oil weapon' – and discouraged by months of long and fruitless negotiations with the oil companies over prices and terms of concessions – the OPEC Ministerial Committee of the Persian Gulf members, assembled in Kuwait, announced an immediate increase in the posted price to $5.12/b – close to the spot market prices. With this historic move, OPEC assumed for the first time in its history the power to consider and set oil prices unilaterally. The unprecedented price hike was defended on the grounds that consumers were willing to pay for it and that some 66 percent of the total retail oil prices went to the consuming country governments while the share of oil countries, whose 'natural heritage' was being depleted, was less than 10 percent.[12]

The voracious global appetite for oil, due to low prices, coupled with the removal of US oil import quotas by the Nixon administration in April 1973, pushed the demand for OPEC oil toward its highest level, and a 'near panic' pushed oil prices 'sky high'. The 20-year oil glut was finally at its end. And the threat of increased oil supply from non-OPEC sources to restrain the oil exporters' 'intransigence' was no longer taken seriously, as the United States and other non-OPEC producers operated at near 100 percent capacity. OPEC, and particularly the Persian Gulf producers, were firmly in the driver's seat. Thus, in an intriguing chronological order, the power to set oil prices began with unilateral decisions by the oil majors from the 1920s on; served as OPEC's *raison d'être* in 1960; became subject to veto by the oil exporters in 1970/1; and was shared by countries and companies between February 1971 and October 1973. By the winter of 1973/4, this power became the sole prerogative of OPEC members. And, the circle was finally closed.[13]

OPEC's power to set oil prices after 1973 resulted in a rapid restructuring of the global petroleum industry. While at the beginning of the 1970s, the seven multi-national majors owned or controlled about two thirds of the world's oil reserves outside the Communist bloc, by the end of the decade, widespread oil nationalisations within OPEC members reduced this control to about one fifth. Iran's hawkish position on the oil price in 1971 emboldened nationalist groups in other OPEC countries to seek further 'independence' from the oil companies. They proceeded first by asking for 'participation' (i.e. partial ownership of oil resources) along with the companies, and later by making full expropriation. Iran itself had nationalised its oil industry in 1951, and the Shah was proudly referring to this fact at every opportune occasion. The revolutionaries who overthrew King Faisal II of Iraq in 1958 had also demanded extensive revisions in the concession of the Iraq Petroleum Company (IPC) and subsequently completed the company's nationalisation by 1972. Algeria and Libya took similar actions against the concessionaire. Kuwait, Venezuela and Saudi Arabia followed suit, so that by 1975 there was no concessional arrangements between producers and companies, and the oil exporters gained full control over their oil resources for the first time since the beginning of the century. From then on, the companies were no longer the 'owners' of the oil in the ground under their concession agreements but merely 'contractors' of the new national oil companies, offering the latter technical services, marketing their production and otherwise acting as their agents.

OPEC's stunning success, after ten long years of helplessness and inactivity, was due to a multitude of forces long in the making: a fundamental change in the world's basic balance of power giving greater independence and influence to the Third World; growing dependence of the industrial world on oil for both military strength and daily lifestyle; a strong solidarity among OPEC members despite differences in ideology and politico-economic philosophy; conflicting interests among major oil-importing countries in spite of their similar political philosophies and economic orientation; an increasing supply of indigenous talents in the oil producing countries who could fully match their Western counterparts in negotiations; and, finally, the declining political influence of the oil majors over both their own governments and those of the producing countries.[14]

As events later turned out, OPEC's victory was, however, short-lived, and was followed by a number of largely self-inflicted setbacks.

CYCLES OF BOOM AND BUST

As in any commodity market where there is an excess of demand over supply, the price of crude oil reacted strongly to the worldwide commodity boom of 1972–3. With oil inventories generally exhausted, the initial decision by the Middle East producers to raise the oil price in October 1973 became self-enforcing. This decision, supported by an Arab oil embargo on 11 October 1973 against countries friendly to Israel, heralded the dawn of the new era. The 'oil weapon', had been bluntly tried twice in 1956 and 1967 – and subsequently resisted by some OPEC members – was now brought into play by the Arab group effectively. The supply cutbacks, although rather minimal, had a disproportionate effect on prices in the prevailing climate of uncertainty about future OPEC moves, amidst panic buying by desperate independent refiners, and fear of the unknown. In November 1973, Iran held an oil auction in the spot market where a small lot was sold for $17.04/b, and rumours later circulated that the buyer had turned around and resold the shipment at $22/b.

Amid subsequent feverish bids for crude oil in the spot market and total disarray in the official 'contract' prices, the oil ministers of six Persian Gulf members met in Tehran in 22–23 December 1973 to discuss oil prices. Delegates from Algeria, Indonesia, Libya, Nigeria and Venezuela attended as observers. Recognising that recent huge auction prices for oil spot delivery were not representative of the long-term supply-demand balance, Iran took the lead in introducing a 'new concept' for oil price determination: the cost of alternative sources of energy (e.g. gas from coal and shale oil). Based on a report prepared by OPEC's own Economic Commission Board, taking into account recent direct spot sales in the market by Iran and Nigeria, and mindful of the rising rate of inflation in industrial countries, the real cost of an alternative energy barrel was estimated at $14–17.

The OPEC ministerial committee convening in Tehran, however, decided to discount this figure, and to set the price at a level which would set the government's share of oil exports at $7/b for Arabian Light 34 API (the marker crude). In line with a previous arrangement under the Geneva II Agreement, the posted price of the marker crude was put at $11.651. The price of a notional 'OPEC basket' (composed of various qualities of crude with different gravity and sulphur content) was set at $10.84. This price was to be effective as of 1 January 1974. The first oil boom thus followed. Curiously enough, while all members were expected to benefit from this spectacular price rise, not everyone (especially Saudi Arabia) was happy about the scale of the increase or Iran's daring role in engineering the price coup.[15]

The 'government take' was further raised, through adjustments in tax and royalty rates. In December 1976, OPEC ministers decided to increase the per barrel price of marker crude to $12.70 as of 1 January 1977, and to $13.30 as of 1 July 1977. Saudi Arabia and the UAE decided to raise their prices by only 5 percent instead of 10 percent (i.e. to $12.09). The two-tier price system was terminated in mid-1977 when the two dissenters agreed to go along with the rest, and the eleven-member majority decided to forego the additional 5 percent price rise previously announced. With these further small upward adjustments, the price of the 'OPEC basket' became $11.46/b in 1975 and $12.70/b at the end of 1977. In December 1978, the oil price was further 'corrected' upward by 10 percent, to be spread over four quarters in 1979, reaching $13.54 by the end of the year.

Despite these seemingly incessant upward revisions, the real (inflation-adjusted) price of oil on the eve of the Iranian revolution in 1979, however, was 10 percent below that of 1974. And while oil-importing countries, particularly the poorer developing economies, were chafing under high oil prices and blaming OPEC for the outcome, OPEC members were pointing to their still 'meagre' share of their depletable fortune. References were frequently made by oil exporters to the fact that in 1975, still about 45 percent of the 'pump price' of gasoline and other oil products in Western Europe belonged to the national treasuries as a consumption tax, while the oil producing governments' take was only 35 percent.

Meanwhile, the six-fold increase in the price of the benchmark crude, from $1.80 in 1970 to $11.65 in December 1973, exerted profound, far-reaching and, as it turned out, permanent effects on the world oil balance. The prevailing relationship between the oil-exporting countries, the major oil companies and the oil-importing world was forever altered. As a result of the massive changes caused by the new terms-of-trade between petroleum and other goods and services, as well as significant dislocations in Western production processes and lifestyles, a deliberate deflationary policy was adopted by the industrial countries to cope with their balance-of-payments deficits and to compensate for the 'OPEC tax'. As a result, the demand for OPEC oil, which had steadily risen from 20.2 mb/d in 1970 to 27.5 mb/d in 1973, changed course, and declined first to 27.2 mb/d in 1974 and then precipitously to 24 mb/d in 1975. Despite higher oil prices, too, OPEC's foreign exchange earnings fell to less than $108 billion in 1975 from more than $120 billion in 1974. This mini-recession in the oil-exporters' economies was only reversed by a subsequent tightening of the oil market (due to world economic recovery) and, ultimately, by the Iranian revolution and the Iran/Iraq war.

The Iranian Revolution of February 1979 reduced Iran's crude exports to a trickle. Revolutionary turmoil, strikes in the oil fields and other disturbances caused oil output to decline from a peak of 6.1 mb/d in September 1978 to virtually nothing in December and to an only partial recovery by March 1979. Overtaken by Iran's upheaval and sharply reduced oil exports, the previously scheduled OPEC price was amended in March 1979, and set at $14.55, with members allowed to add their own 'surcharge' as the market warranted. Meanwhile, global oil markets fell into total disarray. Despite Saudi Arabia's 2 mb/d extra output to compensate for the Iranian shortfall, spot crude prices escalated to upward of $40/b. Panic buying was largely to blame. But widespread anxiety about the outcome of the Iranian turmoil and its possible impact on other Persian Gulf producers had widened the range of uncertainty faced by consumers and producers alike. In June 1979, the basic (marker) price was raised to $18/b. This base price (plus a surcharge of about $3.80 and allowance for quality differentials) was fixed at a maximum of $23.50. By the fifty-fifth OPEC conference in Caracas in December 1979, however, the oil price structure was in total chaos. There were no uniform posted or official prices. Saudi Arabia kept its own price at the $18/b base, while OPEC hawks (Iran, Algeria and Libya) fixed theirs at $28/b, and spot prices ranged from $40 to $50/b on small lots. Iran, in particular, followed a new ideological and revolutionary line in favour of sharp cutbacks in production. The benchmark price was subsequently raised to $24/b while the spot prices still hovered around $30/b. In June 1980, the base price was again raised to $32/b plus an additional $5/b maximum for differentials.

The outbreak of the Iran/Iraq war on 22 September 1980, and a sudden loss of almost 4 mb/d (15 percent of OPEC output and 8 percent of non-Communist world consumption), pushed up spot prices once again beyond $41/b. In December 1980, the official price of the marker crude was set at $32/b, with a maximum of $36/b for the same quality oil. The maximum price of any OPEC crude was fixed at $42/b. Saudi Arabia chose to keep its benchmark crude at $32/b while other members decided to charge $36/b for the same quality oil. However, under domestic political pressures – charges that the Saudi government was dissipating the national wealth and encouraging undesirable rent-seeking activities – Riyadh finally relented, and all members agreed in October 1981 to adopt a unified benchmark price of $34/b.

The second oil boom of 1980–1 marked the zenith of OPEC's glory, power and wealth – with the organisation becoming a dominant force in determining global oil output, having principal price-setting responsibility, and enjoying unprecedented foreign exchange earnings. From then on, however, up until the Persian Gulf war of 1990, OPEC faced a downward trend in both its fame and its fortune, with a sharp and unexpected oil bust in 1986. The second oil boom resulting from the new oil price explosion of 1979–80 produced another global downturn shortly after. As a result of extensive conservation measures adopted by the West after the earlier oil shocks, and weakened demand for oil, OPEC was more than ever becoming a residual producer. In fact, supplies from non-OPEC sources (Alaska, Mexico, Britain and Norway in particular) overtook OPEC's output by 1 mb/d in 1982, and pushed down the demand for OPEC oil to the lowest since 1970.

For the first time since its 1973 triumph, OPEC faced a distressful dilemma. If members were allowed to produce at will in an increasingly competitive world oil market, prices would undoubtedly fall for all, and export revenues, as well as political power and economic influence, would decline. But, if the official price were to be maintained, total OPEC output had to be curtailed (i.e. regulated by cartel-like country allocations). Reduced production, however, meant a loss of OPEC's market share and its 'price-maker' status. Neither option was easy or desirable. One way out of this dilemma was to go for the jugular: engineer a worldwide oil glut, push down prices temporarily, drive high-cost producers off the market and regain a near monopoly position – something akin to what early American oil barons had done in the late nineteenth century. But this strategy was fraught with immense risks since there was every indication – extensively discussed in the Western press and widely suggested by lawmakers – that any such move on OPEC's part was going to be countered by higher import taxes on crude oil and additional excise levies on gasoline in the consuming countries.

After painful soul-searching and wrangling among the major players (and after nearly 22 years of denying that it was a 'cartel,' because it had no quotas or market sharing arrangements) OPEC finally succumbed to the inevitable: if the group wished to sustain cartel prices, it had to behave like a cartel – a procedure that

OPEC's two founding fathers, Alfonzo and Tariki, had in mind from the beginning. Their ideal model was an international authority similar to the US Texas Railroad Commission, to control production in order to maintain price. Meanwhile, with the introduction of oil futures contracts by the New York Mercantile Exchange in 1983, the long-standing system of 'administered prices' for petroleum had come to an end. Oil prices, like those of other commodities, were subjected to simultaneous supply and demand forces in the open market. Thus, in a nearly full circle, the auction pricing of the late nineteenth century in West Pennsylvania oil exchanges was restored by floor traders in the new York and (later) London commodity exchanges. And the saga of the oil price determination took a new turn after a nearly hundred-years interval, during which crude prices were fixed, first, by John D. Rockefeller's Standard Oil Trust, then, by the Texas Railroad Commission (in the United States) and the Seven Sisters (in the rest of the world), later, by negotiations between oil exporters and the oil majors, and, finally, by OPEC alone. After 1983, risk-averse refiners, hedging producers, and paper-barrel speculators took over the oil market in an open, impersonal process. OPEC's oil power from then on resided only in the control of its total supply.

The worldwide slump after 1981–2 turned out to be the harbinger of a tumultuous decade for OPEC and oil. By the end of the short recession, the global demand for oil plummeted to 53.5 mb/d in 1982 from 62.9 mb/d in 1979, i.e. 15 percent. OPEC's output, however, fell to about 18.7 mb/d, down from 30.5 mb/d – a whopping drop of about 40 percent.[16] Some 28 percent of this drastic fall was generally believed to be due to the recession itself, nearly 25 percent to a decrease in commercial energy consumption through conservation and use of oil substitutes, some 17 percent to accelerated drawdown on oil inventories, and the remaining 30 percent to new supplies from non-OPEC sources (e.g. the United Kingdom, Norway, Mexico, the USSR and others). The handwriting thus clearly appeared on the wall. Oil price unity among members – the hallmark of OPEC's success and the backbone of its survival since 1971 – was fast disintegrating. The enforcement of a production/price discipline seemed even more difficult than ever because policy differences among members had become wider and more open.

In addition to clandestine discounts and rebates offered by members to meet the competition, a direct challenge to OPEC's authority and solidarity emerged on 17 February 1983, when the British National Oil Company cut the price on North Sea oil by $3/b to $30.50/b and Norway followed suit immediately. Nigeria, whose light oil was competitive with the North Sea's, saw no choice but to meet the rival's challenge by lowering its 'bonny light' price by $5.5/b to $30/b – or $4 less than OPEC's official base price of $34/b. The USSR also reportedly reduced the price of its Urals oil (of similar quality to Saudi Arabia's) to as low as $27.50/b.

Despite a substantial share loss of the world crude oil supply since the 1973 peak, OPEC members were still unable in 1982 to agree on either a production or a price strategy that could best serve their interests. Algeria, Iran, Iraq, Libya and Nigeria accused Saudi Arabia and other wealthy members of the Gulf Cooperation Council of creating a 'harmful glut' in the oil market at the expense of poorer partners. The Saudis and their GCC allies, in turn, accused their detractors of 'irresponsible' price-cutting and 'misguided' behaviour designed to regain market share.[17] By February 1983, the oil ministers had tried and failed six times within 13 months to coordinate their production and price strategies. The total output ceiling and individual country quotas proposed in earlier ministerial meetings to support the official price had not been accepted. The output target was also torpedoed by a mild winter in the West and by accelerated destocking on the part of the oil majors. Consumers' reaction to escalating oil prices also undermined OPEC's efforts.

In a long and very gruelling series of negotiations in London during 12 consecutive days in early March 1983, OPEC ministers finally resolved their differences, and in that process produced another first. For the first time in OPEC's twenty-three-year history, the price of the benchmark crude was *lowered* by 15 percent, from $34/b to $29/b. An average output ceiling of 17.5 mb/d was established for the group as a whole for the remainder of the year. Individual quotas were allocated to all members except Saudi Arabia who, with a notional allocation of 5 mb/d, was expected to act as a 'swing' producer to supply the balancing quantities to meet market requirements in order to defend the official price. Members were enjoined from giving discounts in any form. Hopes were raised that the new price/output package would restore stability to the market, particularly because some non-OPEC members (e.g. Mexico) were expected to follow OPEC's price lead.

Despite the group's success in averting a price collapse, and despite the collective confidence in the workability of the London Accord, the outcome was far from clear. The new agreement – even if adhered to by all members – gave no assurance that the lower official price could hold. And it did not. Non-OPEC production continued to rise. Oil itself continued to lose market to cheaper alternative sources of energy (particularly coal, natural gas and nuclear power). The quantity of oil needed to produce a unit of output continued to decline worldwide as a result of the adoption of further conservation and efficiency measures. Producing more than the allotted quotas, under-the-table discounts, oil-for-goods barters, and other 'forbidden' practices remained unabated. As it finally turned out, the expected world economic recovery after 1983 was mild, and the demand for OPEC oil continued to decline. Meanwhile the bulk of OPEC exports was sold as much as $2/b below the official price. And, from a peak of some $286 billion in 1980, the value of OPEC's petroleum exports dwindled to less than $153 billion in 1984. The brunt of this enormous shortfall was borne by Saudi Arabia.

In the ministerial conference of December 1985, under Saudi pressure, the four-year effort to prop up the oil price by restricting OPEC's total output was abandoned in favour of a new arrangement. Instead of defending a given price for oil through production cuts (resulting in the loss of market to non-OPEC producers, and loss of earnings due to declining prices) OPEC decided to secure and defend a 'fair share' of the global oil export market consistent with 'necessary incomes' for the member countries' development. OPEC's share in world crude oil exports had fallen to 10.5 mb/d in 1985 (51 percent of global petroleum exports) from the peak of 26.5 mb/d in 1979 (79 percent of the world total). This new audacious strategy meant (1) the end of the Saudis' self-imposed role as a swing producer (2) the end of OPEC as a residual supplier (3) the end of members' daily output quotas and (4) a *de facto* end of unified pricing.

The market-share stratagem, replacing the price-support objective, caused a devastating oil bust. At first, OPEC's crude output rose to more than 18.mb/d in early 1986 (from 14.9 mb/d average in 1985) against a presumed 'fair share' of 17 mb/d–17.5 mb/d. The Saudis had the largest share increase. But, since non-OPEC producers (particularly Britain, Norway and the USSR) ignored OPEC's repeated pleas for 'cooperation,' a glut of some 2–3 mb/d developed in the market, and crude prices crashed to their lowest in seven years. The average OPEC oil price skidded to less than $10/b in mid-1986 – with some small cargoes sold as low as $6/b. At this price, there was no profit in oil production for some members. Thus, the policy which was devised to drive marginal, high-cost, non-OPEC producers off the market turned against OPEC itself. By August 1986, under a severe financial squeeze caused by the oil price free fall, all members (including Saudi Arabia) were eager to seek a way out. Accepting a solution originally championed by Iran and its hawkish allies, the OPEC majority was finally convinced that sustainable oil prices could be supported only by considerable production cuts (particularly by high-volume, rich producers).

Thus, in the December 1986 conference, it was decided to cut OPEC production 'temporarily' by 7.6 percent to about 15.8 mb/d – the lowest level in the organisation's history – in order to move prices up to 'reasonable levels'. Quotas again were set for each member. Concurrently, it was decided to return as promptly as possible to a system of 'fixed pricing'. A level of $18/b for a basket of six OPEC and one non-OPEC crudes was chosen as a 'target' or 'reference' price. OPEC's production for the entire year 1986 ended up in averaging 17.6 mb/d. The total value of OPEC's petroleum exports fell from more than $131 billion in 1985 to less than $79 billion in 1986 (the lowest in 13 years since 1973, and only about one fourth of the 1980 peak). With the value of the US dollar also lower in terms of other major currencies, the blow to OPEC members' finances was unbearably sharp. For all practical purposes, OPEC became a residual producer once again after December 1986, with daily crude oil prices determined basically in London and New York. OPEC's power was limited to keeping members from exceeding their assigned quotas. However, this power was also somewhat asymmetrical: it was much more effective when the oil market was firm than when the demand for OPEC oil was depressed.

By the subsequent conference in June 1987, oil spot prices and the netback values of all OPEC crudes had firmed up significantly, with the benchmark crude selling at over $19/b. But as OPEC's total production began to rise and reached about 19 mb/d by the end of 1987 – about 2.4 mb/d above the new self-imposed ceiling of 16.6 mb/d – prices gradually began to tumble toward $15/b. After the ceasefire between Iran and Iraq in August 1988, oil prices began to drift down toward $12/b, mostly because of over-production by the Arab group (especially Kuwait and the UAE).

Due to excessive production by both OPEC and non-OPEC producers throughout 1988, prices remained weak. In order to legitimise over-production by members, OPEC's total ceiling was repeatedly raised, reaching 19.5 mb/d by June 1989 and 22 mb/d by March 1990, but still trailing behind members' actual total. The target price of $18/b was also renamed the 'reference' price in mid-1989. By March 1990, OPEC output had reached 24 mb/d – the highest daily average in eight years. The old combatants, Iran and Iraq, now joined forces to demand a cut in the 22.1 mb/d total OPEC ceiling and a rise in the 'target price'. This strange alliance, followed by Baghdad's ill-fated adventure into Kuwait, produced the third short-lived mini-boom.

With the ceasefire between Iran and Iraq in August 1988 and a gradual firming of world demand for OPEC oil, prices headed upward again, and averaged around $17.5/b. Under Iraqi pressure, the reference price was raised to $21/b in July 1990 from $18/b, and the output ceiling was set at 22.5 mb/d. Iraq also threatened to enforce adherence to individual quotas by Kuwait and the UAE, if necessary by force. As these countries did not take Baghdad's threat seriously, Iraq invaded Kuwait in August 1990, and the subsequent United Nations embargo on Iraqi oil shipments pushed the oil spot price to nearly $40/b as some 4 mb/d of OPEC oil was abruptly cut off from global supply. To help ease the oil shortage, OPEC ministers, late in August, temporarily suspended all output quotas, and authorised key producers to make up the Iraq/Kuwait shortfall any way they could. Prices subsequently receded to $17.5/b by March 1991, after the end of the Persian Gulf war. In the meantime, the value of OPEC's total exports gradually rose from some $88.4 billion in 1988 to more than $150 billion in 1990, reflecting a third mini-boom. However, while the total revenue for the year was somewhat larger than that of 1977, the overall situation had actually deteriorated drastically in the interval. Per capita oil income within OPEC had fallen from $46 in 1977 to less than $33 in 1990, and the weighted value of the US dollar *vis-à-vis* the major world currencies had shrunk by 14 percent.

In order to raise the market price toward the $21/b 'reference' goal, the total OPEC ceiling was cut to 22.3 mb/d in March 1991 and kept at that level until September when it was again raised to 23.65 mb/d in order to accommodate Soviet supply disruptions. The quota system that had been abandoned in late August 1990 remained suspended. While Iraq and Kuwait ceased exports, and other members (except for Saudi Arabia) were all producing at full capacity, allocation posed no problem. With Iran's production capacity substantially raised after the ceasefire with Iraq, and Kuwait entering the market once again, actual OPEC output reached 24.5 mb/d by mid-1992 – about 1 mb/d above the cap. The ceiling (now called 'market share') was thus raised to 24.2 mb/d. Actual output was 25.3 mb/d, and OPEC's price hovered around $18/b. Once more, the authorised production was raised to 24.5 mb/d in September 1993 – this time with specific allocations for individual countries. In view of the continuing pressure on prices – partly caused by continuing increase in non-OPEC production – OPEC officials decided to maintain the 24.52 mb/d ceiling for the whole of 1994. The price had already fallen to $13.75/b by the end of the year.

PERILS OF MISCALCULATED RISKS

Looking back at the 1974–94 period, with its three oil booms, a major oil bust, and several years in between punctuated by oil price and income volatility, one is tempted to ask why did OPEC members fail to forge a unified OPEC position *vis-à-vis* their customers and competitors? One plausible answer may be that the differences in national interest and domestic absorptive capacity (discussed in Chapter I) were largely responsible for hampering these efforts. And a perpetual clash between price hawks and doves, based mainly on nationalistic considerations, sapped the organisation's solidarity and vigour. Yet it is hard to ignore the thought that OPEC's short-sightedness, and absence of a clear realistic vision, caused it to duplicate the same mistakes in judgment that the oil majors made in the 1950s and 1960s.

OPEC's triumph over the oil majors in acquiring the power to influence prices instantly put an end to the managed oil market where crude supplies, prices and market shares were all, up to then, tightly regulated and administered by the Seven Sisters oligopoly. Once the world oil market was deregulated, crude prices became subject to the interplay of supply and demand. During the 1974–81 period when OPEC was in the driver's seat, oil supply prices were officially set by the Organisation with scant regard to long-term consequences (including reactions from major consumer countries). As a result, it was world oil demand that determined OPEC's total sales. Furthermore, exploration for new oil deposits in non-OPEC areas, the development of alternative sources of energy and the search for new recovery techniques stimulated by higher oil prices were all anti-OPEC developments which were beyond OPEC's control. During the early 1980s, fresh attempts were made by the major players within the Organisation to regulate and ration OPEC's oil supply in order to support the official price. But, by then, the members had lost control of supply as well as prices. Eventually, the many new entrants on the scene, the rise of the futures oil market in London and New York and the preponderance of speculative daily transactions reduced OPEC's price-setting and pro-rationing power, and made the Organisation merely one among many other players. In a word, the oil market became another full-fledged commodity market for the first time ever.

While OPEC's charter called for the maintenance of 'order and stability' in the oil market to ensure 'fair and reasonable prices', subsequent events showed that at no time since the discovery of petroleum in commercial quantities in 1857 were oil prices as volatile as they were in the last quarter of the 20th century while OPEC was presumably in charge. The relatively calm oil market after the end of World War II, mostly due to the oil majors' tacit agreements, suddenly became a nervous, unstable and sensitive market after 1973, in the hands of panic-prone oil traders and over-active 'paper barrel' speculators. Crude prices rose more than fifteen times within seven years, 1973–80, and fell by nearly 80 percent in the subsequent five years, 1981–6. Prices again soared by more than 400 percent between 1987 and 1990; and finally came down by about 50 percent between 1991 and 1994.

These extraordinary market fluctuations had a dramatic impact on OPEC's share of the global oil supply, and the members' oil export earnings. As previously indicated, OPEC's share of the world oil output fell from more than 56 percent in 1973 to about 28 percent in 1985 before rising gradually to 41 percent in 1994. OPEC's export earnings correspondingly soared from $37 billion in 1973 to a peak of nearly $286 billion in 1980; they fell to a record low of less than $79 billion in 1986 before rising again to more than $150 billion in 1990 and finally settling back at $123 billion in 1994.

Responsible for these dramatic changes were, in part, predictable reactions by the major oil consuming countries, which were blindly unforeseen or foolishly underestimated by OPEC's major decision-makers. Part of the blame also had to be attributed to OPEC's own organisational and decision-making weaknesses. The two oil price explosions in 1973 and 1979 brought about four reactions in the major oil consuming countries (1) an increase in non-OPEC oil output previously considered uneconomical (2) rapid shifts to existing alternative energy sources such as coal, gas and nuclear power (3) a search for exotic and renewable substitutes such as solar, thermal, wind and other non-hydrocarbons and (4) a raft of effective energy conservation policies aimed at more efficient use of oil (particularly in vehicular transportation). Soaring oil prices in the mid 1970s and the early 1980s also led to a series of highly successful, defensive monetary and fiscal measures within the OECD countries to deal with their worsening terms of trade.

The surge in crude oil prices in December 1973 to over $11/b resulted in a transfer of an extra $83 billion from the rest of the world to OPEC members. A series of immediate measures were taken by the major consuming countries to deal with the sudden change. Anti-inflationary policies in the major industrial countries – particularly in the United States and Germany – led to the recession of 1974–5 and a fall in the demand for oil. As a consequence, oil prices between 1974 and 1978 rose less than global inflation. Reduced demand for oil adversely affected GDP growth rates in many OPEC member countries. At the same time, inadequate absorptive capacity in many group members in the early years of the first oil boom caused part of the oil windfalls to be left in deposits with multinational commercial banks and international financial institutions (e.g. the IMF and the World Bank) to be recycled to poorer oil importers. Recycling through official and private channels resulted in heavy external debt among developing countries, followed by a global recession – and further drops in oil demand and lower real oil prices.

The second oil price explosion, following Iran's revolution in 1979 and the Iran/Iraq war in 1980, resulted in the transfer of an additional $150 billion from net oil importing to net oil exporting countries. Higher returns on OPEC's petrodollar deposits abroad also added to the members' coffers. This second oil shock triggered another series of strong contractionary policies in the developed countries, and led to an unexpectedly sharp global recession in 1981–2. Short-term interest rates rose to historic heights in the United States and the United Kingdom. Some OPEC member countries with sizeable external debt (Algeria, Ecuador, Nigeria) experienced a cut in their windfall gains as a result of higher interest rates on their floating-rate loans. Reduced demand for oil following the new recession, combined with increased domestic spending financed by higher oil prices (and revenues), led to a worsening of current-account balances in some member countries whose payments for imports fell short of their exports earnings.

The third jump in oil prices after Iraq's invasion of Kuwait in August 1990 was soon reversed, thanks partly to increased output on the part of some OPEC members (particularly Saudi Arabia) but mainly to the cautious policies followed by the major oil importers which resulted in another global mini recession and a subsequent sharp drop in oil prices.

The actions by OPEC members during the 20-year period basically followed from their individual predicaments and specific opportunities. In two areas, however, the members shared a similar fate. First, they all faced a turbulent and unpredictable global petroleum market where relatively small supply changes created enormous price sensitivity.[18] Long-term sales contracts were replaced by spot sales and short-term arrangements; and OPEC members were placed at the mercy of non-OPEC suppliers. The unaffiliated (and uncooperative) non-OPEC competitors (particularly Great Britain and Norway), by deciding to produce at full capacity, were able to capture a large share of the market under OPEC-protected prices but at OPEC's expense. Non-OPEC output thus rose from 44 percent in 1973 to 59 percent in 1994. Second, OPEC members had to adjust to rapid changes in their fortunes in a 'crisis mode'.[19] That is, they were subjected to sudden and significant ups and downs in their foreign exchange revenues; they had to face successive acceleration and deceleration in the rate of oil-induced domestic economic activity; and they experienced severe production bottlenecks during booms and troublesome unemployment and excess capacity during busts.

These exogenous factors were essentially beyond OPEC's reach. Yet the organisation was responsible for its members' price and output decisions which directly or indirectly contributed to the fluctuations in their financial windfalls and shortfalls. While repeated attempts were made to adopt a unified stand against a largely hostile world, OPEC members' behaviour was primarily determined by how well they were endowed with oil reserves and by their specific national interests. These differences among members resulted in the adoption of variable price and output strategies which were, in turn, partially responsible for OPEC's booms and busts.

SEEKING AN UNREACHABLE GOAL

Differences in national oil reserves and domestic absorptive capacity did not bode well for OPEC's unity and solidarity at the outset. And they never did afterwards. Members with very large petroleum reserves, a small population and limited non-oil resources had a different stake and attitude with regard to OPEC's output and price goals than did other members with scant oil deposits, sizeable population, and ample opportunities for domestic development and growth. Saudi Arabia, with colossal oil deposits and relatively limited opportunities for domestic development, could hardly follow the same strategy as Indonesia, Iran, Iraq, Nigeria or Venezuela. Saudi national interest was clearly on the side of prolonging the 'usefulness' of its single valuable commodity – oil – as long as possible. This national imperative called for a policy of price moderation in order to contain the development of other energy alternatives and competition from non-OPEC suppliers.

Expecting an eventual fall in the demand for oil as a rational long-term response to the 1974 price explosion, and anticipating additional supplies from non-OPEC resources (for the same reason), the Saudis, along with Kuwait, Qatar and the UAE, routinely championed the cause of price restraint. Arguing strongly that, burdened by high oil prices, major oil consumers would cut their demand for oil through conservation, increased energy efficiency and shifts to oil substitutes, the oil-rich members counselled in favour of ample supplies and moderate prices. The not-so-rich countries, with greater development opportunities at home and political ambitions abroad – spearheaded by Iran and frequently supported by Algeria and Libya – regularly opted for reduced OPEC output and higher oil prices. Output-optimisers eyed a future for their oil sales extending beyond the 21st century, and wished to pre-empt the rise of oil substitutes for as long as they could. Price-hawks, on the other hand, knew that time was not on their side, and wanted to take advantage of the tight energy market by selling as much oil as they could before viable substitutes became profitable and competitive with oil.

While the defining line between 'price seekers' and 'volume-chasers' was commonly drawn on the basis of petroleum reserves and their life spans, there were certain exceptions to the rule. Occasionally, some non-typical behaviour was shown by members. For example, Iraq, with petroleum deposits perhaps second only to Saudi Arabia, appeared to behave like a price hawk after the Iran/Iraq ceasefire in 1988 when the country was desperate for funds. By contrast, Venezuela, with relatively smaller recoverable reserves, frequently acted as a price dove; it even volunteered to expand production unilaterally, if need be, after Iraq's invasion of Kuwait in 1990. Caracas wished to prevent a new price hike – a position identical with the Saudis'. After the oil price jump in 1974, Kuwait cut its production (from 3.2 mb/d in 1972 to 2 mb/d in 1975) but, in the late 1980s, it routinely strove for higher quotas and was among 'quota cheaters', triggering Iraq's wrath. After the Iraqi withdrawal, Kuwait again began to redevelop output capacity towards 3 mb/d from 2.5 mb/d, and insisted on having the same quota as the UAE's.

Leaving aside such unusual departures from the norms, OPEC members' production/price strategy followed from the magnitude of their petroleum reserves in comparison with their other resources. The larger the oil deposits and the smaller the complementary factors, the sharper was the tendency toward high-volume production and more 'moderate' prices. And vice versa. For example, Iran – except for a very brief period after the 1979 revolution when leftist elements still wielded power within the Revolutionary Council and advocated a reduction in output – always pushed for higher oil prices to be achieved by a cap on total OPEC production at the expense of those who were 'not in need' (e.g. Kuwait, Saudi Arabia and the UAE). Ideally, the organisation as a whole wished to produce oil at or near full capacity, and sought non-OPEC suppliers' 'cooperation' to keep the world price high. Failing this 'ideal' arrangement, small- and medium-reserves holders – the capital-deficit countries – expected the oil-rich, capital-surplus members (namely, Saudi Arabia and its other Gulf Cooperation Council partners) to bear the brunt of production cuts in order to attain high target prices. But the latter had their own agenda, and invariably pressed for appropriate sacrifices on everyone's part.

The same attitude and stance as Iran's were typically shown by Algeria and Libya (and occasionally also by Iraq). They all wished to lower the OPEC ceiling in order to support higher prices but raise their own quota near capacity in order to maximise export income. At the first quota allocation in March 1983, Iran objected to its 2.4 mb/d share, and vowed to maintain its already higher output level. So did Iraq, who wanted a quota equal to Iran's (although its actual output at the time was below that level). In December 1985, when Saudi pressures forced OPEC to abandon its four-year effort to prop up oil prices, and embark on a new 'fair market share' strategy, the three radical members refused to go along – arguing (as it turned out later, correctly) that OPEC stood to lose from such an open price war with non-OPEC suppliers. In January 1987, Iraq again rejected its assigned quota of 1.54 mb/d, renewed its displeasure in the June meeting of the same year and continued over-the-quota production throughout the year.

A concerted effort by the Arab group to produce beyond assigned quotas in order to keep prices down in 1988 and 1989 raised OPEC's total daily output by about 2 mb/d above the official ceiling during the period. The open defiance by Kuwait and the UAE, despite repeated, ominous Iraqi warnings, finally resulted in Iraq's invasion of Kuwait in August 1990. Nevertheless, OPEC's self-imposed periodic ceilings were routinely surpassed by nearly all members, albeit for different reasons: the Arab groups in order to keep the price at a moderate level; others in search of higher revenues. After the end of the Persian Gulf War, while the embargo against Baghdad kept Iraq out of the production loop, Iran repeatedly and single-handedly pushed for lower total ceilings at every ministerial meeting while defending its own quota. Iran's goal was normally supported by Algeria and Libya, and tacitly endorsed by Indonesia, Nigeria and Venezuela among the rest. Lower OPEC output ceilings were supported by all except, of course, the affected Arab members. Iran's stance in favour of higher oil prices was, ironically, the same before and after the 1979 revolution. In December 1973, Iran supported an explosive four-fold increase in the oil price – a decision which was adopted by other OPEC members – with Saudi Arabia the principal silent dissenter. In subsequent OPEC conferences, Saudi representatives repeatedly resisted fresh attempts at further price hikes. In these meetings, Iran persistently advocated a base price for the benchmark crude equal to the cost of the closest energy substitute and also indexed for inflation and currency depreciation in the industrial countries where OPEC members obtained their import needs.

This unconcealed clash of fundamental national interests between the two major groups within OPEC defined the organisation's concern and behaviour. During the twenty-year period 1974–94, only twice did Iran and Saudi Arabia diverge slightly from their chosen paths. In the December 1977 meeting of OPEC ministers, acceding to President Jimmy Carter's request for keeping a lid on oil prices, Iran took a lead against any further price increase. As a result, the 50th conference was not able to reach a consensus on an expected 'price readjustment'. In its turn, it was after the disastrous failure of the December 1985 'market share' strategy that Saudi Arabia agreed to cut its production.

The push for higher prices through production cuts was assiduously pursued by Algeria, Iran and Libya – and steadfastly resisted by Saudi Arabia throughout the period. Thus, in March 1979 when the official OPEC price was fixed at $14.55/b, the three price hawks obtained a concession from the group to add a 'surcharge' on their own sales. In June 1980, the marker crude was priced at $30/b but Saudi Arabia decided to retain the price at $28/b. The radicals blocked the Saudi move in July 1985 to reduce the official price by $1.50/b. In June 1990, under Iraqi pressure and with Iran's support, OPEC raised its target or reference price by $3/b to $21/b (while actual price in the market hovered around $17.5/b). The Saudis again grudgingly relented – fully aware that the new reference price had no more significance for the market than the previous one. In the aftermath of the Persian Gulf War and the restoration of prewar output quotas, the Saudis' large output prevented the average daily price for OPEC's basket from ever reaching the $21/b target. Riyadh adamantly refused to lower the Saudi daily production below 8 mb/d, despite its pre-war quota of only 5.38 mb/d.

ONE AMONG MANY

In assessing OPEC members' overall behaviour, we see that the role played by Saudi Arabia in pursuit of its national interest was paramount. Riyadh, as an OPEC initiator and its kingpin, perpetually tried to strike a balance between several conflicting objectives: its own national interest; the interests of its OPEC partners; maintaining its special relationship within the Gulf Cooperation Council; reaching an accommodation with its own domestic opposition; and, finally, preserving its position *vis-à-vis* the West (particularly the United States). Hoping to maintain such a balance, the Saudis engineered two different strategies during the 1980s, both of which backfired. At the March 1983 London conference – believing that the worldwide recession had run its course and that the global demand for OPEC oil was likely to surge soon – the Saudis volunteered to become OPEC's designated 'swing producer' in order to defend the official price. In such a scenario – with other members being bound by their output quotas – the swing producer could clearly reap all the benefits of a demand surge. As it turned out, however, the world economic recovery was weak and the demand for OPEC oil continued to decline due to non-OPEC competition, which held up better and longer than most experts had anticipated. By mid-1985, Saudi oil exports plummeted to a low of 2.15 mb/d, half of its notional quota, less than one-fourth of the kingdom's average daily output of 9.2 mb/d in 1980, and even less than the output in the British sector of the North Sea. This price-support strategy caused the annual Saudi oil revenue to fall from nearly $119 billion in 1981 to less than $37 billion in 1984. The Saudis took a high-risk gamble and lost.

In a dramatic change of strategy, Saudi Arabia decided to switch from the defence of the oil price to the protection of its own 'rightful' share of the market. After threatening OPEC and non-OPEC members that it would flood the market and drive down prices, and failing to elicit a favourable reaction, the kingdom resorted to a new and ingenious device to reclaim its notional output of 4.35 mb/d. Since increased exports in the depressed market prevailing at the time required higher discounts, and this move was bound to be matched by other sellers in a price war, the Saudis embarked on a new marketing strategy, the so-called netback deals. Under this scheme, there was no fixed price for the purchaser (the refiner), and the price the kingdom received was based on what the refined products fetched in the marketplace minus a predetermined profit per barrel. This mechanism helped minimise the refiner's risk and encouraged sales. The use of this weapon was effective, however, inasmuch as it enabled Saudi Arabia to regain lost sales and even to increase exports beyond its quota, by late 1985. But, as other members learned to use the same arrangement, the benefit to the Saudis proved to be shortlived. In December 1985, in another ill-fated stratagem, the Saudi representative marshalled enough votes to switch OPEC's official stand from the defence of the official price to the recapture of OPEC's exports share in the global market. The new strategy raised OPEC's exports and its world share, and also boosted Saudi exports. But the overall outcome was a disaster for all, and worst for the Saudis themselves.

The Saudi move was intended to punish quota cheaters and discounters within the organisation, drive marginal non-OPEC suppliers out of the market and weaken Iran's military offensive against Iraq – a Saudi ally at the time. As prices subsequently nosedived to below $10/b, some high-cost producers in the United States were wiped out but the bulk of non-OPEC supply stayed put. As a result, the Saudis' oil revenue slid to $18 billion in 1986 from $36 billion in 1984. Riyadh's stratagem failed once again, for itself as well as its OPEC partners.[20]

Since the Kingdom's 'moderate price' policy happened to be also in the US national interest, Riyadh was often accused by OPEC hawks (and its own domestic opposition) of following Washington's line. But, whatever motives the Saudi rulers might have had, their high-output, moderate-price strategy could not be faulted as being against the Kingdom's own perceived national interests.[21] This was the behaviour expected of an 'initiator'. Interestingly, Saudi Arabia's consistency in pursuing a long-term moderate price strategy was questioned by some American analysts because of the kingdom's acquiescence twice in sharp price hikes in 1973 and 1979.[22] But other studies, as well as the Saudis' declared position within OPEC, demonstrated Riyadh's policy to be one of 'shared responsibility' for the maintenance of OPEC prices.[23] A clear example of this policy was the move into the hawks' camp in May 1992, when OPEC faced new European proposals to impose higher taxes on domestic oil consumption. Stung by such threats, and pressed by its own money needs, Riyadh quietly abandoned its anticipated insistence on raising OPEC's ceiling by one million barrels a day, and gamely agreed to extend the prevailing ceilings of 23 mb/d (plus Kuwait's output) in the third quarter of 1992. The Saudi position was that the European Community's proposed carbon tax would not discourage total energy use but simply siphon off OPEC's revenues. In that case, if petroleum consumption were to be cut by higher prices, the benefit should go to the oil exporters in the form of higher crude prices rather than to the European governments through higher excise taxes.[24]

After 1992, the Kingdom's energy policy was focused on 'maintaining the role of oil in the world of energy, and the role of Saudi Arabia in the world of oil'. This policy, in turn, emphasised the volume share rather than the price concern,[25] and was officially declared to focus on (1) stabilising oil prices so that they accurately reflect economic value and help to expand demand in a balanced manner (2) building a strong and competitive oil industry with a high degree of horizontal and vertical integration at home and abroad and (3) maintaining oil's share in global energy consumption and the kingdom's share of the world market at levels that correspond to its reserves and production capacity.[26]

To sum up, after the failure of the second Saudi strategy in 1986, OPEC members were never able to adopt a unified and coherent position between higher prices and higher volumes. Since achieving both of these objectives simultaneously was impossible, OPEC's policy drifted toward *ad hoc* crisis management. With the unexpected availability of non-OPEC capacity to meet slowly rising world demand for oil, OPEC found itself at cross purposes between defending the $21/b 'reference' price and the 55 percent market share. Higher prices called for collective output restraint, 'equitable' quota allocation by some consensual criteria and strict quota discipline – requirements which the organisation was unable to fulfill or enforce. Moreover, raising prices without losing sales required cooperation by non-OPEC producers in keeping production down – a condition which was scarcely assured. A larger market share meant expanding global demand for energy and oil as well as rising OPEC production capacity as residual supplier – neither of which could be easily obtained.

NOTES ON CHAPTER III

1 The intention here is not to chronicle OPEC's history but only to describe specific developments and factors within the organisation which shaped individual members' behaviour and national agenda. For detailed accounts of OPEC's background, its structure and functions, consult the bibliography.

2 For a fascinating story of early global rivalries and clashes over the oil mastery, see Daniel Yergin (1991), Part I.

3 For the content of this agreement and its background, see *The International Petroleum Cartel* (Washington, US Federal Trade Commission, 1952).

4 Since New York was chosen by the oil majors as the 'equalisation point' for the crude brought in from the Persian Gulf, the Gulf of Mexico and the Caribbean (Venezuela), the Middle East posted prices between 1949–59 were deliberately kept low in order to compensate for the higher freight cost.

5 See 'Oil and the Gulf', *The Economist*, 28 July 1984, pp 12, 50.

6 *OPEC At a Glance* (Vienna: OPEC Secretariat, 1993), pp 5–9.

7 *OPEC at a Glance* and *OPEC Bulletin*, June 1994, p 6.

8 For example, discounts of 40–50 cents a barrel from the posted price on the producers' offtake were not unusual, with the result that a barrel of offtake crude by the Persian Gulf producers was often sold at $1.25 in the world markets. The chronic over supply of crude kept market prices routinely below posted prices.

9 For the full text of this and subsequent OPEC declarations, see *OPEC Official Resolutions and Press Releases, 1960–1990* (Vienna: OPEC Secretariat, 1990).

10 While the 1971 posted price in the Persian Gulf barely matched that of 1947, the real (inflation-adjusted) price was actually 15 percent lower still.

11 *The Economist* (London) 26 January 1974, p 87.

12 The claim regarding unequal distribution applied mostly to Western Europe, with a long tradition of high fuel taxes, and not to the United States where gasoline taxes were a rather small fraction of the price at the pump.

13 Meanwhile, the posted price of oil was raised from $1.80/b in 1970 to $2.18 in 1971, $2.90 in mid-1973, $5.12 in October 1973, and finally $11.65 in December 1973.

14 For a fuller discussion of these factors, see Jahangir Amuzegar, 'OPEC in the Context of the Global Power Equation', *Denver Journal of International Law and Policy*, Fall 1974.

15 Some Arab countries even considered the price hike counterproductive to their strategy of oil cutbacks (the embargo) which was devised to penalise foes and reward friends. See *Dialogue on World Oil* (Washington: American Enterprise Institute, 1974), p 21.

16 All figures are from OPEC *Annual Statistical Bulletin*, 1993 and 1996.

17 See *Middle East Economic Survey*, 27 July 1981, p 12 and 18 October 1992, p 2.

18 For example, a less than 9 percent decline in total world oil supply after the Iranian revolution in 1979 caused crude prices to jump by more than 300 percent within a few months. Similarly, a less than 5 percent 'glut' in aggregate oil supply in early 1986 triggered the average price of spot crude to tumble by nearly 80 percent.

19 Ahmad Jazayeri (1988), pp xiii–xiv.

20 Later on, the Saudis blamed the oil debacle on other OPEC members' 'real mistake' in not heeding repeated Saudi warnings against raising oil prices (between 1979 and 1981) to unreasonable levels. They also attributed the situation to a 'Western plan to destroy OPEC and remove it from the scene'. See 'Forum', *OPEC Bulletin*, June 1986, pp 3–4.

21 The Saudis, for example, were accused of giving only lip service to their commitment (under the July 1990 conference) to obtain the $21/b 'reference price' for OPEC oil because the Saudi delegates usually qualified their support by adding 'if the market decides that'. This prompted the opposite camp, led by Algeria, to say, 'If the market dictates the price, why don't we all just go home?' See *Wall Street Journal*, 25 and 27 September 1991.

22 See William Quandt, *Saudi Arabia's Oil Policy* (Washington: The Brookings Institution, 1982).

23 Edward Renshaw, 'Saudi Oil Policy', *OPEC Bulletin*, July–August 1985, p 21.

24 *Wall Street Journal*, 27 May 1992.

25 See 'Saudi Arabia', *Middle East Economic Digest*, Quarterly Report, May 1994, p 10.

26 See, 'Saudi Arabia', *Wall Street Journal*, 2 September 1995.

IV

Windfall Allocation and Adjustment Policies

In all OPEC economies, the adjustment process took place through changes in the level and composition of domestic expenditure in line with the fluctuating levels of petroleum revenues. Development plans or annual budgets allocated available oil revenues (and additional funds) among current public and private consumption, fixed domestic capital formation, build-up of national defence and security, and foreign savings and external assistance. During periods when the global petroleum market was depressed, countries which had over committed themselves to large public outlays (or run into significant cost overruns in development projects) had to supplement their annual oil receipts with foreign loans and grants. Some succeeded in supplementing their domestic investable funds by attracting foreign direct or portfolio investments to finance local projects. A few managed to build a substantial 'nest-egg' abroad.

The specific allocation of windfalls by the government determined the direction, tempo and quality of economic development. Since the state was the sole recipient of petroleum revenues and a major source of wealth accumulation, it felt no need for domestic taxation to finance its favoured projects. For this reason, national socio-economic agenda were set by the ruling elite without being initiated, participated in or scrutinised by the citizenry. As a result, oil windfalls were allocated to priority regions, sectors and projects in a centralised, force-fed and often *ad hoc* manner. This fiscal independence, in turn, not only resulted in increased monopolisation of decision-making by a select few at the top but also led to recurring cost-benefit miscalculations, much waste, a host of rent-seeking activities and widespread corruption. The magnitude and focus of windfalls allocation thus served as the initiator, prime mover and, occasionally, spoiler of domestic development efforts.

By virtue of their exclusive ownership of oil reserves and the monopolistic receipt of oil export revenues, OPEC governments were in a sense neo-patrimonial states whose response to the rise and fall of oil income played a quintessential role in promoting, directing and channelling (or distorting) domestic economic growth and national economic development. Without exception, the government in each OPEC economy played a modernising and industrialising role in the economy by allocating a lion's share of the oil windfalls to the creation and expansion of physical infrastructure, provision of mass access to education, healthcare and social services, and the establishment or enlargement of key industries.

A common thread that ran through nearly all member economies in their windfall allocations was a good deal of optimism. When oil prices were on the rise, the assumption ordinarily was that they would continue uninterrupted. Under this assumption, public and private consumption was allowed to rise above sustainable levels, low-priority capital projects were inadequately appraised, capital outflows were allowed and tolerated as helpful to the adjustment process and, even, additional foreign loans were sought in the hope of repayment out of future oil earnings. And, when the oil market turned sour and prices were on a downslide, some viable schemes were indiscriminately postponed or cancelled in a hurry, various controls were imposed and resort was made to foreign borrowing in the belief that the situation was temporary and higher oil prices would eventually be resumed.

This choice of macroeconomic policies also had a crucial impact on the outcome of windfall allocations by creating new 'belief systems', new modes of behaviour and new vested interests.[1] The need for the appearance of propriety and wisdom invariably suggested some kind of economic planning for both practical and prestige reasons. Unrequited oil receipts tended to introduce a new politics of rising expectations, social welfare largesse and greater state paternalism. Fiscal policy in the form of reduced taxes, increased subsidies, enhanced welfare payments or expanded public employment encouraged profligate consumerism, relaxation of fiscal discipline and living beyond one's means. Monetary policy, particularly in countries with a less-developed capital market and a nationalised banking system, tended to become more or less subservient to fiscal measures. Exchange-rate policy, as the kingpin of other macro policies, not only influenced the sectoral composition of the domestic development model but (and more importantly) invited and reinforced an insidious rent-seeking behaviour. The latter helped create a new class of social parasites, i.e. wheeler-dealers or apparatchiks – which often sabotaged needed timely adjustments.

A combination of 'wrong' policies at the state level often ended up in unrealised targets, a multitude of white-elephant projects and widespread inefficiency. The absence of foresight and prudence in the selection of policy alternatives during oil booms made it doubly difficult to shift gear, correct mistaken policies or adopt counter-acting adjustment measures once booms turned inevitably into busts. The reversal of fortune during periods of depressed oil prices frequently resulted in painful economic deterioration, bankruptcies and debt moratoria.

In short, while unwise allocation and wrong policy choices were by no means the only determining factors in the overall performance of member economies, they were, by all indications, the critical ones. The findings of this chapter will show that despite substantial differences in domestic institutional arrangements, natural resource endowments, the state of technological advancement and absorptive capacity, all countries followed astonishingly similar policies, and fell more or less into the same traps, as through they were under the same spell. A common policy stance among group members, with some rare exceptions, was to miss out on globalisation, i.e. integration with the world economy as an axis for development.[2] Trade in fuels (oil, gas and derivatives) dominated most group economies throughout the period, and energy continued to have a very large share in total exports. Except for the small Persian Gulf economies and Indonesia, which were relatively more integrated in the world economy (with a high ratio of trade to GDP), the group members commonly failed to adopt appropriate policies that could increase their global competitiveness. In addition to continued trade and exchange restrictions, wage/price controls, costly and cumbersome regulations, and a harmful tolerance for unethical business practices, many group members could not provide a hospitable environment for private domestic and foreign investment. The commitment to privatisation adopted by most governments towards the end also proved half-hearted and perfunctory.

ALGERIA[3]

Until late in the 1980s, Algeria chose a socialist and interventionist approach to economic development, and the windfalls' allocation. The national charter, drawn up in 1976, laid down the guidelines for creating a 'socialist society', where large state-owned enterprises were to handle virtually all production, imports, marketing and investment in nearly all major economic activities except small farming, light consumer goods and trade. In 1980, however, under the Chadly Benjadid administration, the government adopted an 'enriched' National Charter, allowing for an increasingly flexible economic policy aimed at reducing the rigidities of the Boumedienne years. Yet the basic culture of state paternalism continued. Algeria's heavy reliance on the public sector was due partly to its revolutionary leaders' long sympathy with socialism. Inasmuch as France symbolised a Western colonialist and imperialist power, Algeria's National Liberation Front, which secured the country's independence, regarded the private sector and multinational businesses as alien to the revolution's objectives.[4]

Algeria's response to the first and, to some extent, the second oil boom was a strategy of deferred consumption whereby the windfalls, in conjunction with foreign credits, were mainly allocated to the expansion of the energy sector and a heavy industrial base – a feature unparalleled in most other member countries. Only after the second oil boom, and the Benjadid administration's slight reduction of Boumedienne's austere regime, did private consumption (including better housing) receive long-neglected allocation. The investment allocated in Algeria's Second Four-year Development Plan (1974–7), hurriedly revised after the onset of the first oil boom, amounted to a total of $26 billion – three times the amount invested under the First (1970–3). The Algerian government viewed the 1974–8 oil boom as a prelude to a continued oil seller's market and steady rises in crude oil prices. In line with this scenario, the ratio of gross domestic capital formation to GDP was raised to about 40 percent, one of the highest in the world.

The Second Plan, underscored by sizeable oil windfalls, emphasised capital-intensive heavy industrialisation at the expense of job-creating light industries and agriculture. The bulk of public investment was concentrated in hydrocarbons, and specifically liquified natural gas (LNG). Banking on its proximity to European markets and the high quality of its gas and oil, Algeria pursued a policy of capacity expansion in oil, gas and gas liquids. The national oil company, SONATRACH, undertook some 27 macroprojects in oil and gas and built the world's largest LNG export capacity and an undersea gas pipeline to Europe.

The preference for heavy industrialisation by Algerian revolutionary leaders was given warm encouragement by the 1973–4 oil windfalls.[5] The slogan of 'sow oil to reap industry' was translated into an expanding programme of investment in steel, fertilisers, petrochemicals, plastics and transport vehicles – the so-called 'industrialising industries'. These activities were expected to be fed and financed by energy inputs and oil export earnings;[6] their products – fertilisers, tractors, steel pipes etc. – were expected to feed self-managed farm units in the agricultural sector in line with the forward-linkage theory. The hydrocarbon industry was given top priority for development in the hope of generating foreign exchange, supplying energy and feedstock for petrochemicals and basic metals, and otherwise filling common development gaps. Modern industries were, in turn, expected to provide a range of processed and finished goods. This strategy put great emphasis on holding down private consumption and sustaining large public investment. Private investment was treated as a residual and given a secondary role.

Popular discontent with this austere strategy appeared shortly after the second oil boom. A variety of social grievances, from a shortage of jobs to poor housing, inadequate transport and shortfalls in farm output led Algerian planners to shift their focus from hydrocarbons to the lagging sectors. Rapid population growth (3.2 percent a year), the rural exodus and increasing demands for urban housing and employment called for a shift of emphasis to agriculture, education, housing, social services and light industry. The 1980–4 plan recommended greater private sector participation in such areas as housing, tourism, transport and trade. However, strategic sectors such as energy, metals, minerals and finance still had to remain under state control. The plan, furthermore, introduced wide-ranging reforms in agriculture including price liberalisation, private marketing of farm products, land tenure and farm management, and availability of bank credit and extension services. In this plan, for the first time, investment in light industry exceeded that for heavy industry, especially in construction materials, food and textiles. And, unlike heavy industry which was entirely state-controlled, light industry had sizeable private participation in a 'controlled way'. But, as the ratio of gross domestic savings to GDP remained persistently below gross investment, a sizeable resource gap developed which had to be filled by foreign borrowing. Total external debt subsequently rose 3.6 times during the second plan, and reached $16 billion by 1985.

The 1986 oil price crash, in the wake of the new four-year plan (1985–9), led the authorities to further redirect domestic planning away from undue reliance on hydro-carbon and heavy-metals toward the development of agriculture, water resources and other sectors, in order to lessen dependence on imports. More reliance on domestic resources at the expense of foreign borrowing was also contemplated. Under pressures from a new middle class of technocrats, the planning ministry was abolished in 1987. A significant deterioration of economic conditions in 1986–9 exposed years of eco-nomic mismanagement and plunged the country into a long period of economic chaos and political instability. The economy's frailties finally forced the authorities to seek help from the International Monetary Fund in 1989 and 1991, whereby fiscal and monetary policies were tightened, interest and exchange rates were adjusted and prices were liberalised.

A deepening political crisis and new social strife, caused by the cancellation of general elections, resulted in a change of government in late 1992 and a slow-down in the pace of reforms. Under the new Abdesslam Belaid administration, and with the reincarnation of Boumedienne's original economic folly, renewed state control and direction of the economy once again became an essential perceived element of domestic strategy. Trade controls were reimposed and foreign exchange was allocated to prioritied sectors and projects. There was a clear change from the previous direction under Benjadid. Austerity again became the dominant economic policy. The transition from state socialist economy to the free market was stalled. However, when the new government proved unable to tackle any of the basic economic challenges, a second reversal got underway. The IMF programme which was suspended in 1992 was later revised. Negotiations on a new standby Compensatory Facility were concluded in 1993 and a new programme of structural reforms was established in 1994. In a series of major policy shifts under IMF programmes, inefficient collectivised state farms (established in the 1970s) were transferred to the private sector, autonomous public enterprises were formed, the central bank (*Banque d'Algerie*) was given greater authority to control money, exchange and credit, a large segment of the economy ('non-strategic' sectors) was opened to foreign investment and joint-ventures, imports were demonopolised and other moves toward privatisation and market-ori-ented solutions were encouraged. Under this agreement, the Algerian dinar was devalued by 40 percent, interest rates were raised, prices were decontrolled and subsidies cut, money supply was reduced, budget deficit was lowered and trade was liberalised. The 1994 IMF agreements opened the way for the drawdown of about half a billion dollar loans negotiated with the World Bank, the European Community and the Japanese Eximbank earlier in 1991 and partly blocked by the creditors pending an IMF deal. A week after the latter, the Paris Club of official creditors agreed to reschedule government-backed debt with up to $14,500 million. A London Club rescheduling and a round of bilateral arrangements followed. Obligated under agreements with the IMF, Algeria gradually relaxed its statist stance and became, grudgingly, market-oriented. With the announcement of its privatisation programme in such sectors as public works, small and medium-size industries, trade, road transport and tourism, Algeria finally discarded its last non-socialist 'economic taboo' by 1994.

Algeria's specific economic policies followed the broad outlines of the developments in national strategic direction. Fiscal policy was, up to the late 1980s, the dominant economic policy. The Treasury served not only as the government's cashier (i.e. revenue collector and money dispenser) but also its banker. All the central government's financial needs were met by the Treasury through revenues, liquid funds deposited by public financial institutions, government bonds and an overdraft facility at the central bank. After the switch from a centrally planned economy toward a decentralised market-based system, the Treasury's economic role was reduced to the financing of infrastructure projects, while banks, public enterprises, financial institutions and the central bank were given substantial autonomy. Taxes on income, corporate profits and consumption were adjusted.

In the monetary field, Algeria's money and credit policy was determined in a centrally planned framework until the late 1980s. Banks remained publicly-owned and highly specialised, with their primary function being meeting public sector financing needs and residually helping the small private sector. Credit was given to public enterprises on the basis of their production plans. This system was totally revised in 1989 in favour of greater market-based credit distribution and private commercial banking. The Central Bank of Algeria, long operating as a *de facto* branch of the Treasury, was given autonomy over monetary policy. Insurance activities, however, remained a state monopoly. Under the 1989–91 IMF conditions, the rates on deposits and loans that were administratively determined and kept artificially low were allowed to move more freely.

Throughout most of the period under review, Algeria maintained tight control over most prices and wages. Price controls varied in rigidity with respect to basic necessities, raw material inputs and other consumer goods – with prices of 'essential' commodities (e.g. basic foods) fixed by the government and kept low through budgetary subsidies. Prices of the other two categories were given greater flexibility. Up to the mid-1980s, most essential prices were administratively regulated and subsidised to protect the purchasing power of low-income groups in an elaborate and cumbersome system. Within the context of the 1989–91 IMF adjustment programmes, most prices were gradually liberalised and market-determined. Yet mandatory ceilings, controlled profit margins and prior notification of price rises on many goods were kept through 1994. The general wage policy was aimed at periodic adjustments of minimum rates in line with price increases, and at narrowing differences among various wage earners. 'Indicative' minimum wages and salaries were determined through agreements between the government and the largest workers' union. There were strict labour laws regarding hiring and firing of workers in the small private sector.

The Bank of Algeria had jurisdiction and control over Algeria's exchange rate policy. This policy sought to keep the value of the Algerian dinar stable relative to a basket of currencies of its major trading partners. After delinking from the French franc in 1974, the rate, announced daily by the central bank, was related to a basket of 14 currencies weighted by their shares in the country's external transactions (including capital transfers). The daily rate was adjusted by the central bank to reflect its domestic economic objectives.

There were elaborate and detailed regulations regarding foreign exchange availability for imports, external financing of imports, non-resident foreign currency accounts and transactions on foreign currencies. Foreign trade was a government monopoly that lasted well into the 1980s. After 1987, and particularly during 1990–4, stringent controls were gradually relaxed. The top tariff rate which in 1992 was cut in half, was still 60 percent, and there were other levies on imports. Also, a large number of goods were prohibited for importation in order to protect home industries. Foreign investment was highly constrained in most of the period. In accordance with the Law of Money and Credit (1990), foreign direct investment was freely permitted except in certain specific sectors. In subsequent years, foreign investment was actively promoted in accordance with standard international protection and guarantees; the investors were also given a number of tax and other incentives.

Algeria's comprehensive welfare system included free education, free healthcare, child support regardless of income for workers in the formal economy, family allowances and retirement benefits. Furthermore, there were veterans' benefits and extra budgetary subsidies, out of the Compensation Fund, for food staples and cash transfers to the poorest segment of the population. After 1992, concerns about the potential social repercussions of necessary macroeconomic stabilisation and reform measures prompted the establishment of a system of unemployment insurance, a social security system and a housing assistance fund – all parts of a new welfare plan called the 'social safety net'. Various benefits under the system benefitted some 15 million people, or 55 percent of the population, in 1994. The combined subsidies from the budget in the mid-1980s was equivalent to about 2 percent of GDP. During the second half of the period, consumer subsidies (food, energy products and public services) by themselves took 2 percent of GDP. Other aspects of the safety net (i.e. family allowance, education allowance, healthcare, unemployment compensation etc.) took another nearly 4.5 percent of GDP. Agriculture continued to be subsidised through both input (low interest rates, tax exemption and price and profit controls) and thus output price supports absorbed another 1 percent of GDP in 1994.

In a move toward intra-sector diversification, Algeria focused its hydrocarbon policy on the development of natural gas to gradually replace its dwindling crude oil exports. By contrast, the country's gas reserves were elevated to the world's seventh largest in 1994, and its gas exports to Europe were the third largest (behind Norway and Russia). With respect to the development of its hydrocarbon deposits, Algeria adhered to its 1971 oil nationalisation policy for a long time and refused to allow foreign share participation. After 1986, this policy was liberalised, and outside participation in production and equity sharing agreements were welcome. By 1994, following changes in the law, a score of foreign companies signed joint-venture agreements with SONATRACH, the country's oil and gas conglomerate.

To sum up, for the largest part of the period under review, Algeria's markets for factors of production, goods and services, exchange and interest rates were all strictly controlled by the state. Private enterprise was relegated to small farming and service activities. Foreign trade was tightly controlled and foreign exchange was allocated administratively. The nationalised banking system and the treasury were used by the government to allocate and direct virtually all major investment funds. The post-1990 commitment to economic liberalisation was still lukewarm and the pursuit of needed reforms rather erratic. Despite the quasi-military character of the regime, Algeria's defence outlays were below 3 percent on the GNP and less than 10 percent of the annual budget.

ECUADOR[7]

Ecuador joined OPEC in 1973 as its thirteenth member only a year after it had discovered oil in commercial quantities. It was also the first member to leave the organisation in January 1993. The oil-price surge was the most unexpected bonanza that followed the membership. Prior to the oil discovery, Ecuador was a relatively poor country with bananas and coffee as its main exports. OPEC's membership helped Quito raise taxes on the US oil companies that developed its oil fields and later wrest the majority ownership from them.

Ecuador's total oil export receipts over the two decades totalled about $25 billion. Although the volume of crude oil exports varied from year to year, its peak in 1986, at 196,000 b/d, was nearly equal to the level back in 1973. Revenue fluctuations thus reflected mainly the changes in the oil price. A sharp drop of 30 percent in 1987 oil production was caused mainly by an earthquake which destroyed pipelines. Ecuador continued to surpass its OPEC quotas, producing an estimated 320,000 b/d in 1992. It finally left the organisation by the end of that year after falling behind in its membership dues for two years. The petroleum sector then became the fastest growing sector in 1993–4 as Ecuador's departure from OPEC allowed it to increase output significantly to an estimated 350,000 b/d. Ecuador was one of the few OPEC members who responded to the oil price rise of 1974 by reducing output to about half of what was previously projected, mostly based on the Lara administration's nationalistic policy of reserves conservation. Also, part of the new windfall was spent on purchasing back Gulf oil's share in the oil consortium that possessed the Ecuadorian oil concession, thus increasing CEPE's (State Petroleum Corp.) share.

To make the best use of the new oil windfalls (and to avoid the pitfalls in previous cocoa and banana bonanzas), Ecuador decided early in the game to launch three fundamental institutional changes: agrarian reform, tax reform and public administration reform. As it turned out, however, the land reform did not change the structure of land holdings or the power of big landlords but did distribute some new land among landless peasants. Tax reforms were virtually ignored; and the public sector actually grew more rapidly than planned.

Except for a brief and unsuccessful period of market orientation, Ecuador followed a policy of extensive state intervention in productive sectors of the economy. Ecuador's strategy for the use of its relatively small windfalls suffered in direction and coherence from two internal factors: political instability and the lack of centralised decision-making.[8] During the first two oil booms, several shifts in military-civilian administrations took place, and no coherent plan for oil revenue allocation could be implemented. Furthermore, the public sector, consisting of more than 40 autonomous institutions and state enterprises, had specially earmarked portions of government revenues, which left very little room for allocative manoeuvering to the central administration. Like Algeria and Venezuela, Ecuador also used its OPEC credit status at the start, soon after 1976, to borrow heavily from foreign commercial banks. As a result, public foreign borrowing – dissaving abroad – financed more than the public sector's deficit. External credit allowed the private sector to thrive on cheap government finance, while the government itself was able not only to finance its domestic investments but also to reduce non-oil taxes. It was only towards the very end of the period that marketisation and privatisation came into vogue. Based on bitter past experience, the modernisation programme instituted in 1992 placed its emphasis on reducing the excessive role of the state, selling-off state-owned companies, improving the investment climate (particularly for foreign investment) and introducing some semblance of fiscal discipline.

Long after Ecuador joined OPEC, the country thus kept its populist, statist tradition. The administration elected in 1984 introduced a free-market approach toward economic liberalisation. With a focus on taming inflation and balancing public finances in the context of an IMF-sanctioned economic stabilisation programme, money supply was brought under control, subsidies were cut back, trade was deregulated, foreign investment was encouraged and the exchange rate was adjusted to maintain international competitiveness. But the painful consequences of these policies, combined with the 1986 oil bust, resulted in a shift of administration to a centre-left government and renewed interventionist policies in 1988. Shortages of basic commodities, high inflation and rising unemployment under restrictive policies made the new government also highly unpopular, and prepared the country for a shift to the centre-right in the 1992 elections and the success of Duran Ballen, leading to the espousal of free-market reforms once again.

Ecuador's oil policy was steadily directed toward expanding exploration, with the aim of doubling the level of proven reserves. In June 1992, the state took control of the country's oil activities by purchasing back the shares of Texaco and Occidental in its joint ventures with them. Facing an increasing need for foreign exchange and having a small crude output, Ecuador regularly exceeded its OPEC quotas, fell behind in its membership fees and finally left the organisation as of January 1993. The 1993 amendments to the Hydrocarbon Law allowed foreign companies to sign production-sharing contracts with the government for the exploration, extraction, refining and marketing of oil and gas. The new legislation abolished the government's monopoly on domestic sales of petroleum derivatives and introduced greater flexibility in setting prices. Ecuador hoped to attract foreign private interest and investment in 13 new blocks of oil put up for auction. Ecuador's industrial policy in non-hydrocarbon fields was to encourage capital-intensive, energy-based industries by offering low-cost fuel, tax incentives and tariff protection under the Industrial Incentive Law. Non-traditional exports were also encouraged through rebates. Ecuadorian public investment out of the windfalls, unlike that of most other members did not embrace large-scale industrial or agricultural projects. During the first boom, for example, only about 28 percent of total public investments went into those two sectors, while nearly 41 percent was channeled into physical infrastructure (power, roads and transportation) and 28 percent to social welfare projects (education, health, water, sewage and housing). The rest involved miscellaneous investments. Part of the public investment expenditures was financed by foreign borrowing.

Ecuador's fiscal policy was uncontrollably expansionist and a bone of contention with the International Monetary Fund under various stabilisation programmes. Unlike in most other group members, petroleum revenues were not a dominant portion of total revenues in Ecuador. Moreover, petroleum receipts were shared by the central government and some 21 different public sector entities according to a complex revenue-sharing arrangement. Faced with large fiscal imbalances at the beginning of the 1980s, the government adopted a series of adjustment policies, including prominently an enhancement of public revenues. Prices of domestic petroleum products and public utilities were raised. The general sales tax and some excise taxes were increased, an import duty surcharge was levied and import duty exemptions were reduced. Subsidies to oil users accounted for a considerable portion of the total windfalls so that the state derived no revenue from home oil consumption. Returns to the state electricity company, INECEL, were also reported to be more or less zero during the first boom and negative during the second, although the company received its fuel oil input at cost. In the various stand-by programmes worked out with the IMF, successive governments tried to achieve a better budget balance by reducing subsidies and enhancing revenues through improving tax collection, particularly on imports. Unlike many other group members, Ecuador periodically adjusted the prices of domestic petroleum products and utilities. Several tax reform packages were also enacted by successive administrations under IMF prodding, particularly in 1988, 1990 and 1993. Nevertheless, non-oil tax revenues represented less than 8 percent of GDP in the 1990s (as compared with an average of 14 percent in South America as a whole). Up to 1982, the main way of financing the public sector deficit was net borrowing from foreign commercial banks. As access to international financial markets became more difficult in the aftermath of the Mexican debt crisis, resort was made to borrowing from the Central Bank. In the eight years ending in 1994, public sector deficits were financed by external borrowing and the rise of external arrears as well as by loans from the domestic banking system and floating domestic debt. Rising interest payments on public debt, combined with subsidisation of credit and consumer goods (e.g. wheat and milk) absorbed a large part of public spending, even more than defence outlays. Public employment, particularly in autonomous entities, nearly doubled between 1973 and 1982. As oil revenues increased, non-oil taxes fell, but considerably faster, mainly because of the tax concessions and exceptions granted to industry.

The thrust of monetary policy was to keep inflation under control by restricting credit and neutralising expansionary fiscal policies. Monetary policy encompassed the standard instruments for controlling domestic money and credit. Rates charged on loans and advances by various monetary institutions, and interest paid on demand and time deposits, were strictly regulated in detail. Following the deterioration of economic conditions in the early 1980s which led to the build-up of arrears, the Central Bank assumed the private sector's debt to foreign commercial banks and made arrangements for their rescheduling. In the second half of the 1980s, a comprehensive and sector-specific credit policy was adopted by the Central Bank. After the mid-1980s, interest rates were further liberalised to try to reverse capital flight, and credit subsidies were reduced.

Price policy was geared to keeping the inflation rate low. Consumer prices were held down by keeping prices of petroleum products and utility rates below cost increases, and by subsidising certain import items (e.g. wheat). Items under price control in 1985–94, however, were limited to only a few (e.g. bananas, coffee and pharmaceuticals). In several successive moves between 1988 and 1994, fuel prices were raised to near those on world markets. Ecuador had an elaborate wage system under which there was a basic minimum wage set by Congress plus minimum wages for different sectors (e.g. agriculture, household etc.). Wages other than minimum wages were determined by tripartite labour-management-government commissions. For some large firms, collective bargaining was in effect. In the mid-1980s, there were nearly 100 sectoral commissions for wage determination.

Ecuador's exchange rate policy was geared to coping with its nearly perennial current account deficits. The national currency, the sucre, operated under a two-tier rate system between 1974 and 1982, when the deficit in the current account reached close to 11 percent of GDP. Between 1982 and 1994, the national currency was adjusted several times following various conditions which the IMF set from time to time. By allowing exporters and importers to trade the currency in the free market, the sucre became effectively subject to floating in 1992. The official rate was abolished in 1994. As part of an anti-inflation policy the government actively intervened in the market to maintain the sucre's strength through liberalised domestic interest rates and sucre purchases.

Ecuador followed a complex trade regime. Imports were divided into permitted and prohibited categories. Permitted imports were divided into several essential and less essential categories. Most imports were subject to tariffs which were generally low in the case of capital goods but up to over 100 percent for certain consumer luxuries. There were other elaborate details regarding countries of imports, origin, specific items needed by state agencies, and others. After gaining access to foreign markets under the Generalised System of Preferences in 1990, Ecuador proceeded to adopt a large number of trade liberalisation measures, particularly in the case of export-oriented manufacturing. The Duran Ballen government joined in free-trade arrangements with Bolivia, Colombia and Venezuela, and a common external tariff was put into effect at the beginning of 1994. Average tariff rate was low but customs procedures were cumbersome, giving rise to much abuse.

In 1993, the State Modernisation Law called for the decentralisation of the state apparatus, and privatisation of the public sector through concessions, private entry into the provision of services and reduction of bureaucratic procedures. Nevertheless, outright privatisation of three main state monopolies, in petroleum, electricity and telecommunications, was still prohibited. A privatisation programme was initiated in 1994 with the sales of assets in the public enterprises dealing with hotels, wholesale companies, cement, dairy products and others. Yet the promised privatisation of major public enterprises was stifled by the Congress, the country's left and centre-left tradition and still-powerful labour unions. Sales of cement, fertiliser and other small public enterprises were also marred by charges of malfeasance and corruption.

Altogether, in terms of general economic policy, it took virtually the entire 20-year period for Ecuador to transform itself from a state-controlled to a market-oriented economy. In March 1980, Ecuador launched a national politico-economic development plan under which some 27 percent of GDP was to be invested in a 40/60 percent ratio between the public and private sectors. Public investment priorities included health, education, housing, energy, agriculture, transportation and industry. Subsequent years' allocations followed no formal planning and were affected by changes in administration. Although Ecuador was ruled by a military government in part of the period and had border disputes with its neighbors, military expenditure for most of the period averaged less than 3 percent of GDP and less than 16 percent of the national budget. Undergoing several changes in its political stance during the period, with mixed results, Ecuador was still groping for a sure and stable road to economic health and clean administration in 1994.

GABON[9]

Gabon was an export-reliant, enclave economy heavily dependent on its natural resources, particularly oil, minerals and forestry products. The petroleum sector accounted for about half GDP in 1974. Oil production in Gabon was in the hands of five major foreign companies (including Shell, Gulf and Elf) in which the Gabonese government had a 25 percent equity participation under a series of specific agreements. Gabon's total oil export earnings between 1974 and 94 amounted to slightly more than $28 billion. Foreign exchange resources during the period were augmented by external loans, foreign aid, some direct foreign investment and World Bank loans. Direct foreign investment occurred on a small scale outside the oil sector from year to year but Gabon proved an unattractive place for private non-oil investment due to high production costs. In 1977, Gabon 'graduated' from the World Bank roster of eligible borrowing countries on account of its rising oil revenues. Following a decline in per capita income and the severe economic and financial crisis in 1985–6, it became eligible again. During the twenty-year period, Gabon received an estimated $2.5 billion in official grants and loans.

Although Gabon is commonly categorised as a non-statist regime, the government, directly or through some 80 public institutions and enterprises, was active in almost all sectors of the Gabonese economy. The state had between 33 and 100 percent stake in more than 50 economic entities; it also pursued an official (albeit *pro forma*) planning process, and was very active in domestic investment aimed to expand the country's main natural resources of oil, timber, manganese and uranium. Oil exploration and development was actively pursued through joint-ventures with some major oil companies.

The oil windfalls had an uncommonly large share in gross fixed investment, which averaged 50 percent of GDP between 1974 and 1978 and divided roughly equally between public and private outlays. Capital formation rose to an uncommonly high rate of 60 percent in 1976, and remained high at about 35 percent of GDP 1981–5. As a percentage of GDP, aggregate consumption absorbed only about 36 percent on average for 1974–8. Private consumption fell to as low as 20 percent in 1976 – the lowest among all members in that or any other year. Between 1985 and 1994, however, aggregate consumption took an increasing share of GDP, reaching about 75 percent in 1994 and averaging 65 percent in 1986–94. Private consumption averaged more than twice as much as public consumption.

Under the first oil boom, Gabon launched an ambitious development programme, centred primarily on the construction of the Trans-Gabon Railway and other infrastructural facilities, mostly around three major coastal cities, especially the capital, Libreville. The construction of prestige buildings for the 1977 summit meeting in Libreville of the Organisation of African Unity was another principal oil-financed project. Another major allocation priority was the reduction of the external public debt both in relative and absolute terms. This policy was followed through 1983 as amortisation exceeded new borrowing. In the mid 1980s, however, foreign borrowing became necessary in order to finish on-going projects (especially the Trans-Gabon railway). Defence also absorbed a not insignificant portion of the windfalls. With a population of only about one million and a salaried employment force of about 120,000 in the 1990s, Gabon's central and local governments employed about 35,000, of which 13,000 were in the armed forces.

Due to the small and subsistence nature of its agriculture, and its heavy dependence on food imports, the Gabonese government placed increased emphasis on promoting farm output. To cope with underlying constraints (e.g. manpower shortage, inadequate transport facilities and limited public assistance), the government paid attention to improving the quality of rural life, slowing down the migration of young active farmers to the cities, making markets more accessible to the general population and upgrading cultivation techniques. During the 1980s, the government raised producer prices, provided other incentives and promoted agro-industrial projects. Due to its immense forestry resources, timber regulation constituted one of Gabon's major economic policies. Conservation considerations required licences for the felling and marketing of timber and payment of concession taxes.

Industrial policy was directed to increase Gabonese participation in domestic industrial ventures (e.g. oil refining, chemicals, food and wood processing) that were mainly owned by foreign enterprises. To this end, the government began to acquire substantial equity holdings in those firms. New foreign enterprises after 1975 were required to give the government, without charge, 10 percent of their equity, set aside 3 percent of their capital for subscription by Gabonese nationals and train local staff for senior managerial positions. In 1994, these requirements were eased in order to attract private investment.

Gabon had arguably the most detailed and complex fiscal policy structure of all group members. The tax system was the most elaborate of all. There were taxes on the net income and profits of individuals and companies, payroll taxes, taxes on property, taxes on sales of goods and services, taxes on international trade, royalties on mining and other taxes – each with several detailed categories and rates. Throughout the period under review, the fiscal policy stance was central to the economy's behaviour because of the role of the government in channeling part of its main income (oil) to the rest of the economy. The government's fiscal position was also chiefly influenced by changes in oil output, prices or the exchange rate *vis-à-vis* the US dollar. Despite gradual increases in non-oil revenues, petroleum income still accounted for nearly half of the government revenue at the period's end.

Gabon's monetary policy was largely determined by its membership obligations in the French Franc zone. Members shared a common currency, the CFA franc, issued by a common central bank, BEAC. Credit policy was directed to the promotion of development objectives and priority sectors. After a runaway rise in bank credit for both public and private sectors in 1974–7, monetary policy generally gave way to fiscal, and played no pivotal role in Gabon's development. Part of the reason for the secondary role of monetary policy was that interest rates were not actively used as an instrument of credit control.

Gabon followed a very complex system of price controls and subsidisation. Both domestically produced items and imports were subject to various regulations and supervisions. In practice, only the prices of essential consumer goods were monitored and controlled by a government agency, and adjusted from time to time when justified by changes in costs. Imported products were subject to profit-margin limitations. Prices of basic commodities were subsidised through a special government fund which received transfers from the Treasury and had its own source of revenues through taxes on some import items. Prices of gasoil and kerosene were subsidised by another fund through a tax on gasoline. Other funds for stabilising cocoa and coffee were financed through the budget. Prices of several other products were 'equalised' throughout Gabon by subsidising differences in transport costs. This process was self-financed through taxes on producers.

Gabon pursued a statutory wage policy throughout the period. Workers in farm and non-farm employment were guaranteed a uniform minimum wage, which was adjusted from year to year. Actual wages in the private sector were determined by collective bargaining and generally exceeded the minimum wage by large margins, even without the various allowances paid during the labour shortage. All private sector employees were covered under a social security system financed by both employers and workers (mainly by the former). Wages and salaries were steadily raised during the 1980s and 1990s. Gabon's wages (including national bonuses) were by far the highest in equatorial Africa. Almost half of total public expenditure was typically earmarked for the wages and salaries of public workers.

Gabon continued to maintain a rather liberal system of trade and payments until the mid-1980s. Imports from other partners in the Franc Zone were free of restrictions. After 1986, growing pressure on the balance of payments began to intensify restrictions on both exchange and trade. By 1994, except for France and countries within the Franc Zone, there were administrative controls on exchange transactions, particularly with respect to payments for invisibles. Imports, exports, borrowing and lending abroad, capital movements and dealings in gold were all subject to statutory or administrative controls of one kind or another. The exchange regime was virtually free of restrictions on payments and transfers for current transactions. The national currency, (the CFA franc), was fixed to the French franc at CFAF50=FF1 from 1948 until its 50 percent devaluation in January 1994. Gabon opposed devaluation for a long time, despite IMF-World Bank recommendations, on the grounds that its principal exports were dollar determined, that it relied on imports for most of its food needs and that the domestic economy was too small to allow substantial growth of non-energy exports. The ultimate acceptance of the devaluation was prompted by international organisations' refusal to finance its external resource gap.

During the two decades in question, Gabon's economy remained under government domination, and economic reforms followed a slow and cautious pace as the authoritarian political regime enjoyed power, backed by its former colonist. Although the government officially followed a pro-business, pro-private-enterprise policy, and declared itself ready to grant tax exemption, tariff protection, public credit and contracts to private interests, the regime's highly bureaucratised nature proved a formidable obstacle to such undertakings.

INDONESIA[10]

On the eve of the first oil boom, Indonesia was a poor rural economy, with the lowest per capita income, largest population and smallest per capita oil wealth of all OPEC members. The one significant difference from the rest of the OPEC membership was that Indonesian crude oil exports accounted for only 40 percent of total exports. Indonesia's economic direction and strategy were stipulated in the 'Guidelines of State Policy', sanctioning a mixed economy with a largely private agriculture and a predominantly public energy and heavy industry sector. The government's role was to help the private sector mobilise and develop the country's resources under state guidance. The government retained ownership of certain national resources and exercised extensive regulation over many economic activities.

The means of achieving national objectives were a series of five-year indicative plans (*Repelita*), designed to establish national priorities and sectoral growth targets ,subject to annual review in the yearly budgets. Between 1974 and 1994, four *Repelitas* were enacted. Each of these development plans had a specific target for the average annual GDP growth rate and a set of qualitative goals (e.g. higher standard of living, greater social welfare benefits, increased job opportunities). A large part of the total investment projections in each plan was to be provided by the private sector. As in the case of other group members, these plans were not based on detailed input-output calculations and mandates but, instead, formulated as pragmatic guidelines for investment. Adherence was flexible and pragmatic rather than strict and rigid.

Indonesia's total earnings from oil/gas exports between 1974 and 1994, amounting to an estimated $166 billion, were augmented by continued substantial capital inflows in the form of grants, loans, transfers and foreign investment. Between 1979 and 1986, Indonesia received commitments for nearly $16 billion in bilateral and multilateral aid from Japan, the European Community and the United States, as well as the World Bank, the Asian Development Bank and others. Another $5 billion in new commitments was obtained during 1994. Foreign private investment inflows rose steadily from around $140 million in 1981 to about $2 billion in 1994. Offsetting the initial oil receipts was the government bailout of its giant national oil company. Prior to the oil price rise of 1974, Indonesia's national oil company,

Pertamina, had engaged in substantial short-term borrowing to build a conglomerate of energy, steel, air and sea transport, construction, real estate, chemical, agricultural and tourist enterprises. Failing to meet its obligations in these far-flung activities in 1974, it had to be bailed out by the government, and some two-fifths of the first year oil windfall was allocated for repayment of Pertamina's debt. During the second oil boom when the windfalls rose again and were widely expected to endure, Indonesia moved very cautiously, partly due to the experience gained in the Pertamina case. Part of the new income was 'sterilised' by creating a budget surplus and doubling the international reserves.

The country's diversified resource base was well suited to take advantage of the oil windfalls. Accordingly, the light manufacturing industries that operated in the pre-boom years were supplemented, under state tutelage, by such heavy industries as petroleum refining, LNG, basic metals and cement. By introducing the Green Revolution, the government spurred agricultural production and productivity. Funds were also allocated to modern and sophisticated services in telecommunications, banking and insurance. Oil windfalls were also directed toward public investment in social and economic projects, primarily in the formation of human capital and improvement of infrastructure. In line with the basic objectives of alleviating poverty and promoting a fairer income distribution, a larger part of development outlays was allocated to the less-developed rural areas.

The key objective of diversification was served by focusing public investment on those import-substituting industries in which Indonesia was believed to possess a long-run comparative advantage. Under *Repelita III* (1979–83), development of basic industries, non-oil export manufacturing sectors and labour-intensive activities were emphasised. The government took the lead in launching industrial projects (basic metals, fertiliser, cement, paper and hydrocarbons), as well as infrastructure (transport and power). Oil production capacity was to be raised to 2 mb/d by the mid 1980s. Resource-based industrialisation in Indonesia was largely centred on LNG development (and export). By 1983, net LNG exports – which had started with less than $100 million in 1977/8 – reached $1,355 million. Part of public investment was also allocated to metal processing (steel and aluminum), oil refining and petrochemicals. Steel production, and manufacturing in general, were oriented toward the domestic market under considerable import protection and with various subsidies.

In the early 1990s, a group of 'economic nationalists' backed by President Suharto saw the new vision of Indonesia's future in anything but agricultural and low-tech industries. The group, which had introduced helicopter and light aircraft manufacturing into Indonesia, was aiming at such high-tech endeavors as shipbuilding, communications satellites, nuclear power plants and the like.[11] Despite the apparent non-viability of such projects in the medium-term, President Suharto and his ambitious technocrats defended them as an effort to 'leapfrog technology into the next century'. The Indonesian leadership supported the scheme as 'the consummate national morale builder' and 'part of the national pride'.[12]

Throughout the years, a larger share of the windfalls was allocated to primary education, sanitation and health facilities, roads and rural electrification, i.e. social and physical infrastructure. In addition to direct public outlays on these sectors, Indonesia allocated a substantial portion of the oil income to subsidies. Fertilisers, pesticides and other farm inputs were heavily subsidised to expand the production of rice, rubber and oil palm, where Indonesia had a comparative advantage, (along with coconuts and coffee).

Despite the prominent role assumed by the military, since Indonesia's independence in 1945, in guiding and managing the economy, defence spending was kept at relatively modest levels. During the three five-year plans between 1979 and 1994, total annual expenditure on defence typically averaged about 2.5 percent of GDP and about 9 percent of central government expenditures. The main reasons for this frugality were Indonesia's non-alignment with any multilateral or bilateral defence treaties, an almost total lack of a credible external threat, the absence of any arms race in most of the region and, most significantly, the engagement of the defence forces in a variety of business enterprises (e.g. banks, industrial firms and commercial ventures) that helped finance part of their needs.

Encouraged by rising public sector investment and expenditure, private sector projects approved by the government rose several times during the period. Three quarters of these outlays were in manufacturing (including chemical, mechanical, electrical, vehicles and cement projects). The major part of foreign private investments was in mining and manufacturing (metals and automobiles). The extremely heavy reliance of the private sector on bank credits, and equally alarming dependence of the Indonesian banking system on short-term foreign borrowing, exposed the economy to possible external shocks.

At the core of Indonesia's adjustment process was a basic and steady strategy of agricultural promotion and industrial diversification. However, as large industrial projects offered only a small share of employment opportunities, the government was duly concerned about the rapid increase in the labour force. Five-year development plans routinely encouraged an increasing portion of the new job seekers to be absorbed by small manufacturing, trade and service businesses in the private sector. Private investors were invited to undertake appropriate projects in exchange for import monopoly rights, protection from abroad and restrictions on other potential domestic rivals. In exchange for undertaking large-scale investment projects within the national plans, private conglomerates were given monopoly privileges on the production and imports of key industrial items (e.g. steel and plastics). While these giant corporations were responsible for the country's steady growth, they were 'high cost' enterprises that often added considerably to the domestic price of goods bought by Indonesian users. Political patronage was always a crucial factor in private business success: firms that possessed import licences were highly profitable.

Up until the early 1980s, a steady increase in oil and gas revenues supported domestic development efforts, and the pre-1974 strategy was followed. During the first two oil booms, high development expenditure was partly financed by external borrowing on the strength of future oil earnings. However, the debt-service ratio began to rise as oil/gas exports were increasingly compressed. The collapse of the Indonesian average crude price from $25/b in 1985 to $13/b in 1986, combined with the worsening terms of trade in other export items (coffee, rubber, tin), called for the adoption of urgent remedial measures. To deal with the emerging problems, a comprehensive adjustment effort in the areas of fiscal, monetary, exchange, trade and foreign investment policies was initiated in mid-1986. President Suharto's 'New Order' reflected a clear departure from Sukarno's 'Indonesian-style socialism'. The salient feature of this new order was fiscal and monetary conservatism, or a new 'Indonesian-style free-market system'.

On fiscal policy, Indonesia followed the so-called 'dynamic balanced-budget' principle which, in contrast to all other OPEC members, it strictly maintained throughout the period. Under this sensible creed, public expenditure was divided into 'routine' (i.e. current) and 'development' (i.e. capital) items. Routine spending was essentially financed by domestic tax revenues, and development outlays were covered by oil receipts and foreign assistance. Government borrowing in the domestic market was ruled out. Short-run fiscal flexibility was achieved through the Treasury's accumulated stock of deposits, with the domestic banking system to be drawn upon during budgetary shortfalls.

The post-1983 downturn in oil revenues and the worsening of the trade terms led the state to adopt a series of comprehensive tax reforms, broadening the non-oil tax base and raising its yield. At the core of this reform was the introduction of a value-added tax (VAT) and a new income tax system. After 1985, the country's customs service was thoroughly reformed to raise non-oil revenues. In 1988, the tax on dividend income was extended to income from bank deposits, in order to encourage the growth of the capital market. Indonesia was unique among group members for taking far-reaching reforms in the trade and tax systems after the oil bust, and doubling its non-energy taxes as a share of GDP, thus shielding the budget from oil-price fluctuations. The balanced-budget rule and the prohibition of domestic deficit financing helped the authorities to keep inflation under control.

Monetary policy carried much of the burden of demand management in the Indonesian economy. However, the country's open capital market, and its financial sector's increasing integration into the world money markets, gradually reduced the policy's effectiveness as changes in domestic real interest rates often resulted in disequilibrating capital flows in and out of the economy. Monetary developments, as well as money and credit policy, were mainly influenced by changes in the external payments position. During oil booms, a part of foreign exchange earnings was 'sterilised' in the form of increased government deposits with Bank Indonesia (the central bank). Nevertheless, the sizeable increase in net foreign assets, together with a continued large expansion in credit, resulted in a steady rise in liquidity. Since 1974, Indonesia's credit policy was designed to maintain 'stable' prices, i.e. an annual inflation rate below 20 percent (and since 1991, not more than 6 percent). This policy was implemented through a comprehensive, albeit flexible, system of monetary controls – credit ceilings, selective credit allocation, subsidised low interest loans to small *pribumi* (indigenous) businesses etc. Surveillance was to be exercised by the central bank over the five commercial state and development banks. The objective was to promote economic growth in disadvantaged sectors and to help state enterprises. However, a combination of controlled domestic interest rates and a highly liberal foreign exchange policy prompted occasional capital outflows.

Despite its basically market-oriented stance, Indonesia exercised various controls over prices and wages. After the 1978 exchange devaluation, temporary price ceilings were decreed on rice, petroleum products and utility rates. A temporary freeze was also declared in government salaries. After the second major rupiah devaluation in 1983, prices of non-traded goods (particularly administered prices) were frozen. Price adjustments were made for certain items (e.g. cement, fertiliser and steel) to reflect cost increases caused by the devaluation. Petroleum products, domestic transport fares, electricity and water supply and other public services had their prices determined administratively by state enterprises. Like most other OPEC members, Indonesia kept domestic oil prices below international levels. Oil price subsidies formed a rising proportion of total budgetary expenditures, and relatively low energy prices unduly encouraged energy-intensive production methods.

The Indonesian government set legal minimum wages for some 27 regions of the country on the basis of periodic price surveys. The statutory minimum wage set by the Manpower Department was usually at an extremely low level and even occasionally below the official subsistence level, in order to stimulate employment. Considerable wage differentials existed between areas (urban *vs* rural) and occupations (manufacturing and commerce *vs* agriculture). Wage adjustments for both private and public workers in the 1990s were well above the rate of inflation, in contrast to the deterioration that occurred during the last part of the 1980s. The objective was to bring minimum wages to a level commensurate with a 'minimum quality of life' by 1994.

Indonesia followed a relatively free-exchange regime throughout the period, by following a system of managed float. Policies on foreign exchange market operations were taken by Bank Indonesia. There were no restrictions on foreign currency accounts in authorised domestic banks. Unrestricted transactions in foreign exchange by Indonesian residents was a unique feature of the regime. Indonesia maintained a realistic exchange-rate policy throughout the period. In 1978–94, it continued to follow a very sensible stance aimed at maintaining the economy's international competitiveness in non-oil exports, safeguarding the health of domestic industries and raising the rupiah's worth of the oil export dollars during depressed oil markets. In line with this strategy, the rupiah was devalued from time to time during the twenty-year period – from R415=$1 in 1974 to R2150=$1 in 1994 – under a system of managed float. By allowing repeated exchange-rate adjustments, Indonesia was able to maintain a foreign exchange regime with remarkably few and simple regulatory measures. Exchange-rate policy also played a supportive role to both fiscal and monetary policies in that the rupiah's value was always kept on a path of depreciation to offset domestic/foreign inflation differentials, and speculative pressure for faster devaluation was resisted through monetary and fiscal restraints. In 1988, Indonesia joined the developed member countries of the International Monetary Fund by accepting the obligations of Article VIII of the Fund Agreement.

Unlike its exchange regime, Indonesia's trade system was highly regulated. Foreign trade was governed by regulation, protection, subsidies, quantitative quotas and monopolies. Goods for which an adequate domestic supply was available were either prohibited or available only through approved importers. Allowable imports were subject to licences and classified into several categories from essential to luxury.

Tariffs and/or surcharges for most items ranged from zero to 200 percent. Export taxes and quotas were also imposed from time to time to safeguard domestic supplies, as were export incentives for some items. Beginning in the late 1980s, a number of import liberalisation measures were adopted with a view to improving domestic competitiveness. Non-oil exports were further encouraged through a variety of mechanisms such as duty-free zones for export assembly activities, export credit facilities and technical aid. With accession to GATT, the government agreed to phase out the subsidised export credit system in stages. Following the shift in trade policy from import-substitution to export promotion, a new package of trade reforms was announced virtually every year, the main features of which were to further liberalise tariff and non-tariff barriers on manufactured items and to facilitate non-oil exports.

Foreign direct investment was subject to the Foreign Capital Investment Law of 1967 and its subsequent amendments. Foreign investors were welcomed to invest in projects approved by the government, in certain approved sectors and under certain criteria. Investors were guaranteed the right to repatriate capital, transfer profits after taxes and make transfers in respect of depreciation allowances and other business expenses. From 1992, full foreign ownership in the form of foreign direct investments was allowed in certain sectors and under certain criteria. Foreign investment had to begin divestiture after some years of operation. Incentives for foreign investment included exemption from duties on imported inputs, generous annual depreciation allowances on machinery and certain other tax benefits. The Jakarta Stock Exchange was to serve as a vehicle for facilitating private investment.

Privatisation was another feature of Indonesian strategy towards the end. Under state guidance, public enterprises had begun to expand after 1974 and remained large, accounting for 15–20 percent of GDP and dominating such major industries as basic metals, chemicals and cement. With the 1986/7 monetary, trade, tax and investment liberalisation, the emphasis for development was shifted to the private sector. A new banking law in 1992 allowed state banks to sell up to 49 percent of their equity to domestic and foreign investors. By some estimates, the private sector in 1994 accounted for three quarters of non-energy economic activity and about two-thirds of fixed capital formation.

In sum, Indonesia based its development strategy on lowering barriers to trade, opening the economy to direct foreign investment, keeping taxes low, curbing inflation, allowing timely exchange rate adjustments, obtaining concessional foreign grants and loans and avoiding budgetary deficit.

IRAN[13]

Of all the group members Iran, between 1974 and 1994, experienced by far the most profound and far-reaching transformation in its system of government, its socio-political institutions and its economic fortunes. The 1979 revolution resulted not only in the regime's change from a constitutional monarchy to a theocratic republic, it also led to rising uncertainties regarding private property and free enterprise, the rule of law and the direction of the economy.

Between 1974 and 1979, Iran had a mixed economy where both public and private sectors had their roles to play: it was a free-enterprise economy guided by the state, without a sharp ideological slant. The post-revolution economy (1979–94) was still a mixed regime but guided by an avowedly populist and statist tendency, without a solid rudder. The emphasis of economic policy was directed toward self-sufficiency, redistributive justice and Islamic values. For part of the latter period, 1982–8, coinciding with the management of a war economy, the economic model was the old Indian-Soviet variety characterised by economic insularity, domination of the state over the economy, repudiation of 'consumerism' and reliance on import-substitution industrialisation. In 1989–94, a nationalist/populist stance was adopted, in a Saudi-Chinese fashion, whereby a tightly managed capitalist economy was to coexist with a non-capitalist, non-democratic political ideology and a strong social welfare orientation.

Immediately after the 1979 revolution – coinciding with the second oil boom – when the Iranian government faced the exodus of many businessmen, property owners and skilled workers, a shaky financial system and labour unrest, the bulk of the country's manufacturing, banking, insurance and foreign trade was taken over by the government. As a result of the wholesale nationalisation of banking, insurance and, *de-facto*, foreign trade, and following the confiscation of businesses and properties belonging to the former regime's officials and associates, the regulatory role of the government in the economy expanded rapidly during the 1980s, even though the state's monetary share steadily declined. In addition to the interventionist bias of the left-of-centre government in most of the 1980s, war exigencies prompted the authorities to extend economic regulation and control into almost all aspects of the economy.

With the rise and proliferation of parastatal foundations (*bonyads*), total public control over the economy steadily advanced both in monetary and regulatory spheres, so that by the late 1980s, no more than 30 percent of the economy was effectively in private hands. At the same time, due to the fragmentation of political power among various clerical groups, some 20 percent of the economy was estimated to have fallen into the so-called 'informal' category, i.e. outside formal national accounts. The largely independent *bonyads* – run by the clerics and responsible only to the country's supreme leader, with special access to cheap foreign exchange, low-cost credit, import priveleges, tax exemptions and lucrative contracts – had a very high stake in the public economy.

In the first oil boom, Iran had a prescriptive plan for the public sector and an indicative plan for the private side. The bulk of oil revenues was earmarked for direct, public, fixed capital formation in specific fields, and the private sector was induced through incentives and assistance to invest in other fields. With the exception of infrastructural and defence-related activities, public investment policy was to select enterprises which were unattractive to private interests by virtue of size, capital needs or risks. Encouraged by the successes of its previous plans up to 1972, state planners thus began with a 'big-push' industrialisation strategy during the first oil boom. The strategy called for the purchase or lease of foreign technology, capital equipment, management know-how and even skilled manpower, while simultaneously presumably intent on preserving the country's heritage and strengthening its traditional values.

In the post-revolution era of populism and left-leaning distributive stance, the necessity of development planning to ensure 'social justice' was inscribed in the new 1979 Constitution. Several attempts at the preparation of a national plan, in 1981, 1983 and 1986 proved fruitless due to the exigencies of the Iran/Iraq war, internal bickering among ruling factions in the *Majles* (parliament) and the oil price collapse. The first post-ceasefire plan (1989–94) was approved in January 1990. But as with the similar plans of other group members, results deviated from planned targets by wide margins, and the exercise turned out to be largely intellectual and academic.

Iran's total receipts from oil and gas exports during the 20-year period was upward of $315 billion. But this enormous sum was very unevenly earned. Except for the latter part of the 1970s and the earlier part of the 1990s, Iran did not make much use of foreign credit. Due to the revolutionary change of the regime, as well as marked fluctuations in annual oil earnings and the 1980–8 war with Iraq, the oil revenue allocation followed several different paths. Iran's Fifth Five-Year Plan (1973–8) of about $37 billion was already three and a half times that of the Fourth. With the nearly quadrupling of the oil exports revenue in 1974, planned outlays were revised upward to nearly $70 billion. By all indications, the original plan was revised with haste and in an atmosphere of unbridled euphoria. Several new projects in direct-reduction steel, nuclear energy, satellite telecommunications and equity investment in foreign firms were added to the list of those already set aside due to lack of funds. On the whole, the revised plan exceeded Iran's absorptive capacity economically; it also made some social promises that it could hardly keep.

Between 1973 and 1978, most of the windfall was spent domestically, and the rest was earmarked for foreign loans, investment and aid. The adjustment process was thus carried out through domestic absorption plus a reduction in some import levies. During the period, private consumption in real terms was allowed to rise by more than 40 percent, public consumption by 84 percent, gross domestic fixed capital formation by 100 percent and merchandise imports by 350 percent. There was clearly an attempt by the government to reach its three-pronged objectives of raising the people's standard of living, diversifying the oil-based economy and meeting national defence needs. As a clear priority in promoting economic diversification, the government favoured capital intensive industries that depended heavily on imports of raw materials, semi-processed goods and equipment. At the same time, dependence on imported inputs made the manufacturing sector imprudently vulnerable to changes in oil prices and government revenues. Insufficiently promoted also were smaller industrial enterprises that depended on renewable domestic resources and technology, and were thus relatively immune to external stimuli beyond national control. The government's policy toward agriculture was one of secondary attention.[14]

The post-revolution chaos of 1979/80, the Iran/Iraq war starting in 1980 and the post-1982 oil bust played havoc with development planning. Although faint efforts were made by the government in the early 1980s to construct a viable plan, considerable internal political bickerings between various radical and conservative factions within the theocratic oligarchy brought these efforts to naught. Although the second oil boom was caused by the abrupt decline in the global oil supply as a result of the

Iranian revolution, Iran itself did not become a major beneficiary. A combination of political disturbances, economic disruptions and output dislocation in 1979–80 – plus a deliberate decision by the new revolutionary government to limit oil extraction and exports – denied the country another oil bonanza. The post-1979 allocation strategy was akin to an Algerian-Indian model, rejecting 'economic dependency' in favour of self-reliance. As in the Algerian case, there was a strong bias against consumerism and export promotion, and much sympathy toward domestic production of basic necessities, import-substitution and the so-called 'mother' industries. A 'consumption model', emphasising frugality and avoidance of luxury consumption items, was to be formulated by the government to guide the people's daily life. After the ceasefire with Iraq, and the beginning of a more market-oriented stance, sectorial priorities became a bit more balanced.

After a hasty and short-lived decision to limit oil extraction to meet 'no more than essential national needs' (i.e. to reduce oil output), the Oil Ministry in 1983 announced a comprehensive, 20-year plan for the expansion of the hydrocarbon sector. After the ceasefire in the Iran/Iraq war in 1988, the government announced its intention to expand daily oil production, both onshore and offshore, to 5 mb/d by the year 2000. Of significance in resource allocation under the second oil boom (and post-revolution priorities) was the concentration of state economic activities in the rural areas as part of the government's 'social justice' promises as well as a means of preventing rural depopulation. To these ends, substantial public investments were earmarked for rural road building, rural electrification, extension of telephone lines and provision of safe drinking water and health care. Chastised by the precipitous fall in oil revenues following the 1986 oil bust, the government raised the self-sufficiency banner once again and announced an ambitious new plan to make Iran 'dependent on its internal resources and domestic production'. Agriculture once again was to become self-sufficient in five years. In industry, the priority was given to the expansion of steel, development of gas and oil fields, construction of power plants and downstream utilisation of the existing copper complex.

With industry operating at or below 40 percent capacity by the end of the Iran/Iraq war, unemployment at least 16 percent and the cost of living index running at nearly 30 percent a year, the new government, formed under the revised constitution, in 1989 adopted a peacetime strategy of reconstruction and economic restructuring. This strategy, while significantly market-oriented and liberal compared to the statist and interventionist strategy of the previous seven years, followed more or less the same priorities as before. The new First (post ceasefire) Five-Year Plan for 1989–93 was ratified in January 1990. Priorities were now slightly changed in favour of social rather than economic development. In the public sector investments, however, the largest share was given to oil and gas, water and power, transport and communications, industry and mines, and agriculture, in that order. The third oil boom, 1990–1, was a short-lived bonanza for Iran. Believing the oil price rise would continue, the government opened the imports gate, and allowed both the private sector and public enterprises to satisfy their pent-up demand by ordering goods from abroad through short-term export credits (usance) of a year or so in duration. As a result, imports almost doubled in the next two years, and the country, which had conducted a costly

8-year war with Iraq without foreign borrowing, went into heavy short-term debt and substantial arrears.

In both the pre- and post-revolution periods, Iran pursued a series of moderately extensive social welfare programmes designed to improve the national levels of education and healthcare and to alleviate poverty. Between 1974 and 1978, these programmes under the banner of the 'White Revolution' could be easily financed and, except for continued housing shortages, were fairly successful. The post-1979 attention to the poor (in whose name the revolution had taken place) resulted in a significant allocation of resources to the depressed rural areas and underdeveloped urban sectors. Food subsidies, below-cost sales of public goods and services, and the quantitative expansion of educational and health facilities absorbed a significant portion of the national budget.

Military expenditures also took a large share of the oil receipts. Iran followed an extensive programme of defence build-up and military modernisation before the revolution. During the eight-year war with Iraq, a sum of $4–6 billion a year was reportedly spent on defence (out of a total defence budget that accounted for 30 percent of annual expenditures), with an estimated $2 billion on foreign arms purchases, mostly on international black markets. After the ceasefire in 1988, defence expenditure seemingly declined but the very low official figures were suspect. According to informed estimates, annual military expenditure in the 1990s averaged about 11 percent of total government expenditure and about 4.5 percent of GDP. Iran had a modest foreign assistance and foreign savings programme during the first oil boom and prior to the 1979 revolution. Grants and loans were offered to a number of developing countries. Contributions were made to the IMF Oil Facility and the World Bank lending funds. The government also lent to some European countries and purchased manufacturing assets on the continent. Some joint oil refineries were also built with India, South Africa and others.

The policies adopted to implement these windfall allocations followed the conventional lines. Fiscal policy, particularly in the first two years of the initial oil boom, was exceedingly expansionary. There was a deliberate attempt to spend the windfalls as soon as they were received. The spending spree was naturally accompanied by the approval of many sub-optimal projects that escaped careful scrutiny. And, while the state's regulatory power did not change, its fiscal and financial roles were considerably enhanced as public consumption expenditures rose into the double-digit range. Oil-financed public expenditures raised aggregate demand for both traded and non-traded goods but put its heaviest price pressures on the latter as domestic supply could not increase in the short-run. The post-revolution government followed a rather conservative fiscal policy. The ratio of public expenditure to GDP fell from more than 40 percent just before the revolution to less than 20 percent by the end of 1988, despite the war raging between 1980 and 1988. The decline in public spending, caused mainly by falling oil revenues during the period, affected mainly the capital side of the budget.

As a percentage of GDP, public capital formation was only a third of the pre-revolution level in real terms. No attempt was made to raise non-oil revenues through taxes. In fact, the tax/GDP ratio declined by more than 55 percent in the 1980–90 period. During the post-war reconstruction period (1989–94), total public expenditure rose to around 30 percent of GDP, of which capital outlays accounted for about 7 percent in 1994. Persistent annual budget deficits (5 percent of GDP in 1994) were covered mainly by borrowing from the central bank. Oil revenues continued to account for the lion's share of the total government budget – as high as 70 percent in 1994. The ratio of (non-oil) taxes to GDP was about 7 percent. Even with consumption and other taxes, the ratio did not exceed 12 percent – one of the lowest among countries in Iran's income group.

Monetary policy, 1974–9, was largely accommodative to fiscal policy due to the absence of a developed capital market and the small volume of monetary assets. While the central bank could effectively control private domestic assets through reserve requirements and credit ceilings, the major component of domestic liquidity, i.e. changes in the net foreign assets (accrued to the treasury and converted into local currency), was beyond its control. The treasury's use of oil revenues in the domestic market made the task of monetary control most difficult. Domestic public expenditures were thus the major determinant of the rate of monetary expansion. The central bank used its two key policy instruments, raising the cost of funds (interest rate), and regulating the money supply (reserve requirements and credit ceilings). In deference to the Islamic prohibition of interest (usury) payments on deposits, the nationalised banks were required in 1984 to bring their operations into line with 'reba free' (no interest) practices. Under the 1984 law, depositors were bank partners and were thus paid on an equity-based profit sharing system out of bank profits. Borrowers also shared risks with the banks, for which they had to pay the bank part of their business profits. In practice, this type of Islamic banking had several expected (and not altogether salutary) consequences. Concerned about borrowers' choosing riskier projects and exposing loan portfolios to possible large losses, the banks shied away from long-term investment lendings, and placed an increasing portion of their assets into short-term commercial, trade and services projects. In this type of quick-returns operation, partnership and profit-sharing lost much of their significance. Depositors were routinely paid various rates of interest called 'profit on capital', and borrowers were charged various interest rates based on type of activity, called 'service charges'. Furthermore, finding public sector enterprises safer bets for lending, banks steadily diverted their credits away from the private sector in favour of public entities. Still further, since the structure of depositors' 'profits' and borrowers' 'charges' was fixed by the central bank at any given time, many depositors and borrowers stayed away from the formal banking system, and preferred to deal with a host of other private, informal and unregulated financial institutions in order to escape close scrutiny, earn higher returns and engage in riskier undertakings. Finally, since rates, credit ceilings and sectoral allocation of bank credits were all regulated by legislation or the central bank – and deposit interest rates were always below the rate of inflation – depositors routinely lost part of their wealth and selected borrowers routinely reaped windfall 'rents'.

Price control was a permanent part of Iran's national economic policy. With a surge of prices during the first oil boom, despite expanding domestic output and rising imports, the government was forced to institute price controls. Price and wage controls continued throughout the post-revolution period. So did public subsidies on fuel, food and other essential consumer goods. Rationing was in place during the war years, and the sale of certain items against coupons at lower (subsidised) prices in government-supported stores lasted through 1994. Minimum daily wages were set by the government and changed periodically, but the changes barely kept up with inflation. The large informal economy also eluded wage or price enforcement. Various government attempts to reduce food and fuel subsidies were resisted by the *Majles* or grudgingly allowed in the case of petroleum products after 1994, albeit still at a fraction of international prices.

During the first oil boom, the government tried to maintain a relatively stable nominal exchange rate to the US dollar. By the end of the first oil boom, the real exchange rate in terms of the US dollar appreciated by 36 percent. Yet since the government routinely supplied foreign exchange at the official rate, there was no black market for foreign currencies, and the official rate remained between Rls 70–71 to the US dollar. During the second oil boom after the 1979 revolution, the demand for foreign exchange suddenly surged as most well-to-do Iranians tried to leave the country and transfer their assets abroad, and as economic disturbances intensified inflationary pressures. In 1980, the Iranian rial which had, since 1976, been pegged to the Special Drawing Rights with large margins for fluctuations, was officially delinked from the US dollar and valued at the SDR/dollar equivalent of Rls 93. As capital flight continued, the revolutionary government put restrictions on exchange transactions and rationed the available supply of foreign exchange (mostly dollar revenues from oil exports). Military imports, capital flight and the freezing of Iranian foreign assets by the United States after 1980 prompted the government to impose capital controls and introduce a system of multiple exchange rates. Western sanctions on Iran precluded any meaningful access to foreign capital markets (while the government was also not disposed to use them). The result was the intensification of black market activity. The government maintained the official exchange rate at the (artificially low) pre-revolution level and adopted a very complex multiple-rate exchange regime, which eventually encompassed 12 separate rates. This policy, in turn, contributed to massive cost/price distortions, high budget deficits and the worsening of the non-oil balance of payments. Available foreign exchange (mainly oil revenues) was allocated among claimants under a special foreign exchange allocation budget. With the black market premiums steadily climbing toward 2000 percent by the end of the 1980s, the mere receipt of a foreign exchange quota by select importers became an active (rent-seeking) preoccupation in itself.

After the end of hostilities in 1989, the government began to chart a new strategy aimed at reconstruction and the restructuring of the war-ravaged economy. In line with these objectives specified in the 1989–94 plan, the authorities proceeded over the 1990–2 period to lower the budget deficit, liberalise imports and exports, contain inflation, reduce the number of exchange rates to three, remove controls from many prices, eliminate quantitative credit ceilings, raise public utility rates and energy prices, and set forth a privatisation plan to sell money-losing state enterprises. The pre-revolution law for attraction of foreign private investments was reactivated. The foreign exchange budget was eliminated. Nevertheless, the government until 1993 refused to lower the official rial rate, which would have been an easy way of eliminating the budgetary gap by selling petrodollar revenues at higher rates. Concerns about rising inflation and falling real wages were among the reasons for maintaining the artificially low exchange rate. In some influential government circles, devaluation was also regarded as a sign of weakness and failure. In early 1993, the rial was finally devalued, unified for nearly all transactions and declared officially 'convertible'.

Even limited amounts of capital transfers were authorised at the new unified rate of Rls 1600=$1. Failing to adopt supplementary monetary and fiscal policies to support exchange devaluation, the liberalisation scheme soon backfired. Liquidity expanded at a rate of 20–25 percent a year, producing similar rates of inflation. A sixty percent jump in imports not only failed to effectively dampen price pressures but added to higher external indebtedness and difficulties in the short-term debt service. Budget deficits which had been temporarily tamed began to rise again. The unified exchange rate was ultimately abandoned in December 1993.

Confronted with severe macroeconomic imbalances, e.g. some $23 billion of external debt – mostly short-term and in arrears – depressed oil prices, declining reserves, new American trade sanctions, rising prices, external deficits and decreasing access to the world capital market (including the World Bank credit), Iran was not able to inaugurate its Second Five-Year Development Plan in March 1994 as scheduled. During the one-year hiatus, the economy was subjected to renewed economic restrictions and re-regulations. As part of its bilateral debt rescheduling efforts, the government introduced a new dual exchange rate system, reimposed import and export controls, reinstituted non-oil exporters' obligation to surrender their foreign exchange proceeds to domestic banks, banned all exchange transactions outside the official channels, established strict allocation of oil revenues at the new official exchange rate of Rls 1750=$1 and set the rate for all other transactions at Rls 3000=$1. The pace of privatisation was also considerably slowed down, and the attraction of foreign private investment faced new resistance.

In sum, during this 20-year period, Iran moved from a progressive mixed economy where private investment and activity were to increase, back to a regressive variety where the state and its parastatal appendices dominated the scene without greater national resources or managerial competence. The Iran/Iraq war also had its share in intensifying the state's interventionist tendency.

IRAQ[15]

Iraq's economy throughout the period, and under the Baath military dictatorship, was a centralised socialist system whereby the state dominated and controlled nearly all aspects of economic activity with the exceptions of small farming, retail trade and personal services.[16] With the nationalisation of the last foreign oil company (Basrah Petroleum) in 1975, the entire oil sector was taken over by the state. In a Baath Socialist party congress in 1982, the 'legitimacy and importance' of the private sector in the economy was formally recognised. However, the ultimate control over private business activity was still reserved for the state, and official admonishing of 'exorbitant profit' kept private entrepreneurs largely at bay. The public, or 'socialist', sector included the central and local governments plus hundreds of public enterprises operating in the agricultural, industrial, transport, communications and services sectors. Other public corporations or agencies also operated in the banking and insurance sectors. The state continued to be the largest employer of labour.

Iraq's oil export receipts during the two decades had the most volatile path, due to fluctuations in the volume of crude sales. Petroleum revenues ranged from a record $26 billion in 1980 to a paltry $420 million a year on average in 1991–4. The total for the period amounted to about $190 billion. With oil reserves estimated to be the second largest in the group, an abundant supply of water and some 8 million hectares of cultivable land, Iraq presented an ideal model for successful oil-based development. The nationalisation of foreign companies' oil assets in 1972 and the sudden jump in oil prices offered the government a rare opportunity for growth through diversification. Iraq's strategy for the use of the growing oil windfalls was to proceed with full-scale domestic investment, especially in capital-intensive infrastructure and energy. Basic development agenda emphasised economic self-reliance, rapid growth and diversification. Development priorities focused on energy, water and irrigation, land reclamation, power, transport, communication, health and education. Priority in industrial investment was given to the oil sector (with a target of 6 mb/d output by 1980) and the chemical, electricity, textile, food, and metal industries, in that order. After the onset of the first oil boom, attention was focused on petrochemicals, iron and steel works, sulphur, phosphate and fertiliser projects. In agriculture, the aim was self-sufficiency in food production by 1980. Agriculture was particularly emphasised as 'permanent oil'.

Except for direct involvement in the energy and military sectors, the state's principal engagement was in the expansion and improvement of the physical and human infrastructure. With some 99 percent of total export earnings derived from oil and accruing directly to the government, and 65 percent of GDP emanating from the public sector, the state was the instigator, planner, investor and executor of the country's economic development. With the nearly nine times increase in total state revenues between 1973 and 1980, government expenditure rose eight times and development outlays more than 10 times. In this initial period, planning was concerned more with the subsidisation and promotion of social welfare than with economic efficiency.[17] Iraq's relative manpower shortages, and Saddam Hussein's determination to break out of 'third-world status' led the authorities to seek projects with up-to-date technology, including forays into the nuclear field. Under the impact of the war with Iran, the oil sector was virtually devastated, and detailed planned allocations became irrelevant. Most of the projects, except those contributing to the war effort, had to be suspended. And a rush to secure foreign credits to finance the projects already under way began in earnest in 1983. During the war with Iran, long range planning was replaced by tentative and flexible annual budgeting. And only a small fraction of development allocations was actually spent. Although annual budgets contained allocations for investment projects, the latter were rarely followed. Development funds were routinely diverted to the ordinary budget for current expenditures. Agriculture suffered most by such diversions.

As the war with Iran wore on, the priorities gradually shifted toward defence, followed by the allocation of oil revenues to an irreducible minimum of imports of staple foodstuffs, raw materials and spare parts. Agricultural self-sufficiency (twice postponed in its deadline) and exports diversification occupied the back burners for a while. Baghdad was later accused of diverting a significant portion of oil export receipts into the development of nuclear, chemical and biological warfare.

Since the majority of development projects initiated by the state were effectively carried out by private (both domestic and foreign) companies, the private sector continued to exist, and the market played its role in economic policy determination. To be sure, those in the private sector who benefited most from the state largesse often enjoyed privileged access to the regime's leadership. But the overwhelming majority of the work force earned their living from the private sector, and effective wage levels were set by that sector. The government also provided the private sector with low-cost credit, subsidised machinery and raw materials, and advances on contracts.

In line with socialist principles, private land holdings were limited, and large private estates broken up in the 1970s to encourage small owner/operator farms. As excessive fragmentation precluded mechanisation and economies of scale, the government resorted to collectivisation. However, state and collective farms, as well as state-directed farm cooperatives, admittedly proved 'worthless' and impeded the growth of farm output. After party officials' criticism of agricultural performance, certain measures for farm privatisation were adopted after 1983. In order to encourage farm output, prices paid to farms for some staples were periodically raised, and other incentives were offered. These included subsidised inputs tax exemptions, decontrol of retail farm prices, low interest loans and, after 1987, leasing or selling state farms to private investors.

Industry was concentrated in the so-called 'socialist sector' under state tutelage, receiving large government funds. Most public investments were at first oriented toward the domestic market to encourage import substitutes. Priority was accorded to industries using domestic raw materials i.e. petrochemicals, fertilisers, power generation, steel, cement and construction materials. In order to encourage private participation, a number of incentives such as exemption from profit and import taxes, low-cost credit, low-cost industrial sites and infrastructure, and technical training were offered by the government. Furthermore, the private sector was allowed to import raw materials and foreign labour; and privately produced goods were priced high enough to ensure an adequate return on capital. As a result, the private sector expanded its activity in light consumer industries and agrobusiness.

Fiscal policy was largely focused on the distribution of oil revenues. During the first oil boom, gratified with the sudden windfalls, the Treasury decided to reduce the tax burden of both public and private enterprises. Customs duties on the 'socialist sector's' imports were eliminated; so was its income tax. Expenditures were divided between current and development outlays and kept up with the fluctuations in oil receipts. Most economic activities derived from these public spendings. The war with Iran, cutting sharply into Iraq's oil revenues and combined with the oil price collapse of the mid-1980s, made non-oil revenues relatively more significant – although information on fiscal trends was not published until 1989. The Persian Gulf war and the UN-imposed sanctions again made information on budgetary operation fragmentary. Due to the strong government fiscal position, there was no noticeable change in the internal national debt during the 1970s. In the 1980s, the central bank sold some bonds to the public, mainly to encourage savings and to absorb excess liquidity. But the shortfalls in oil revenues were made up mainly through foreign loans and grants.

Monetary policy was easily handled by the government since the entire banking system – the central bank, the single commercial bank (Rafidain) and three specialised banks in agriculture, industry and real estate – was state-owned. Domestic liquidity fluctuated with the ups and downs in government domestic expenditures. Rapid expansion of credit reflected partly the rapid increases in borrowing by public enterprises as well as the private sector. Iraq's three specialised banks in the housing, agriculture and industry sectors were given the responsibility to offer low-cost credit for investment in their specific areas. In order to encourage homeownership, real estate loans were kept in the 0–3 percentage range depending on location and the amount of the loan.

Prices were subject to an extensive system of controls and subsidies. Domestically produced commodities were priced at cost plus a certain mark-up for overheads and some margin for return on investment. Imports were priced on c.i.f. invoices plus similar mark-ups and margins. Price changes were made by administrative decrees issued by the Trade Regulating Committee. In order to encourage farm output, prices paid to farmers for certain special staples were raised periodically. Salaries and wages in the public and mixed sectors were set by the TRC in a range of upper and lower limits in each category of labour. The private sector was expected to observe the same scales. In the aftermath of the Persian Gulf war, rationing of all essential consumer goods became a norm.

Exchange and trade policies followed more or less the socialist line. Foreign exchange was controlled and regulated by the central bank. The Iraqi dinar (ID) was pegged to the US dollar, the intervention currency, between 1973 and 1982 at ID1=$3.38. As of 1976, all foreign exchange transactions had to take place through a licenced dealer (e.g. branches of the sole commercial bank, Rafidain). Foreign exchange allocations for imports were made, according to an annual import programme, among major economic sectors. All payments for invisibles required authorisation. Export proceeds had to be surrendered through a licenced dealer. In October 1982, the dinar was devalued by 5 percent, while the nominal effective exchange rate for the currency had appreciated by 42 percent. Under the 1987 liberalisation programme, state and mixed-sector companies were allowed to keep a large portion of their export proceeds for the purchase of their own import needs, thus effectively opening up a semi-official parallel foreign exchange market, in addition to the flourishing black market. While the official rate was retained past the Persian Gulf war, the post-1991 hyper-inflation and freezing of Iraqi foreign assets abroad under UN sanctions made the rate unreal and, for all practical purposes, meaningless.

Foreign trade – essentially imports – was controlled by the annual state imports budget, drawn each year according to the availability of foreign exchange. Permitted and prohibited imports were listed on the import programme for each year. The bulk of imports was actually carried out by government departments and public enterprises. The private sector's share was marginal. Certain specified goods were prohibited for export when domestic supply was short. The export of certain items was reserved for a state agency. Non-oil exports were subsidised through taxes levied on import licences.

Overburdened by rising war costs, domestic subsidies and debt, Iraq in February 1987 launched a liberalisation and privatisation drive aimed primarily at increasing domestic efficiency. Breaking with two decades of socialist ideology, the reform programme emphasised domestic price adjustments, mobilisation of private savings, reduction of external deficit and improved access to foreign credit. As part of the reforms, the old, socialist-based labour law, guaranteeing lifetime employment to workers but maintaining obligatory job assignment, was abolished, and private sector employees were allowed to form labour unions. The goal was to reduce the labour costs of production, estimated to be two to four times higher than in non-socialist countries. The hope also was to reduce the number of administrative and service staff in typical state enterprises (which sometimes reportedly equalled 50 times more than in developed countries) and cut the labour cost of output from 20–40 percent to 10-15 percent, the norm in similar non-socialist economies. The 1987 privatisation drive envisaged outright sale of public enterprises (bus companies, gas stations, farm enterprises, department stores and, most of all, the national airline). Restrictions on foreign direct investment were also to be reduced; and joint ventures, which, as late as 1980, were considered an infringement on national sovereignty, were to be encouraged. In actuality, very little came of these visionary reforms.

After the end of the war with Iran, the Iraqi economy – plagued by stagnant oil revenues, reduced output and living standards, and rising foreign debt – had still nearly two-thirds of its GDP in state hands. The government controlled nearly every segment of the economy, affecting the private sector directly and indirectly. Although the compatibility of private enterprise with Baath's brand of socialism was acknowledged by Saddam Hussein as early as 1979, and reaffirmed by him in 1984 and 1986, the policy had not been translated into concrete action. The Persian Gulf war and its aftermath further intensified the state role in the economy, and put an end to any semblance of liberalisation. In sum, Iraq's development strategy and economic policy were determined by its Baath socialist ideology and immensely influenced by its debilitating war with Iran in 1980–8, the disastrous 1991 Persian Gulf war and continuing UN sanctions.

Iraq's virtual 'war economy' for nearly two thirds of the period under review naturally absorbed a disproportionate part of available resources. Although data in this category are least available or reliable, there are some informal estimates. In the first four and a half years of the war with Iran, annual military expenditure reportedly averaged $24 billion a year ($5 billion on foreign procurements). In the last three and a half, the figure was down to $14.5 billion. With total oil revenue of about $121.5 billion between 1980 and 1990, Iraq is estimated to have spent nearly $205 billion on military mobilisation and warfare. Given Iraq's total gross national product during the period of about $558 billion, military expenditure gobbled up close to 37 percent of the total.[18]

The total cost of the Iran/Iraq war for Iraq has been the subject of various un-official estimates. At the low end, the direct actual cost in terms of destroyed war material and damage to domestic non-oil economic installation was probably $50 billion. Indirect costs in terms of lost oil revenues, losses in non-oil GDP and unrealised capital investment were partially estimated at $175 billion.[19] At the high end – including military expenditure and arms imports, replacement cost of destroyed assets, loss of oil revenues, extra cost of trade routing and potential losses in GNP – the material damage has been estimated at more than $450 billion.[20] The value of assets destroyed during the invasion of Kuwait and its aftermath was esti-mated to be nearly $235 billion or several times the level of Iraq GDP at the time of the invasion.[21]

By all indications, Iraq's welfare programme in the 1970s and 1980s was among the most comprehensive in the Arab world. In addition to generous outlays for edu-cation and health, food and other commodities were made available at subsidised prices through the state distribution system. By one estimate, Iraqis had the highest calorie consumption per capita in the region by 1990. In 1974, Iraq established the Iraqi Fund for External Development (IFED) with an initial capital of $190 million to channel aid to developing countries, mainly to help meet their energy needs. Between 1974 and 1979 some $700 million of financial assistance was given out by IFED. Another $220 million was loaned interest-free for 20 years to 12 countries to help them pay for their highest imported energy cost after 1979. Bilateral foreign aid was stopped by the end of 1982. Aid through international organisations was also cut in the same year. From then on, Iraq itself became a major recipient of foreign aid from its Arab neighbours and other benefactors.

KUWAIT[22]

The three distinct features of the Kuwaiti economy were the crucial role of oil and gas in GDP, the dominance of the public over the private sector and the heavy depen-dence of the city-state on expatriate labour. Petroleum production and refining remained the mainstay of the economy. Kuwait was one of the few OPEC members with oil operations worldwide. It produced oil in several regions of the globe and operated oil refineries in Europe and elsewhere. And Kuwait Petroleum Corp. had thousands of service stations in Europe under its Q8 brand. Kuwait welcomed the par-ticipation of foreign investors in oil projects in order to benefit from foreign expertise, know-how and capital, especially in the construction of petrochemical complexes.

The government owned and operated the oil industry, supplied all public utilities, provided most community and social services and part of communications, trans-portation, and construction facilities and ran a large number of enterprises in petrochemicals, manufacturing and the retail sector. The state contributed more than 80 percent of GDP, was responsible for more than half of domestic expenditure, and was the largest buyer of goods and services (including defence).

Kuwait received a total of just about $200 billion for the 20-year period, albeit in a highly volatile fashion. A tiny city-state with a very small population, sitting on about 10 percent of the world's known oil reserves, Kuwait in 1974 was a defenceless country surrounded by ambitious neighbors. For this reason, the country's allocation strategy was geared to both domestic economies and foreign politics. This strategy was to make prudent and effective use of oil reserves not only for national defence but also for the achievement of sustained and diversified growth by (a) channeling a large portion of oil income into investment in infrastructure and industry at home and (b) the acquisition of diversified income-producing assets abroad. A major objective of this strategy was the development of a fully integrated oil sector under national ownership and control. To this end, high priority was given to increased domestic refining of oil output and the development of downstream operations abroad. The development of non-energy industries – mainly construction material and transport equipment – were relegated to the private sector, aided by government financing and infrastructural facilities.

Conservation played an important role in Kuwait's oil production strategy for nearly a decade. Oil production which had reached 3.3 mb/d in 1972 was placed under a ceiling of 2 mb/d as a conservation move. The ceiling was further lowered to 1.25 mb/d in 1981. In the aftermath of the 1985/6 oil price slump, Kuwait reversed that policy and adopted an increasingly hard line with respect to its quota *vis-à-vis* other members. (Proven reserves had been steadily revised upward from 67 billion barrels in 1983 to 94 bn barrels in 1992.) Kuwaiti delegates in OPEC meetings routinely demanded higher quotas, and when such requests were denied, the government chose to ignore its assigned quota of 1.5 mb/d and produce close to 2.2 mb/d. The latter was one of Iraq's excuses for invading Kuwait in 1990.

Due to the country's limited productive base, a principal objective of investment strategy was to increase the domestic value added of its petroleum exports. This strategy called for substantial investment in domestic oil refining and the production of petrochemicals (starting from fertilisers and moving into more sophisticated products). Again, due to limited domestic absorptive capacity, opportunities for petroleum refining, petrochemicals and retail gasoline marketing abroad were actively pursued.

Kuwait had both a ministry of planning and a series of five-year development plans. Its public sector exercised a considerable degree of centralised control over financial operations. The central government, public institutions and commercial and public enterprises constituted the public sector. The main focus of Kuwait's planning was on the use of oil windfalls for strengthening and broadening the economy's production base, increased Kuwaitisation of the labour force through training and education programmes, an emphasis on technical and vocational studies, and raising its citizens' economic welfare. Kuwaiti nationals normally comprised less than half of total population and less than 25 percent of the labour force. Some sectors (e.g. construction, agriculture and trade) were almost totally (99 percent) dependent on foreign labour. Kuwait's civil service traditionally employed about 25 percent of the labour force and about half of all working Kuwaiti nationals. The state as a whole (the civil service plus public enterprises) was the predominant source of employment for Kuwaitis, accounting for more than 93 percent of working citizens.

Kuwait's non-oil industries were essentially in private hands or were joint ventures with the government, enjoying such advantages as a solid infrastructure, long-term loans at concessional interest rates, subsidised land leases, low-cost energy, import protection and priority in contract awards. Most manufacturing activity involved building materials and consumer goods. During the mid-1980s, the government bought equity interest in local private companies and banks, as well as providing grants to private creditors affected by the collapse of the unofficial stock market in 1982.

The most significant economic policy in Kuwait was fiscal policy, as the net domestic expenditure by the government was the single most important determinant of economic activity, rendering public finance the main instrument of demand management. The treasury's near total dependence on foreign-sourced income was highly unusual not only among OPEC members but also within the Organisation of Arab Petroleum Exporting Countries (OAPEC). Between 1974 and1984, revenues from oil and gas exports plus income from previous investment abroad generally accounted for nearly 97 percent of total government revenue, with only 2 to 3 percent generated by more traditional sources (e.g. custom duties and services fees). Kuwait had neither a general income tax nor a sales tax. In the absence of a significant source of non-hydrocarbon revenue, a major decline in oil export receipts usually resulted in a decrease in the level of transfer payments and occasionally an increase in the fees charged for public services.

The absence of taxation on either income or consumption (except for minimal tariff), combined with a fairly inflexible interest rate policy and the genuine shortage of financial instruments, effectively constrained government efforts to conduct an effective monetary policy. Monetary policy thus had a limited role in Kuwait, due, additionally, to the openness of the economy and the dominance of the government in both aggregate expenditure and domestic lending. As in all other countries where government revenues were largely earned in foreign exchange, monetary policy followed the level of government net domestic expenditure plus credit supply to the private sector and the differential between domestic and international interest rates. Interest rates on Kuwaiti dinar's deposits and loans were initially subject to various ceilings mandated by the central bank. There was no ceiling on foreign currency deposits. Later, the rates became market-determined. Kuwait's interest rate policy favoured relatively low lending rates, officially defended as a means of containing inflation, keeping the value of the Kuwaiti dinar stable and ensuring an adequate supply of funds for private investment.

Kuwait followed a moderating price policy. As in most other group members, prices for oil products, electricity and other energy sources were kept low, and well below international levels. Price controls were in effect on a wide range of commodities. There was also a chain of cooperatives in the retail sector which kept prices low. Subsidies, too, exerted a highly moderating influence on both foodstuff and basic utility prices. Wages were normally set by the government in the public sector; they were freely determined by the management in the private sector. Collective bargaining was rare. There were typically wide gaps between wage levels for Kuwaitis and guest labourers and between remunerations in the public and private sectors (except at the executive level).

Kuwait had a liberal exchange and trade system, with no restrictions on current account transactions and virtually none on imports. All transactions took place in convertible currencies and at a unified rate, except for a brief period in 1984 when a two-tiered currency rates system existed for current and capital transactions. After 1975, the Kuwaiti dinar was linked to a weighted (undisclosed) basket of the currencies of its major trading partners with the largest weighting for the US dollar. There was no exchange control for residents or non-residents. Sales and purchases of the Kuwaiti dinar or other currencies were free. So were capital movements. But at least 51 percent of ownership of new Kuwaiti companies had to be held by nationals. Imports were subject to general and individual licences (except for food). Certain imports (e.g. pork, alcoholic beverages and used cars) were prohibited. A limited control over exports was administered by the customs office, particularly on food during periods of shortage. By the end of the second oil boom, the Kuwait Investment Authority had vast holdings in dozens of countries around the world covering real estate, banking, shipping, manufacturing and, of course, oil refining and marketing. Although official information on the size of the government foreign assets was not usually disclosed, private estimates placed Kuwait's total foreign assets before the Iraqi invasion in 1990 at about $100 billion, divided between the General Reserve ($40 billion) and the Reserve for Future Generations ($60 billion) producing an annual income of some $8-10 billion, i.e. larger than the income from oil exports. In 1994, the total is said to have shrunk to $65 billion.

Kuwait's well-publicised 'cradle-to-grave' welfare system, which was established shortly after independence from Britain in 1961, was vastly expanded following the 1970s oil windfalls. Acting as a 'model welfare state' – and a consummate rentier society – until the mid-1980s, Kuwait gave its residents subsidised (mostly free) social services, education and health care. Citizens (limited to those with ancestry proven to 1920 and those formally naturalised) were entitled to free medical treatment at home and overseas and cash allowances to defray the costs of weddings and funerals. Public utilities (water, power, telephone) were made available to all at negligible prices. Petroleum products (gasoline, kerosene and gas) were offered at prices far below international levels.

Beginning in 1985, with the onset of the austerity era, the government introduced user charges for certain services, raised fees and stamp duties and increased domestic hydrocarbon prices. After the expulsion of Iraq, and the rising costs of reconstruction, non-Kuwaiti residents were largely excluded from some of the generous government benefits, while benefits and compensation for Kuwaitis were notably increased. Nevertheless, Kuwait preserved its reputation as the number one welfare state. Consumer production subsidies constituted another aspect of Kuwait's welfare state. By 1985, state subsidies identified in the budget reached 21 percent of total expenditures, covering such categories as food and commodities, petroleum products, public utilities, housing and direct welfare payments. Subsidies and transfers between 1980 and 1985 averaged 9 percent of GDP. Reduced incomes from oil exports (and, later, from foreign investments) continued to affect both the scope and size of public subsidies. Total domestic transfers and subsidies between 1986 and 1990 were down to about 6 percent of GDP. But after the liberation from Iraqi occupation, these transfers went up again, and averaged about 15.5 percent of GDP and 22 percent of total government outlays.

A very significant portion of Kuwait's oil income was appropriated for defence. According to outsiders' estimates, some 6 percent of GDP and 15 percent of government expenditure was on average spent on defence up to 1990. By the Kuwaiti government's own account, defence dominated all public spending in 1994, claiming one third of total state expenditure. Underway also was a 10-year, $11 billion strategic defence programme, including the most advanced hardware in the arms sellers' arsenals. Kuwait's rationale for such stress on security was mainly rooted in the threat from Iraq, both current and potential. Despite repeated declarations of commitment by Washington and other allies to guaranteeing Kuwait's sovereignty and security, the government insisted on investing heavily in up-to-date defence technology. Drawing a parallel with Israel as a small country where manpower is a major constraint on conventional defence, the Kuwaiti government justified its state-of-the-art defence build-up on the grounds that smart weaponry means fewer people are needed to do the same job. Left out of this equation, however, was the fact that smart weapons need smarter people to operate them. And while sophisticated gear could be purchased with oil money, the highly professional people needed to operate it could neither be easily trained nor safely hired from abroad.

Kuwait had a substantial foreign assistance programme, with a wide geographic coverage, designed largely to gain friends and supporters in dealing with an envious and covetous external world. During the two oil booms 1974–81, Kuwait's aid transfers amounted to about $1.6 billion, including assistance to such multinational development institutions such as OAPEC, OPEC Special Fund, Arab Fund for Social and Economic Development and others (e.g. African Development Bank). After 1982 the aid package gradually shrank, but even after the 1991 liberation, Kuwait expected to maintain its annual loan disbursement. About two-thirds of the Kuwaiti aid programme was in the form of grants and the rest highly concessional loans, with interest rates averaging 3.5 percent and a maturity period of 18 years with a generous grace period. During the Iran/Iraq war of 1980–8 Kuwait reportedly gave Iraq some $15 billion to finance its war machine. Through the large number of expatriate workers – as large as three-fourths of the total labour force at times – Kuwait also transferred part of its oil revenues to other developing countries via workers' remittances abroad.

Following the world-wide trend in the 1990s toward the privatisation of money-losing public enterprises in the developing and newly liberated countries, Kuwait also declared its intention, after the Iraqi withdrawal from its territory, to join the movement as part of its comprehensive post-1991 economic restructuring. A World Bank mission in 1993 recommended that about 75 money-losing public enterprises, wholly or partially owned by the government, be sold to local or foreign investors.[23] These inefficient enterprises included telecommunications, health services, public utilities and certain downstream oil operations.

In 1994, Kuwait announced a $12 billion, 5-year plan to privatise state-owned enterprises, including some 64 local entities wholly or partially owned by the Kuwait Investment Authority and listed on the Kuwait Stock Market. However, this was hampered by such factors as existing state subsidies to economic enterprises, overmanning, job guarantees and wage-maintenance legislation. There was also foot-dragging on the part of the enterprises themselves. Further shadows hanging over the programme were the memory of the 1982 Souk al-Manakh stock market crash and the well-publicised troubles of the Kuwait Investment Office in London, which managed a large part of Kuwaiti overseas assets.

Another domestic policy initiative after the 1990 crisis was the so-called 'offset' obligations on the part of foreign companies engaged in the country's reconstruction. For political reasons, the post-liberation Kuwaiti government tried to reduce the onus of 'foreign dependence' by adopting a policy that required major foreign contractors working in Kuwait to reinvest at least 30 percent of the value of their contracts in local joint venture companies within an eight-year period, and against a 6 percent non-fulfillment penalty. The results remained to be seen. Apart from clever ways and means of getting around these offset obligations, the limitations of the local economy and the dearth of joint investment projects caused new headaches for the authorities.

In sum, Kuwait, followed a relatively open-economy policy despite undergoing several major economic shocks in 1982, 1986 and 1990. Trade barriers were kept to a minimum; foreign investment was welcome in selected sectors, private-sector wages and prices were set mainly by the market except for subsidised goods and sources, and regulations were reasonable and fairly evenly enforced.

LIBYA[24]

The Socialist People's Libyan Arab Jamahiriya – the name adopted in 1977 – functioned as probably the world's most unique political regime, organisationally very different from other OPEC members. With a single political party – the Arab Socialist Union – since 1972, and a mixture of quasi-Marxism and Islamic theology, the Libyan regime was to be ruled by the people's 'direct' participation in local congresses. In reality, the country was a one-man military dictatorship, not dissimilar to Iraq's or Nigeria's. All sources of power in various formal political institutions emanated from the top. The country's effective constitution was Qadhafi's Green Book. Between 1971 and 1990, Libya engaged on more than one occasion in open warfare with some of its former federation partners (e.g. Egypt and Tunisia).

Throughout the period under review, the Libyan economy operated under a system of comprehensive controls and elaborate regulations in ownership, price control, credit allocation, foreign trade and currency exchange. Under the ideology expounded in the Green Book, private ownership was to be limited to certain basic necessities (e.g. home, means of transport, work tools). Farmers had user rights to the fruits of their own labour only and not to the land itself. Production was to be organised on a self-employment or collective basis where workers would be 'partners' instead of being mere wage earners. A 1978 law allowed every citizen to own a single residence. The law disallowed the renting of any privately-owned property. In turn, the state Real Estate Bank provided 25-year interest-free loans to eligible citizens to build individual residences. All rental properties were confiscated by the state. Hired employment was allowed only in the public sector. The social product was to be 'equitably' distributed according to the individual's basic material needs, and no more. All foreign trading activity was nationalised in 1978, and all imports were assigned to various public sector trading organisations. Retail trade remained in private hands, but the state indicated its plan to nationalise that sector as well. Throughout the 1980s and early 1990s, the Libyan public sector was the dominant sector, with public enterprises engaged in the production and the distribution of a variety of goods, as well as in construction, transportation, banking and public utilities.

Domestic production was organised under state tutelage, particularly in the modern sector of the economy (i.e. petroleum, industry, telecommunications and finance). Libyan oil was produced and marketed by several national and foreign companies with various equity-participation arrangements. The state had 80 percent overall equity rights in the oil industry. Libya's National Oil Corp., founded in 1968, represented the country's share. Libya was another OPEC member who opted for an oil conservation policy after the oil price rise in 1973 and deliberately reduced its daily extraction (from a peak of 3.32 mb/d in 1970) to 1.48 mb/d in 1975. Production further dropped to a record low of less than 1 mb/d in 1984 before picking up gradually to an average of 1.42 mb/d in 1991–4, in line with the OPEC quota and constrained by capacity limitations

During the 20-year period, total oil income amounted to $214 billion. Dramatic changes in annual revenues were caused partly by variations in the volume of crude exports and partly by variations in the price of Libya's high-quality light crude. The allocation of oil earnings, as both the single largest component of the national budget and virtually the sole exchange earner, was decided upon by the country's supreme leadership and carried out through medium-term planning and annual budgeting. Although separate from the development budget and not officially revealed, the country's military commitment took priority in windfall allocations, at the expense of other, productive sectors.

Under the first two oil booms Libya's oil and gas revenues almost tripled and aggregate consumption doubled. The consumption/GDP ratio hovered around an average of 53 percent and the investment/GDP ratio averaged at 25 percent, most of the investment being by the public sector. In 1981–6, due to a significant retrenchment in public expenditures, the investment/GDP ratio dropped from the 33 percent peak in 1981 to about 20 percent in 1985. Specific data on domestic expenditure and its components are not published for post-1985 years. The investment/GDP ratio, however, is estimated to have fallen to as low as 5 percent in the 1990s.

Prior to oil discovery, farming was Libya's economic mainstay. With revenues from oil, the Libyan government, through various farm policies, encouraged agricultural sector activity toward increasing self-sufficiency in food production. In earlier years the sector was assisted by the provision of irrigation facilities, land reclamation, housing, education, input subsidies, guaranteed producer prices and interest-free loans. In later years, the incentive structure was changed from an emphasis on subsidies to stimulating productivity. Farmers were encouraged to establish partnerships and farmer families. State-owned cooperatives were given greater autonomy. Credit was made available by the Agricultural Bank at zero or low cost. Extension services were provided in the area of land and water utilisation. With the exception of some retail services, all other non-farm economic sectors were in the hands of the government and publicly run.

Despite the fact that nine-tenths of Libya was sand and rock desert – with less than 4 million hectares of cultivable land, of which only 120,000 hectares were irrigated – agriculture received special attention in Libyan plans second only to industry. This strategy reflected the government's desire to broaden the production base of the economy and to attain self-sufficiency in basic food. The emphasis on this sector was also prompted by the desire to improve the farmers' living conditions and prevent rapid migration to urban areas. Of paramount significance in the development of agricultural infrastructure was the controversial Great Man-Made River Project (GMR), the country's largest single development scheme. Dubbed 'the biggest civil engineering project ever undertaken anywhere in the world',[25] the 4000-km pipelines, launched in 1984, were designed to transport water from underground aquifers in the southern desert to the coastal farmland. The first stage of the scheme was to transport 2 million cubic meters of water a day from some 250 wells for a minimum of 50 years, and to provide irrigation for up to 250,000 hectares. The second stage, with 450 wells, was started in 1990. The project was to be financed by special fees, funds and loans.[26] The execution of this project, which accounted for about 2–3 percent of GDP in later years, continued almost uninterrupted, despite the decline in earmarked revenues, through recourse to bank borrowing. The project, which reportedly cost $25 billion by 1994 (and was not yet fully operational after 13 years), had a disastrous impact on the environment and could barely provide cheaper water than a desalination plant (although originally calculated to cost one-fifth). For this reason, perhaps, the regime also planned to build a desalination plant in Tripoli in 1994. Allocation to the GMR, however, continued uninterrupted in each year's annual budget ($1.6 billion in 1995) despite Libya's tight foreign exchange earnings.[27]

Under three successive plans the existing light industry's capacity was expanded and new mega projects were launched in petrochemicals, iron and steel, and building materials. Public investments in housing and public facilities made the construction sector the second largest contributor to real GDP after oil, employing more than 20 percent of the labour force in the early years. The 1986–90 plan did not go into effect largely due to uncertainties in the world oil market and the disappointing results of the preceding plans. A three-year plan for 1994–6 was announced in January 1994.

Like most other members, Libya allocated part of its oil revenue to social welfare in the form of consumer and producer, subsidies in order to reduce domestic discontent. The net cost of commodity subsidies fluctuated from year to year in accordance with international prices. According to foreign observers, however, subsidised food and agricultural equipment did not always reach their destinations. Like some other Arab members, Libya also allocated part of its foreign oil receipts to foreign investment (i.e. 'savings abroad'). The Libyan Arab Foreign Bank and the Libyan Arab Foreign Investment Co. had the task of undertaking investment in agriculture, industry, tourism and communication outside Libya, either directly or through joint ventures and syndicated loans. Libya also expanded its downstream oil operations through the state-holding firm Oilinvest which had equity participation in oil refineries and retail stations in several European countries. Returns from such investments (including interest income in official reserves) reached about half a billion dollars a year on average in 1989–94.

Under strong public pressure for a better distribution of oil earnings, the Libyan leader in 1991 promised the General People's Congress that each of the 600,000 poorest Libyan families would receive $7,000 to $10,000 a year of oil money directly, starting January 1992. The amount was later reduced to $5,000 annually and limited to only 100,000 of the poorest families, when oil revenues declined. By the end of 1994 the money had not been distributed.

The portion of windfalls going into defence expenditures was not made public. Military expenditures in some years were estimated to be as much as 20 percent of GDP, and as high as 40 percent of government expenditure. According to private estimates, Libya bought foreign arms at an average of $2 billion a year between 1983 and 1992, when the UN banned such purchases. A direct and immediate effect of increased financial burdens, caused in part by UN sanctions, was a cut-back in defence-related expenditure (along with the freezing of a number of development projects).

A host of conventional and country-specific policies were adopted to implement and support the national allocation strategy. Manpower policy, as in other sparsely-populated OPEC members, was a major concern for Libya. Attempts to diversify and develop the economy were hindered by a small population, shortages of skills and a low level of education. By 1979, nearly 14 percent of Libya's population, and a much larger percentage of the labour force, consisted of foreign workers. Moreover, the priority given to military service (5-year active duty) and the policy of 'militarising the nation' held back the development of technical and managerial skills.

Due to the very large share of oil revenue in total revenues, Libya's fiscal policy was always deeply impacted by the behaviour of oil receipts. The thrust of fiscal policy was to adjust both current and development expenditures to projected oil income. Libya for the most part lacked any coherent fiscal policy or strategy. In some years (e.g. 1992–4) no budget document was published. In other years, 'supplementary expenditures' (comprising of domestic and foreign transfers, subsidies and defence) were not part of the published formal budget. Some outlays (e.g. the Great Man-Made River) also were off-budget items. The approach was often *ad hoc* and erratic. Monetary policy relied on direct central bank control of the volume of credit allocated to each sector, with limits on credit provision established for each bank. As with many other OPEC members, Libya followed a low-interest policy in both saving and investment. All banks were nationalised and under state control.

During the entire period under review, Libya had a comprehensive system of price controls and regulations. Prices were set by the Central Prices Committee for a specific list of basic consumer goods, all manufactured and imported goods, and for a few farm commodities. A few minor items were left to fetch their own prices in the market. Basic items (e.g. food staples and beverages) were marketed by the National Supply Corporation which procured these items and sold them to the public at fixed prices determined by social objectives and involving considerable subsidies. Trade at the domestic retail level was still carried out mainly by the private sector, although there were public supermarkets established after 1979. Libya was somewhat unusual among OPEC members in adjusting the domestic prices of petroleum products more closely to their costs. In early 1994, at the official exchange rate, the domestic prices of most oil derivatives in Libya exceeded international market prices, reflecting an element of taxation over and above their full costs.

Libya's trade policy was strictly controlled by the General People's Congress and executed by the Secretariat of Planning, Trade and Finance. Since the transformation of the economy required a very large volume of annual imports, external trade policy acquired great importance. After 1978 foreign trade was nationalised, and carried by special state enterprises specialising in one or more commodities, according to annual quantitative quotas in the commodity budget. Imports of some consumer and luxury goods were prohibited, and others were subject to custom duties and surcharges. Under the 1988 'liberalisation' policy, imports by private companies and partnerships were permitted for their own use; and, later, private individuals were permitted to import limited amounts of goods from Arab countries. An Export Promotion Council was established to help promote non-oil exports. Due to rising economic difficulties under US and UN sanctions, a range of export products was banned in 1994.

Libya's foreign exchange policy was in the hands of the central bank. The Libyan currency, the dinar, was from 1973 to March 1986 tied to the US dollar at the rate of LD1=$3.38. After that it was pegged to the SDR at a rate of LD1=SDR2.80. All transactions were effected through state banks. There were small travel allowances for pilgrimage to Mecca and other personal trips. All remittances abroad were subject to central bank approval. Starting in 1985, all foreign exchange transfers abroad attracted certain fees as part of the financing of the Great River project. The nominal rate was officially devalued in March 1992 to LD1=SDR2.52. Further devaluations during 1993 and 1994 reduced the currency's official value to LD1=SDR1.90 (LD1=$2.82) by November 1994, and created an official two-tier exchange market where private importers could legally buy foreign currencies at a reduced rate. By late 1994, the black (parallel) market value of the US dollar in terms of the Libyan dinar was reportedly more than 10 times the official rate.

Under the so-called Qadhafi's 'green perestroika' of early 1988, a move towards liberalisation and privatisation got underway. Private activity in trade, light industry and farming was given greater scope. State trading enterprises in consumer goods were abolished. Steps were also taken to broaden the concept of private ownership of residential dwellings by allowing each family to have additional residences for their children. In 1992, a privatisation law was enacted, calling for the sale of public assets to private interests and for greater private sector participation in the economy. Incentives in the form of various tax exemptions and access to low-cost credit from the Development Bank were offered to the private sector for it to acquire state industrial enterprises.

The subsequent UN sanctions, however, put an effective stop to the implementation of the 1992 privatisation law and any other liberalisation move, as the government tried to tighten its control over the economy to deal with the new situation. Capital needed by private entrepreneurs became scarce, and the rising deficits of public enterprises made them less attractive to the private sector. The country thus continued as OPEC's most economically repressive regime.

NIGERIA [28]

Nigeria's economy was more than that of many group members dependent on oil extraction and exports, both directly and indirectly. Although the petroleum sector constituted a relatively small share of GDP, oil operations influenced virtually all government institutions. Some 70–80 percent of the federal budget revenue was derived directly from the oil sector. Non-oil revenues were also affected by oil activities as many other levies and incomes themselves depended on the foreign exchange made available by oil exports.

Nigeria's income from oil exports 1974–94 was one of the largest among top exporters, amounting to more than $237 billion. Annual oil revenues were also among the most volatile. The record high receipts in 1980, at $24.9 billion, were 5.2 times larger than the record low of only about $4.8 billion in 1986. Nigeria was the largest beneficiary of the second and third oil booms. Mainly responsible for these drastic fluctuations were changes in the price of Nigeria's high quality crude which competed with North Sea oil although export volume also declined by 50 percent between 1974 and 1988 from 2.2 mb/d to 1.1 mb/d.

On the eve of the first oil boom, Nigeria – the most populous and oil-rich country in Africa – had a wealth of natural resources, an active private sector and a large labour force. This considerable development potential was backed by certain civil traditions and administrative institutions inherited from British colonial rule. At the same time, its multiple cultures, 250 languages and three ethnic regions – also a legacy of the colonial design of its territory – harboured a great deal of ethnic and regional tensions which had caused a civil war and a continued struggle for power among leaders from different regions.[29] Unlike many other group members, Nigeria continued to extract and market its oil not through outright nationalisation but via equity participation with foreign oil majors. Nevertheless, the formulation of the country's petroleum policies and strategies remained with the federal government (both civilian and military).

Despite its rich resources and potentials in 1974, Nigeria was in dire need of larger private consumption as well as improved public services. In response to strong pressures from the public and as a result of the crucial decision to spend all the new oil gains at home, both aggregate domestic consumption and investment increased rapidly after 1973/4. As in most other member countries, the main instrument of adjustment was increased levels of government spending which fuelled higher levels of domestic aggregate demand and expenditure. Since the huge balance-of-payments surplus, equal to over 16 percent of GDP could not be readily absorbed domestically, the Nigerian government opted to provide the International Monetary Fund with a total of SDR300 million to finance the Fund's Oil Facility. This prudent and proper decision, however, turned out to be merely a brief postponement of domestic spending. As soon as the size of new affluence became widely known, political pressures to use the windfalls to relieve domestic poverty and deprivation became irresistible. Accelerated spending within a short time absorbed more than the entire windfalls, with the gap being financed by drawing down accumulated reserves, expanding budget deficits and, finally, by external borrowing.

Nigeria's revised Third National Development Plan (1975/6–1979/80) was launched soon after the first oil boom. Priorities in the plan were given to the expansion of physical infrastructure (e.g. transport and communications), development of mining and manufacturing, agriculture and water supply, and improvement in health, education and housing, in that order. In manufacturing, light industries (food and textiles), oriented to satisfying local demand, were given a modest share. Other industries – petrochemicals, steel, cement and motor vehicles – were all essentially import substituting projects and highly capital intensive. The new oil wealth was also used to placate the restive population through a generous dispensation of petrodollars among influential claimants.

In addition to this emphasis on physical infrastructure, investment in human capital focused on education – especially through massive primary education expenditure. In the non-oil traded sectors agriculture was given mostly lip service through such schemes as the National Accelerated Food Production Programme (1974) and Operation Feed the Nation (1976), designed to ensure self-sufficiency. A vast number of the investment projects were quickly started, with evidently scant regard to domestic absorptive capacity, project coordination or sequencing, and with few safeguards against waste and corruption.[30] The second oil boom provided a temporary respite for the country's rising financial constraints towards the end of the Third Plan. Buoyed by an oil price jump from \$19/b to \$34/b between April 1979 and February 1980, a new spending spree began with increased salaries and new investment projects – including the new inland federal capital, Abuja, on top of large irrigation outlays, housing schemes and steel plants. Faced with three consecutive years of decline in oil revenues from \$25 billion in 1980 to less than \$10 billion in 1983, the government faced an intolerable long-term fiscal crisis. In any event, the Fourth Plan remained largely unfulfilled. No other plan followed the fourth, and investment allocations were subsequently made through annual federal and state budgets on a three-year rollover basis.

The choice of sectors of the economy for public investment during the first two oil booms was partly responsible for those shortfalls. The pattern of investment exemplified the tendency among windfall recipient governments to direct resources toward the non-tradeable sector – a trend which in the Nigerian case was also followed by private investors. Furthermore, the lion's share of investment in agriculture and industry went into a few large, high-cost projects (e.g. steel, irrigation and fertiliser production).

Despite a succession of military governments in Nigeria, both the size of the armed forces and total military expenditures were surprisingly moderate. With a military force of about 110,000 for a population of some 108 million in 1994, Nigeria's annual military expenditure was estimated at less than 1 percent of GNP, and less than 5 percent of total central government expenditure (Table 15).

The standard and specific policies chosen to direct the economy during the two decades also left a lot to be desired. Fiscal policy had the largest impact on the Nigerian economy, determining the stance of other macroeconomic policies, particularly monetary and exchange measures. The upward momentum of public spending shortly after the first oil boom, in the face of domestic supply constraints and low to moderate taxation of individual income, corporate profits and capital gains, led to accelerated domestic inflation and exchange rate appreciation. To deal with both internal and external imbalances, the Nigerian government began to take drastic fiscal measures to 'restore sanity' to the economy. The military coup that toppled the civilian Shagari government in 1983 refused to seek help from the International Monetary Fund, preferring to manage the economic slump through 'war against indiscipline'. A programme of fiscal austerity, including across-the-board budget cuts, reduced imports and foreign exchange rationing, was initiated. Yet, facing a build-up of payment arrears, and the oil shortfalls – while imports-dependent industries were becoming idle for lack of raw materials – Nigeria was forced to cut development programmes and frantically seek foreign loans. Since the wage bill was untouchable for legal and political reasons, reduced public revenues resulted in cuts in materials, equipment and maintenance, leaving many projects unfinished.

Continued high fiscal deficits led the government finally to seek help from the IMF and adopt a comprehensive structural adjustment programme in 1986. The programme emphasised downsizing the public sector and improving fiscal administration. However, management of public expenditures proved politically difficult. A tight budget policy was observed in 1986/7 but in 1988, as soon as the expected effects of downward adjustment and austerity showed their harsh realities, fiscal policy turned expansionary. Increased oil revenues in the third oil boom (1990) led to further large-scale public spending, this time through the so-called 'dedication accounts' and other devices outside the regular budget. Like most other OPEC members, Nigeria kept the domestic prices of petroleum products far below the international levels. Although, under IMF pressure, fuel prices were raised four times after 1988, the retail gasoline price was still one fifth of the European level. The policy not only cost the Nigerian Treasury nearly $400 million in annual subsidy but also resulted in a widespread black market, smuggling to neighbouring countries, neglect of refineries' maintenance and inequitable fuel distribution. Persistent annual budgetary and external resource gaps from 1985 to 1994 were financed by incurring external debt and by external capital borrowing, plus some small foreign direct investment. As a result, domestic public debt rose to over $12 billion, and external debt and arrears to nearly $34 billion, by 1994.

Monetary policy was largely influenced by protracted budget deficits and fiscal needs, rising private sector demand for credit, a populist desire to keep interest rates low and the exigency of combating inflation. The Central Bank of Nigeria used the standard monetary instruments to regulate credit to the private sector, ceilings on interest rates and overall liquidity. During most of the period, interest rates were kept moderately low to fight inflation. Savings rates were occasionally positive when measured against the official inflation rates but mostly negative (as much as 30 percent below the consumer price index) in other years. Required by the government to finance the fiscal deficit, monetary policy followed an increasingly expansionary and inflationary course. The 1986 Structural Adjustment Programme sought to reform the financial sector by deregulating interest rates and spreads. But the central bank used its moral suasion to deter upward pressures on interest rates after the 1988 reflation. Rates were re-regulated in 1994. Meanwhile, the mushrooming of sectoral banks and other non-bank financial institutions outside the supervision of CBN complicated the task of monetary control. Until 1987 Nigeria chose to cope with its soaring inflation by turning increasingly to a web of price controls. Wages were determined by negotiations between employers and unions subject to government approval. There was a minimum wage.

Exchange rate policy was determined at the highest rung of political or military power. The central bank of Nigeria was the principal administrator of foreign exchange regulations: it allocated the available foreign exchange on the basis of the type of industry, product and transactions. All proceeds from the export of goods and services had to be surrendered to the bank. The ministry of finance was in charge of exchange control (e.g. transfer of capital and remittances, dealing with securities etc.). At the outset of the first oil boom, Nigeria adopted an independent policy committed to keeping the naira (which had replaced the Nigerian pound in 1973) stable against the US dollar with no obligation to observing specific margins. Later in 1978, the naira's exchange rate was formally pegged to a basket of the currencies of Nigeria's seven principal trading partners, although the actual official naira/dollar rate was subject to daily quotations by the central bank, based on several factors. Even when oil and commodity prices started a decline in the early 1980s, Nigeria refused to devalue the naira and continued to suffer economically. As Nigeria's trade creditors refused to reschedule the country's rising payment arrears without a prior agreement with the IMF, the mid-1985 military government was forced to come to the Fund for help under a new slogan: from 'austerity alone to austerity with structural adjustment'.[31] Under the IMF agreement, the naira was devalued to N4.6=$1, and a two-tier exchange rate system was established through which a second rate would be set by market forces in auctions held by the CBN. An abortive attempt was made at exchange rate unification in July 1987, but daily auctions were resumed in 1989. Finally, in 1994 foreign currency was again administratively allocated by the CBN at a single exchange rate of N21.9=$1,[32] and free market transactions were prohibited.

As indicated by these crisis-induced decisions, Nigeria's trade arrangements were continually subject to elaborate restrictions and regulation to the very end of the period. Trade regulations were administered by the federal ministry of trade. Trade policy was focused on diversification and protection of infant industries, support of external balances and curtailment of luxury imports. Licences and quotas were used to implement this policy. With the fall of oil exports and foreign exchange reserves in 1981–2, Nigerian trade policies became more restrictive. Hoping to discourage foreign exchange payments, import-substitution policies were promoted, often without much regard for comparative advantages, resulting in an excessive reliance of these industries on foreign raw materials and other inputs. Periodic changes announced among the categories were designed to save foreign exchange with respect to imports of 'non-essential' goods. In addition to import licences, relatively high tariffs on certain imports helped to protect new local industries which faced stiff competition from imports, due to inefficiencies and the overvaluation of the naira.

With respect to industry-specific policies, an indigenisation decree in 1972 (later revised in 1974, 1976 and 1977) set the limits of foreign ownership of Nigerian enterprises. Retail trade, transport and distribution were reserved entirely for

Nigerians. Other activities required various percentages of Nigerian ownership. Under a new decree announced in December 1989, foreign companies could hold a 100 percent stake in any new Nigerian venture (except in banking, insurance, mining and oil prospecting) with a minimum initial capital of N20 million. In July 1988, the guidelines for external debt conversion into the local currency allowed the purchase of selected Nigerian foreign debt instruments for use in the expansion or recapitalisation of privatised enterprises. However, political uncertainty, regulatory hurdles, unreliable provision of public infrastructural services (e.g. power and telecommunications) and other impediments raised the cost of doing business and discouraged potential investors. In early 1988, 96 state-owned companies were listed as candidates for either commercialisation or sale to the private sector. By 1993 some 58 small companies on the list were privatised through the Nigerian Stock Exchange. The privatisation of large enterprises, however, was delayed. The commercialisation of giant parastatals was not very successful and the sale of others was postponed. Privatisation efforts were strongly resisted by labour unions.

In sum, the quality of macroeconomic management in Nigeria, never very outstanding, progressively deteriorated after 1990. Trade and price distortions that had been partially rectified by the 1986 Structural Adjustment Programme reemerged. Rates of exchange and interest were again determined administratively and by non-market criteria. Economic policy was epitomised by a complex bevy of restrictions and regulations. Business activity was affected by personal ties and endemic corruption.[33]

QATAR[34]

Despite the essentially private enterprise nature of the Qatar economy, the government wholly owned and operated a large number of domestic corporations, including the Qatar General Petroleum Corp., iron and steel, fishing and poultry companies. The state also owned telecommunications, insurance, hotels and banks. All agricultural land was owned by the government. Unlike many other members, Qatari public enterprises were, for the most part, profitable and net contributors to the budget. Total government expenditures averaged about 47 percent of GDP in most years. The government also directly imported an extensive range of foodstuffs for domestic sales, below private sector import prices.

Qatar's overall development strategy was aimed at enhancing the value added of its energy sector with a view to industrial diversification. This strategy focused on export-oriented, energy-intensive industries. Privatisation and private sector growth was also emphasised. Relatively inexpensive expatriate workers, low-energy costs, virtually no corporate taxes and an open economy provided the country with a substantial comparative advantage. As a result, Qatar was successful in attracting foreign capital and technology.

Qatar was one of OPEC's smallest oil producers, with less than 0.4 mb/d in 1994. In July 1976, Qatar took full ownership of foreign companies, giving them service contracts to operate the oil fields at various fees. After the steep rise in oil prices in 1974, the government imposed a ceiling on production to safeguard the long-term productivity of the oil fields, particularly as the ratio of associated gas to oil output increased. As of the mid-1980s, Qatar entered into production-sharing arrangements with major foreign oil companies to improve efficiency in extraction and to fully exploit oil reserves. Between 1990 and 1994, as crude oil reserves gradually declined, the energy policy shifted toward accelerated oil exploration. Equity-sharing agreements were offered to foreign companies for finding new reserves and enhancing recovery from existing fields.

Qatar's oil revenues during the 20-year period amounted to about $60 billion, with a peak of $5.5 billion in 1981 and a low of $1.7 billion in 1988. After 1991, a good part of domestic energy consumption was met by LNG supplied from the giant offshore North Field. Qatar was the third lowest OPEC oil producer after Gabon and Ecuador. Reliance was thus increasingly placed on natural gas, in which the country had the world's largest reserves after Russia and Iran.[35] Following the rise in oil prices in 1973/4, Qatar's accelerated economic development was underwritten almost totally by the government, and proceeded on two parallel routes: oil and gas, and non-oil. Domestic government development expenditures typically averaged about 12 percent of total public expenditure. Foreign capital was also used in joint ventures. The bulk of development outlays was directed to housing and construction, land reclamation, water desalination, electric generation and other physical infrastructure including roads and communications.

Qatar began its post-1973 development programme with the construction of a small oil refinery, two LNG plants, a fertiliser plant and a petrochemical complex – all using as feedstock the associated gas from the oil fields or other energy operations. Non-oil industrial policy stressed the expansion of the industrial base. The manufacturing sector was dominated by steel, cement and a host of small, light consumer plants. An Industrial Development Technical Centre, initially set up to monitor the country's industrial development, became the Ministry of Industry in 1989. While large, capital-intensive industries were export-oriented, small and medium-size undertakings were promoted for import substitution. Incentives offered to the latter included tariff protection, exemption from import duties, land leases, tax holidays, subsidised utilities, direct financial assistance and the availability of expatriate workers. In 1993, a specialised industrial bank was established to promote further industrialisation.

Outside the hydrocarbon-related industries, the Qatar Steel Company (a 70-30 percent Qatari-Japanese joint venture), established in 1978, was the largest venture, followed by cement. Steel production used natural gas and imported iron ore to make semi-finished steel pellet and steel bars, eighty percent of which was exported to the GCC member countries. The 600,000 ton/y capacity company had its first profitable year in 1988. The cement company was a joint-venture between the government and the private sector. The government was also the source of several major capital projects in the construction industry.

Qatar's farm policy was to assist private land ownership through the provision of seeds, pesticides, water pumps and fertilisers, free of charge. Water rationing, extension services and financial aid were also part of the agricultural policy. The fishing industry was jointly operated by private fishermen and the state fisheries. However, constrained by the scarcity of water and arable land (less than 3 percent of the total area), farming did not attract much private investment despite extensive government incentives.

Manpower policy was a major concern for Qatar. Qatar, more than any other Arab oil producer, was dependent on foreign labour. By necessity, the government followed a very liberal immigration policy. Some 85 percent of private employment consisted of expatriate workers, as did more than half of public sector employment. All Qatari citizens with higher education were guaranteed a position in the public sector, with generous salaries and fringe benefits. Due to the higher wages prevailing in the public sector, the Qatarisation of workers in the private sector was difficult. In 1994, Qatar's population was estimated at about 500,000 with an estimated labour force of about 100,000, of which less than 20 percent were Qatari workers.

Income from oil and gas was the main source of Qatar's government revenues, averaging over 90 percent of the total in the five years ending 1979. Non-oil revenues consisted mainly of returns on foreign investments and about 2 percent of the total from corporate profits, customs duties and public utility fees. There was no income tax for Qatari nationals, but a progressive tax on foreigners. For the years when information is available, government consumption typically absorbed 35 percent of GDP, private consumption about 30 percent, and gross, fixed, domestic capital formation about 18 percent. The rest was invested in foreign assets. In fiscal matters, Qatar was described as a 'current account' economy where cash flows basically determined disbursements. Thus, in certain years (e.g. 1983), despite the government's massive accumulated reserves, payments to contractors were delayed and projects stopped for lack of cash flow. Fiscal policy followed the fluctuations in oil and gas income. The brunt of the adjustment applied to separable items such as domestic lending, foreign equity participation and, finally, development outlays. Budgetary developments followed the government's attempts to contain public expenditures in order to adjust to declining oil and gas revenues, and to avoid an excessive drawdown of foreign assets. Annual fluctuations in oil and gas receipts were made up by transfers to the budget from the government's investment income. Rigidities in such outlays as subventions and subsidies continued to put pressure on the budget balance.

Qatar's monetary policy was principally determined by the government's total expenditure and net lending. The policy was aimed at regulating domestic liquidity in order to maintain a stable exchange rate and ensure an open trade and payments system. The principal instrument was the regulation of interest rates for riyal-dominated deposits and bank loans. Reserve requirements and discounting were adopted, beginning in 1986. In the absence of domestic treasury bills or bonds, the policy did not have open-market operations capability. A low interest rates policy was designed to facilitate the establishment and expansion of local businesses. Until 1991, there were fixed legal limits on riyal interest rates, causing speculative capital flows.

Qatar's exchange and trade policy followed an open, liberal and market-oriented regime. The Qatar riyal was unpegged from the dollar and pegged to the SDR at the rate of QR 1=SDR 0.21, effective March 1975. In practice, however, the riyal followed the US dollar at a relatively stable rate of QR3.64=$1. During the entire period, Qatar had no exchange control, but some imports were subject to licences.

With some minor exceptions, there were no limitations on capital or current payments or receipts. Foreign investment in Qatari enterprises was subject to certain controls with respect to equity participation.

With the exception of certain merchandise subject to licensing, for reasons of health or public safety, all imports were unrestricted. With some exceptions, all exports were freely allowed. Export proceeds could be disposed of freely regardless of the foreign currency involved. A general customs duty at the rate of 4 percent was applied to most goods, with selectively higher rates on steel, tobacco and alcoholic beverages. Goods imported from other GCC countries were exempt from duties under reciprocal agreements. Most foodstuffs were also duty-free.

Qatar's current balance of payments surplus was invested abroad in major foreign capital markets. Investment policy emphasised protection of asset values and maximisation of safe yields. Government balance of payments surpluses were under the direction of an investment board which set general guidelines on foreign investment policy. Part of the increasing foreign reserves was generously spent on foreign grants, loans and private capital flows, until the end of the second oil boom when such transfers were drastically reduced. Instead defence and security began to absorb a very large share, reaching 47 percent of current expenditure in 1985. In the 1990s, expenditures on defence and security and general administration made up to 70 percent of current outlays, and education, health and public utilities absorbed another 20 percent.

Qatar established one of the most modern and comprehensive welfare systems in the Persian Gulf. Education was free but not compulsory. The national health service was free to Qatari nationals. The bulk of development expenditure was financed through loans and equity participation. Resort to foreign borrowing for financing the budget was also had, in order to avoid the crowding out of the private sector. Oil revenues were also directly distributed among the population at large through subsidies. The Qatari government had a subsidy programme for basic food (rice and wheat flour). While direct budgetary outlays for subsidies were relatively small as a percentage of GDP, implicit subsidies in the provision of public goods and services were considerable. Water and electricity were provided free of charge. Gasoline was sold to nationals at considerable discounts from world prices. Utilities were sold at heavily subsidised rates to non-Qatar consumers (including commercial entities), as were telecommunications and other public services.

In sum, Qatar represented a small, energy-based, neo-patrimonial state with both limited resources and limited developmental potentials beyond oil and gas. Economic activity and policy were both influenced by an oil and gas income which allowed the pursuit of a fairly liberal strategy.

SAUDI ARABIA [36]

Saudi Arabia was governed during the period under review as a conservative, traditional, quasi-tribal state, where the al-Saud ruling family had a monopoly on high public offices and exerted unparalleled influence over all major economic decisions through business partnerships and blood relations with the local business community. Despite a genuine commitment to free enterprise, from an ideological standpoint the economic regime remained effectively statist, as the private sector continued to depend on government for both contracts and subsidies. As in other group members, the government played its development role both directly, through extensive publicly assisted investments and welfare transfers, and indirectly through macroeconomic policies.[37]

Saudi Arabia's total oil revenues between 1974 and 1994 amounted to an estimated $930 billion, of which almost half was received during the first two oil booms, and nearly a third in only three years, 1979–81. Public sector spending was by far the most significant determinant of developments in the non-oil sector. Moreover, public allocation of oil revenue influenced aggregate demand, domestic liquidity and incomes in the private sector. The Kingdom used its massive oil revenue to finance a vast and continued programme of infrastructural, agricultural and industrial development, while allocating substantial funds for the modernisation of its educational and health facilities and for the build up of its armed forces. Through all this, the Kingdom also maintained an elaborate welfare system, matched only by a few other oil-based economies in the Persian Gulf. The Saudi government also carried out a large foreign assistance programme. By official estimates, Saudi Arabia spent nearly $900 billion over the twenty-year period 1974–94 on roads, schools, hospitals, power plants and other infrastructural facilities. Another $100 billion or so was invested in oil refining, petrochemical, metal and other key industries.

Saudi Arabia's development planning, in the aftermath of new oil windfalls in 1974, was carried out, in the context of 10-year and 20-year planning perspectives, through four five-year plans running up to 1994. Altogether Saudi planning in the 1974–94 period showed a tendency toward lower overall civilian spending. A major priority in allocation was for physical infrastructure – roads, seaports, airports, housing, communications. However, expenditure on human resources (including all levels of education) was the only category of non-defence outlays which increased with each succeeding plan. Health and social services also absorbed a sizeable portion of development expenditures.

The main thrust of Saudi development policies was toward economic diversification. As part of this strategy, the Saudi private sector was encouraged to establish secondary industries through a number of incentives and facilities. Saudi Arabia's industrial policy covered a very generous programme, including infrastructural facilities (good transport and communications, low industrial park rents, cheap utilities), tax holidays for foreign joint venture participants, tariff exemptions on imported equipment and materials, low-cost credit and interest-free long-term loans, low wage foreign workers, adequate protection against competitive imports, government purchases of local wares, a fairly open export market within the Gulf Cooperation Council area, and assistance in the preparation of feasibility studies. The agency charged with the task of providing credit for industrial ventures was the Saudi Industrial Development Fund (SIDF). By 1994, SIDF had financed nearly 1700 factories. Other agencies set up to promote industrial development included the government-backed, private National Industrialisation Company, the Saudi Venture

Capital Group, the Saudi Advanced Industries Corporation and the Saudi Industrial Development Company.

Saudi agricultural policy was avowedly based on political and social engineering considerations. Agricultural production was assisted by a comprehensive system of subsidies including interest-free, long-term financing, free water, free fallow land, below-cost inputs and machinery, guaranteed above-market prices for products, subsidised fuel and electricity, tax holidays for joint ventures – all on top of improved infrastructural facilities. Under the banner of 'prudent self-sufficiency', this policy was designed to guard against a possible retaliatory 'grain embargo' by wheat growing nations. The government set aside large sums each year to buy wheat grown on well-water irrigated Saudi desert land at several times the cost of imports.[38] This expensive farm policy was adopted for a three-pronged objective: to sustain rural incomes, to encourage Bedouin settlement and to slow the population drift to major cities.

Basic Saudi industrial investments were undertaken directly by the government, and concentrated mainly in large-scale projects. Public investment in industry was implemented through two main organisations, Petromin (in oil, gas and minerals) and the Saudi Basic Industries Corporation (SABIC). Petromin established a tanker fleet for exporting Saudi oil, expanded the country's oil refining capacity at home and abroad, built a trans-peninsula pipeline to carry crude oil from fields to a new export terminal, Yanbu, on the Red Sea, and promoted exploration of Saudi Arabia's other minerals. SABIC was established in 1976 as a joint-stock company to develop the Kingdom's petrochemicals industry. The feedstocks were to be provided by national gases (ethane and methane) associated with oil production, which had been previously flared. It was a joint-stock company with 30 percent private participation. The bulk of Saudi Arabia's petrochemicals industry became located in the twin industrial cities of Jubail on the east coast and Yanbu on the Red Sea, under the supervision of a Special Royal Commission (RCJY) established in 1975. By 1994, SABIC, with a marked capitalisation of nearly $10 billion, had joint partnership in 15 different local companies, in joint ventures with foreign conglomerates including Exxon, Hoechst-Celanese, Mobil and Panhandle Eastern.

Saudi Arabia was one of the world's most generous welfare states, dedicated to providing its growing native population with life's most common amenities virtually free of charge. The Saudis' cradle-to-grave, domestic welfare system of subsidies, grants and free services – with virtually no taxes – created a make-believe economy where nearly everyone depended on some form of government generosity. Social welfare subsidies touched every aspect of Saudi life: free healthcare, free education at all levels, plus monthly allowances for college students, interest-free home loans, interest-free loans to businesses, energy prices at a fraction of world prices, water, electricity, telephone and domestic travel rates at a fraction of the actual cost. Saudi Arabia's welfare policy was rooted in several essential considerations. First and foremost, there was the need to raise the national standard of living from the low levels of the early 1970s (e.g. less than $500 in per capita GDP). Second, from the standpoint of equity, there was the imperative of distributing the oil windfalls among as many people in the Kingdom as rapidly as possible. Third, human resources had to be developed and enhanced in quality through better health and education, for the sake of improved efficiency. Fourth, subsidies on foodstuff and utilities acted as part of an anti-inflation policy. And, finally, the spread of the benefits of oil windfalls to all social strata was motivated by the royal family's desire to preserve political contentment throughout the Kingdom.

To these ends, Saudi Arabia turned into a full-fledged rentier state. Existing road-user charges, customs duties, and individual and company taxes were reduced or eliminated. Consumption was subsidised through the low prices charged by state enterprises and government agencies, frequently below cost. In particular, domestic oil prices were kept low and fixed for long stretches of time. They remained virtually unchanged through early 1984 when the fiscal deficit finally led to a new tax on oil products sold domestically. Nevertheless, gasoline prices remained less than one third of US prices to the very end. Prices charged for electricity were one fourth of the production cost, with the balance covered by budgetary transfers to regional power companies. Water bills for non-agricultural use were well below cost, and sometimes one tenth of what might be expected elsewhere. Domestic airfares in the state-owned Saudi Airline were the lowest in the world.

Although the security of the Saudi Kingdom was expected to be effectively guaranteed by US and allied military and diplomatic forces, as long as the Kingdom produced oil at levels (and prices) consistent with its protectors' wishes,[39] the Saudi government was clearly determined to follow a highly modern and high-cost defence policy. By spending over a third of its annual budget on defence, Saudi Arabia had acquired an extensive and costly military infrastructure equipped with the most sophisticated hardware available at almost any price. Between 1974 and 1985, the cost of military infrastructure and weapons procurements was estimated minimally at $150 billion. In the decade ending in 1994, the Kingdom reportedly bought or placed orders for well over $50 billion in warplanes, tanks and other weapons, most of them from the US alone. The 1993/4 budget of some $52 billion had a 'defence and security' contingent of $16 billion. In 1994, Saudi Arabia reportedly had about $30 billion worth of US military equipment on order. By some estimates, Saudi defence spending, in proportion to the size of its economy, was as much as any country's in the world.

Saudi Arabia also allocated substantial and increasing amounts of financial assistance to developing countries as well as multilateral financial institutions. Loans at concessional terms and bilateral grants, mainly to neighbouring countries, amounted to $3–4 billion a year during the first boom. The Kingdom made a number of loans to African, Asian and Latin American countries during the boom years. Official foreign aid from 1979 to 1983 averaged nearly $5 billion a year, making the Kingdom by far the largest aid donor among OPEC members. Despite the sizeable current and overall balance of payment deficits in later years, Saudi Arabia's concessional economic assistance to the Third World countries remained substantial (Table 16). The Saudi Kingdom was also a major contributor to various regional and international financial agencies such as the Islamic Development Bank, the OPEC Special Fund and the Britton Woods institutions, particularly the IMF in which the Kingdom was given a seat on the Executive Board.

Saudi Arabia's large and extensive foreign assistance programmes were motivated by several factors including self-defence, regional security objectives, solidarity with

Islamic nations seeking moral support within the Third World, and humanitarian considerations. The Saudi government reportedly spent billions for foreign policy reasons in the form of assistance to such nations as Iraq, Jordan and Lebanon. Iraq is reported to have received some $25 billion from Riyadh during its eight-year war with Iran. Syria, which supported Iran against Iraq during their 1980–8 war and was doing business with Libya – another Saudi foe – was nevertheless also routinely subsidised by Riyadh.

In sharp contrast to Kuwait's strategy, Saudi Arabia elected to invest its rising oil revenues largely at home (rather than place them in foreign assets). An exception to the domestic investment focus was the development of an integrated oil industry. Saudi Arabia's Aramco signed its first foreign joint venture with Texaco in 1988 under the name of Star Enterprises, with the intention of refining up to 600,000 b/d of Saudi crude oil in Texaco refineries. Later, in 1991 and 1993, Aramco bought shares in refining and marketing companies in South Korea and the Philippines.

Saudi Arabia's indirect intervention strategy followed the familiar route, with fiscal policy playing the dominant role. Income from oil was the main source of government revenue, averaging about 90 percent of the total during the first oil boom and gradually falling to 70–75 percent towards the end. Return on foreign deposits and investments provided another 8 percent on average in early years. There were no personal or corporate taxes on Saudi citizens and no income tax on individual foreigners after May 1975. Direct taxes included the *Zakat* (an Islamic tax on wealth). Foreign companies were subject to tax on their non-exempt profits. Few indirect taxes included customs duties, road taxes or various fees and licences. On the expenditure side, normally one third of the budget was spent on defence and security, and half on salaries. Interest payments on domestic debt also claimed a rising share. The rest was spent on welfare outlays and development projects. In addition to financing current and development expenditures, a substantial portion of the budget (as much as 20 percent in some years) involved transfers to public sector entities (municipalities, public utilities and various funds, commissions and organisations).

The Kingdom's budget deficit that emerged after 1982 and continued through 1994 was dealt with not through enhancement of non-oil revenues (e.g. income and sales taxes) but by budget cuts, mostly on development projects. The deficits were financed in various ways, e.g. sales of treasury bills, government development bonds, direct placements with the autonomous state institutions, and domestic and foreign bank borrowings. Between 1987 and 1994, a private loan was obtained to cover part of an arms purchase from the United Kingdom; the Treasury borrowed from local banks directly for the first time; a $4,500 million sovereign loan was raised to meet commitments made during the Iraq/Kuwait crisis; and another general liquidity loan of $1,300 million for the Finance Ministry was provided by the Gulf International Bank.

Saudi monetary policy was exercised by the Saudia Arabian Monetary Agency (SAMA), the central bank. Credit to the private and public sectors, however, was provided by SAMA, as well as six other public institutions and a number of commercial banks. SAMA's control over the economy was limited as monetary expansion was determined essentially by changes in government net domestic expenditure and the external payment deficit of the private sector, over which the agency had little direct control. As Islamic *Sharia* disallowed interest, SAMA's main instruments of monetary policy were limited to the legal reserve requirements and moral suasion. A thriving private money market also existed outside SAMA's control. The focus of policy was to keep inflation in check.

Saudi Arabia followed liberal exchange and trade policies throughout the period, accepting the obligations of Article VIII of the IMF statute. The Saudi Arabian riyal was pegged in 1975 to the Fund's Special Drawing Rights at SR4.28255=SDR1 with margins of 7.25 percent. The US dollar served as the intervention currency. Ryadh's exchange rate policy was broadly aimed to maintain a relatively stable relationship with the US dollar and the currencies of the other GCC nations, to minimise exchange risks and to maintain domestic competitiveness in non-oil export markets. The riyal was freely convertible, and remittance in any currency into and out of the Kingdom was permitted. The riyal/dollar rate was frequently adjusted to conform to the pegged SR/SDR rate.

Purchases and sales as well as import and export of the Saudi currency were free of restrictions. There was no exchange control on capital movements by residents or non-residents. Following June 1986, the exchange rate remained fixed at $1=SR3.75 through 1994. Although the Saudi currency came under periodic selling pressure in 1993 and 1994, SAMA stood ready to defend the currency. Explanations presented for the reluctance to follow a floating or flexible exchange rate policy included: the adverse impact of high-priced imports (i.e. the bulk of household consumption); reduction of importers' profits as a result of cuts in imports; capital losses for Saudi citizens and others who had repatriated their funds in search of higher Saudi returns; the traditional prestige associated with a strong and stable currency; and perhaps also the greater political clout of the private importing business community compared with that of state-run, non-oil export enterprises. The policy also reflected a public welfare posture whereby the state was willing to forego its own benefits for the sake of helping with the income and savings of Saudi households and private businesses.[40]

Saudi Arabia also remained an open economy with minimal trade restrictions. However, imports of a limited number of goods were subject to protective tariff duties. After the second oil boom, there being a need for revenues, tariffs on most items were raised in 1985 and 1988 with exemptions for the foodstuff and development inputs. Within these limited restrictions, imports were freely allowed into the Kingdom and foreign exchange readily obtained for their payments. Except for a few commodities that were prohibited for special reasons (e.g. Islamic religion, health or national security), all other items were only subject to duties. Exports had no controls nor required licences (except for the re-export of subsidised imports).

Capital movements were free of all restrictions (except to or from certain countries). Foreign investment was encouraged, although a joint venture with Saudi nationals was always necessary. The Foreign Capital Investment Law allowed certain specified benefits to foreign investors. There was a tendency toward major Saudi control over joint ventures. There were also preferences in official purchases and contracts for national products and for products originating from GCC countries. Also, foreign contractors were required to subcontract 30 percent of government contracts to Saudi firms and to procure machinery, equipment and services from Saudi companies.

Employment was not a major policy problem for Saudi Arabia for some time. The very fast rate of GDP growth after the first oil boom required a very substantial rise in employment which had to be achieved largely through a large inflow of foreign workers by means of liberal immigration policies. Manpower policy gradually became a thorny issue. Saudi nationals often shunned non-administrative and non-managerial jobs, and preferred trade and commerce to industry and agriculture. The manpower policy, as of the start of the Third Plan in 1980, became the so-called Saudisation of the labour force, i.e. the replacing of several million expatriate workers with Saudi nationals at all levels of activity.[41] During the post-1982 period of fiscal stringency, job guarantees to Saudi citizens were gradually eliminated. And, with the population growing at about 3–4 percent a year, a gradually increasing number of young Saudis entered the labour force each year, threatening rising unemployment. The use of sophisticated high technology in the Kingdom also directly contributed to the unemployment problem.[42]

Another facet of Saudi policy was a move toward the privatisation of public companies. The issue of privatisation was brought into sharper focus in the Fourth Development Plan (1985–90). The plan envisaged privatising the country's two largest, immense, state enterprises – Petromin and Saudi Basic Industries Corp. (SABIC) –and following them with the state airline, Saudia, and electricity utilities.[43] No serious action took place in the subsequent decade. In 1994, the government again announced its intention to sell 45 percent of SABIC to the private sector and reduce the state's holdings from 70 percent to 25 percent. As in some other group members, however, the programme remained mostly on paper by the end of 1994.

Some of the public policies adopted by the Saudi government were in conflict with one another, others had unintended, unfavourable side effects and a few were not sustainable over time. For example, Saudi agricultural policy encouraged unneeded surpluses in some crops and enriched urban farm entrepreneurs and mega-farm owners more than small farmers. The Saudi farm policy, while perhaps defensible for a number of reasons (including economic diversification), was also in conflict with the Kingdom's water development and conservation policy. Agriculture accounted for nearly 90 percent of water demand in the late 1980s. Domestic consumption and industry absorbed the rest equally. Water was drawn from wells that dried fast and required increasingly deeper drilling and costly purification. Despite all this, only about 2 percent of the Kingdom's land area was cultivated. Both agricultural and industrial policies were also inimical to the Kingdom's environment policy. And the welfare policy was not compatible with a sustainable growth policy.

THE UNITED ARAB EMIRATES[44]

The United Arab Emirates – a federation of seven small shahdoms in the Persian Gulf – achieved independence in 1971.[45] Through the Emirate of Abu Dhabi, it had joined OPEC in 1967. Before the oil exploitation, it was one of the poorest Arab countries. Of the seven emirates, the leading three – Abu Dhabi, Dubai, and Sharjah – operated jointly under the federal umbrella while pursuing independent development policies. The three also made up the largest part of the federal budget, and Abu Dhabi subsidise four out of seven. Abu Dhabi and Dubai alone accounted for more than four fifths of the total GDP.

During the 20-year period, UAE's total oil and gas export receipts are estimated at about $247 billion. Starting with less than $7 billion in 1974, the country's oil and gas revenue peaked at nearly $20 billion in 1980 and ended up at about $12 billion in 1994. The vast increase in oil export earnings allowed a sharp increase in government expenditures on physical infrastructure, health and education facilities and social welfare programmes, in addition to the development of oil and gas processing industries. The private sector, benefitting from the increased state spending, also experienced significant expansion in construction, banking and trade areas. Development planning and expenditures were undertaken by the Federation Government and the individual Emirates, largely within the context of annual investment budgeting. The Federation Government gradually assumed the responsibility for developing the infrastructure (including water and power, transportation, public works, education and health). The individual Emirates focused on developing industry, ports, local roads and municipal services.

Between 1974 and 1986, all components of nominal GDP rose dramatically. From 1987 to the end of 1994, GDP components continued their annual growth but at a much slower pace. Throughout the period, total investment hovered around 27 percent of GDP. Welfare expenditures absorbed a large part of the oil windfalls. Education was free and compulsory at primary level, and free at all levels for both male and female students. The Emirates also had a comprehensive public health care system, free for all citizens. A sizeable portion of oil windfalls was earmarked for a comprehensive system of subsidies. By its own description, the UAE was a 'model welfare state', where all citizens were assured of government support from cradle to grave. In addition to the provision of free education, health, rural housing, local telephone calls and other things, there were subsidies on water, electricity and other public services.

The major part of the UAE's external payments surplus was held abroad by the public and private non-bank sectors, including the Abu Dhabi Investment Authority, established in 1977 to manage the country's foreign investments. The net foreign assets of the banking system (both central and commercial) were also on the increase each year, with a few exceptions. Limited domestic investment opportunities, differentials between domestic and foreign interest rates, a desire for asset diversification and the uncertain political climate in the region led the private sector to place funds abroad. The country's liberal exchange arrangements, in turn, facilitated such transfers. The Bank of Commerce and Credit International (BCCI), in which the Abu Dhabi Ruler was a major shareholder, was one of the conduits for asset placement abroad.

The UAE had an impressive record of foreign assistance to other Third World countries. Its annual aid ran at about $1 billion a year between 1974 and 1979. The official grants and loans gradually declined, due to falling oil revenues and domestic fiscal constraints. The UAE had sizeable contributions or capital participation in several international or multilateral institutions such as the Islamic Development Bank, the OPEC Special Fund and the Arab Monetary Fund. Abu Dhabi contributed to the IMF Oil Facility and the Supplementary Financing Facility.

Defence also loomed large in the UAE's oil windfall allocation. Although a member of the GCC defence pact, and ultimately protected by the West against any real threat to its security (due to its possession of enormous oil reserves), the UAE had a larger armed forces and defence expenditure than many other Arab countries, as each Emirate insisted on having its own security identity. The UAE was also believed to have paid some $6 billion toward the cost of the Persian Gulf War.

Of all OPEC members, the UAE was probably the closest to a free market economy. Government ownership and operation were limited to the energy sector and a few energy-based basic industries. Internal economic policies were shared by the federal government and individual Emirates. On the federal level, intervention in the economy was narrow-based and conventional. Policies with respect to the petroleum sector were determined by the individual Emirates, mainly Abu Dhabi and Dubai which possessed the bulk of the federation's oil resources. Abu Dhabi was one of the very few OPEC members which did not have 100 percent state ownership of its oil industry. Ownership was shared with BP, Shell, Mobil, Total, and Exxon and others under different corporate arrangements. Dubai acquired full ownership of its offshore facilities in 1975 but allowed foreign-owned companies to operate onshore under certain tax and royalty arrangements. The UAE followed an oil conservation policy in the late 1970s, in order to safeguard the long-term productivity of its existing oil fields. At the same time, the federation pursued a vigorous policy of oil capacity expansion (and sought higher OPEC quotas) toward a 2.5 mb/d output.

Economic diversification, as one of the Emirates' major objectives, was the focus of almost all economic policies, particularly industry-specific policies. Industrial development was the UAE government's top priority. The state concentrated on large-scale, capital-intensive, basically export-oriented enterprises, leaving small and medium-size undertakings to the private sector. Large enterprises, mainly located in Ruwais and Jebel Ali, included gas and petroleum processing plants, petrochemicals, aluminum smelter and water desalination plants. Private sector industries, aided by the state, included construction materials, steel pipes and rods, tires, foodstuff and others. State assistance consisted of providing, in addition to the basic infrastructure, equity, working capital and low-cost loans through the Emirates Industrial Bank, as well as research to identify potential areas for new projects.

The UAE's agricultural policy was focused on reducing dependence on food imports, broadening the country's production base and providing income and employment for the rural population. With less than 5 percent of the land suitable for normal cultivation, the UAE government played a very active role in the development of the farm sector. State assistance included the establishment of the necessary infrastructure (land distribution, land reclamation, irrigation, drainage and rural roads), as well as direct aid to farmers. The range of state services included the provision of seeds, fertilisers and pesticides as well as credit and extension services.

Manpower policy received the greatest state attention as the UAE was more dependent on foreign labour than most other countries in the region. Only about 10–20 percent of the total population, and 10-12 percent of the labour force, consisted of

UAE nationals. The percentage of citizens in the private sector workforce in 1994 was probably no more than 1 percent. For this reason, a major policy concern was the orderly management of the expatriate labour flows. As the expansion of the economy over the second decade was mostly based on the non-oil sector, and the latter's expansion was in such labour-intensive activities as textiles, electronics, construction and tourism, the need for expatriate workers increased faster than in the early 1980s. The Federal Labour Law of 1980 sought to promote the employment of native and other Arab nationals through a number of regulatory and administrative measures. There was also a move to make the educational system more responsive to local labour force requirements.

In view of the crucial role of the oil earnings received by the Emirates in the development of the economy, fiscal policy had an overriding influence on all major decisions. Oil revenue was the main component of government revenues, accounting for 90–5 percent of the total during the 1970s. Customs revenues, certain administrative charges and fees, returns on foreign investment and municipal taxes on property and liquor were others. There were no federal income taxes in the UAE but there were some limited taxes on business entities at the emirate level, which were frequently evaded (except in the case of foreign companies in the oil and gas sector). The share of oil income in total federal revenue gradually declined toward the end, but remained crucial.

Each Emirate's oil income was divided between the Ruler and the government ,according to individual formulae. For the smaller emirates, the division was somewhat blurred. Richer emirates carried the largest burden of federal expenditure. As already indicated, government spending, at both the federal and emirate level, was a major source of aggregate demand for consumption and investment, and set the pace of economic activity. A federal budget deficit served as a mainstay of the comprehensive welfare system and a stimulus to private endeavours. To augment their oil revenues, individual Emirates borrowed from international capital markets in the early years for the financing of their own development projects. Dubai was particularly involved in such borrowings, to finance major projects in the Jebel Ali industrial area. But Sharjah and Ras al-Khaimah also obtained foreign loans.

Monetary policy was mainly shaped by the net domestic expenditure of the public sector and the expansion of credit to the private sector. The rate of interest was regulated by the Currency Board and kept moderate (5–5.5 percent) during the first oil boom. The Currency Board was converted into a central bank in 1980 with enhanced powers of bank regulation. After 1982, all deposit and loan rates in the commercial banking system became free of official restrictions. The Emirates Industrial Bank, established in 1982 to help private industries, normally charged less than the commercial banks, as it drew most of its funds from the federation's budget.

There was no incomes policy within the federation. The UAE had no formal wage and price controls. There was no minimum wage law, and private sector wages were fully market-determined. Labour, however, was not permitted to organise, strike or engage in collective bargaining. The Ministry of Labour was in charge of settling the grievances of industrial workers. Prices were market-determined. Unlike in most other OPEC members, all petroleum products refined internally were marketed at international prices plus distribution costs. There were no energy price subsidies.

The UAE had the freest exchange and trade policy of all members, subject to only minimal controls. The currency of the UAE, the *dirham* effective January 1978 became pegged to the SDR at Dh 4.762=SDR1, within large margins. The dirham/US dollar rate, however, was kept unchanged at $1=Dh 3.673. The Currency Board bought and sold US dollars at the rates set by it within the official margins. The country accepted the IMF's Article VIII obligations as far back as February 1974. There was no exchange control and no restrictions on flows of funds. There were no limitations on the availability of foreign exchange for imports. No control requirements were imposed on capital receipts or payments. Foreign trade was nearly free. Imports were freely allowed, with the exception of a few commodities for health or security reasons. The UAE had the lowest tariff levels of all GCC countries. Virtually all commodities were allowed to be exported or reexported freely, and the proceeds disposed of with no restrictions. Foreign private investments were subject to the Company Law No. 8 of 1984. Except in the free-trade zones, where 100 percent foreign ownership was allowed, and except for GCC, investors who were treated more favourably, all foreign investors were limited to no more than 51 percent of equity ownership of joint venture companies. Foreigners could not own land or buy stock. Some sectors, including oil and gas, petrochemicals and utilities, were closed to foreign investment.

In sum, while the emirate and federal governments owned full or partial shares of agricultural, financial and industrial enterprises and certain services, the UAE was fundamentally a modern market, economy where private property was protected by an effective judicial system and intellectual property rights were recognised. Government regulations were extensive but not inimical to business enterprise.

VENEZUELA [46]

Venezuela's basic economic policy was based on traditional populism and state socialism. The two major political parties, which held the presidency alternately between 1974 and 1994, both subscribed to the same social-democratic ideology. While the political system remained democratic, the state continued to occupy a predominant role in the economy.

Venezuela's total receipts from oil exports during the 20-year period amounted to more than $245 billion but the oil revenue in 1994 was only slightly more than in 1974. Annual income volatility was mainly attributable to oil price fluctuations. Unlike some newcomers to the oil scene, Venezuela's dependence on oil went back to the late 1920s. In 1974, petroleum accounted for more than 40 percent of GDP (at current prices), over 70 percent of central government revenues and about 95 percent of total exports.[47] Investment and growth were both heavily influenced and sustained by developments in world oil prices. Not surprisingly, the quadrupling of international oil prices in 1973/4 provided the socialist-leaning Accion Democratica (AD) administration of Carlos Andres Perez (1974–8) with enough resources to expand and diversify the economy according to the AD party's populist agenda. The allocation of the new wealth (beyond oil conservation) included savings abroad, increased domestic consumption and an ambitious public investment programme designed to diversify the economy.

Contrary to most other group members, the Venezuelan government responded to the first oil boom with initial caution. In a gesture uncommon in most members, Venezuela's oil production and exports were reduced after 1973, because the government thought the gain in export revenues made conservation both possible and affordable. The nationalisation of the oil industry in 1976 favoured and facilitated this policy. This move toward oil conservation was followed by another (albeit temporary) prudent measure in favour of foreign currency sterilisation. As the windfalls surpassed the country's domestic absorptive capacity, some of the new revenues were converted into foreign assets through the Venezuelan Investment Fund, created for this purpose in 1974. After 1977, however, while the VIF and the national oil company were still increasing their overseas investments, Venezuelan public enterprises in the non-oil sector became net borrowers from the international commercial banks, and accumulated massive short-term debts. By 1979, in aggregate terms, there were actually no net savings abroad, as the country's $15 billion foreign investment was matched by $16 billion of external debt overall in the public sector and another $10 billion of private sector obligations.

In accordance with the Perez government's commitment to have the country's wealth reach all social classes, in order to ensure socio-economic democracy, the oil windfall was used to raise wages, workforce and output in the public sector. Between 1974 and 1978, current public expenditures rose annually at double-digit rates. Direct transfers to households during the first oil boom were also part of the windfall use. A similar expansive platform was adopted with respect to capital investment at home. The nationalisation of the oil industry in 1976 allowed public enterprises to double their investment outlays, particularly in the expansion of steel, chemical, and aluminum plants, absorbing one-third of the total non-oil, public investments. This large-scale resource-based industrialisation drive was designed to promote new exports, substitute some intermediate imports and to establish a domestic capital-goods industry. The 1976–80 plan also emphasised the improvement of physical infrastructure, the revival of the lagging agricultural sector and increased easy credit to the private sector. Much of the initial capital expenditure went into office towers, apartment buildings, multi-level highways and elaborate cloverleafs, underground subways, public works and lavish facilities similar to outlays in the Arab oil capitals. A near doubling of gross domestic investment, in addition to net foreign investment and combined with rising aggregate consumption, created a large domestic resource gap that had to be made up with imports, which rose by more than four times during the 1974–8 period, and was mainly financed through short-term borrowing of as much as $30 billion from private international banks.

The opposition political party, COPEI, that came to power in 1979, while accusing the former AD administration of wasting the oil fortune on too many expensive public projects, runaway foreign borrowing, growing inefficiencies and corruption, actually fell into the same trap once oil prices soared in the second oil boom.

The new government embarked on a spending spree of its own, using the fresh windfalls and borrowing billions in short-term, high-interest loans to finance new expenditures. In the second oil boom, 1979–82, Venezuela received a larger windfall, as a percentage of GDP, than in the first. As before, consumption of oil products at subsidised prices increased, and other transfers to households also went up. Investment in public non-oil companies continued to rise. But substantial allocations had to be made to public enterprises and private firms in order to enable them to service their past debts. Altogether, more of the second windfall was spent in the domestic economy than the first. But in the aggregate, the government became a net debtor to the rest of the world and the private sector a net creditor, as some $9 billion left the country by the latter's agency between 1980 and 1982. As a result, much of the second windfall was used to make up for the fall in national savings or was exported by private capital flights.

The economic stagnation and serious recession which developed in the first half of the 1980s was caused by the precipitous fall in gross domestic investment from about 36 percent of GDP in 1974–9 to 21 percent in 1980–5. The government initiated an expansionary policy that led to an upswing of economic activity between 1985 and 1988, but the recovery was achieved at the cost of the exhaustion of international reserves, a widening of the budget deficit and a renewed upsurge in inflation. Swallowing its nationalistic pride, in 1989 the government finally asked the International Monetary Fund for assistance, under which new funds were received from the IMF, the World Bank and the Inter-American Development Bank. Nearly $20 billion (or 75 percent of the total medium and long-term) external debt was also successfully rescheduled. Between 1989 and 1994, annual growth of domestic investment averaged less than 20 percent of GDP, of which public investment formed about half. The relatively low-rate of current public expenditure in the face of a rising population indicated a relative deterioration of public services.

As a consequence of such protracted extravagance, Venezuela's foreign debt increased steadily during the late 1970s and 1980s, reaching a record $38 billion in 1983, when a moratorium had to be declared. A rescheduling worked out in 1986 fell apart due to the collapse of oil prices later in that year. Part of the debt was reduced under the so-called Brady Plan after 1989. Between 1989 and 1994, Venezuelan external debt increased to $37 billion. Of this, about $35 million constituted medium and long-term debt. Total debt accounted for nearly 65 percent of GDP.

Venezuela carried out an elaborate welfare programme. The central government played the crucial role in redistributing the oil windfalls. About two-thirds of central government expenditures in a typical year consisted of current and capital transfers to the rest of the economy. These transfers took the form of grants, loans, interest payments and subsidies. Transfers and subsidies (excluding interest payments on debt) were broadly equivalent to central government revenues derived from the hydrocarbon sector. On the global plane, too, Venezuela, with the highest per-capita income in Latin America in 1974, adopted an elaborate programme of international economic cooperation during the first oil boom. This programme included a series of commitments to the International Monetary Fund, the World Bank, the Inter-American Development Bank, the OPEC Fund and the UN Emergency Fund, as well as bilateral loans and grants. This generosity was partly supported by the fact that Venezuela had one of the smallest defence forces for its size in Latin America, and one of the smallest defence spending (Table 15).

The policies adopted by the Venezuelan government to carry out its allocation strategy were both flawed and inconsistent. In general, under both political parties during the 20-year period, Venezuela's fiscal, monetary, trade and investment policies were all influenced, if not dictated, by two main factors: annual oil export earnings and irresistible populist welfare considerations. Fiscal policy was the initial hostage to the rise in petroleum revenues. During the first oil boom, in the Venezuelan public sector, the rate of growth of domestic expenditures outran that of GDP, and the public sector's deficit reached more than 9 percent of domestic output. The new administration that came to power in 1979 took corrective measures to cool off the overheated economy, adopting a programme of price decontrol on a wide range of consumer goods, two-thirds reduction in import duties, notable cuts in subsidies, a sharp reduction in public spending and some increase in interest rates. The second sharp oil price hike and the attendant revenue increases made such efforts highly unpopular, and the authorities switched to an expansionist fiscal policy once again. A drop in oil income during 1982 and 1983 (equivalent to 10 percent of GDP) worsened the government's finances, and the budget deficit surpassed 11 percent of GDP in 1982. Accordingly, capital expenditures were significantly reduced, particularly by state enterprises. A dramatic increase in non-tax revenues turned the fiscal balance into surplus and the 1984 bolivar devaluation raised the local currency value of the government's oil income.

The oil-bubble bust in 1986, and the adoption of IMF's austerity programme – considered 'shock therapy' – was followed by deep budget cuts, sharp reductions in domestic liquidity, an increase in interest rates, higher energy prices, more taxes, an end to price controls and subsidies, new incentives for private investment, sale of money-losing state enterprises and a freer exchange regime. Some reforms, such as privatisation of seven major state enterprises, reduction of transfers to public entities and sharp cuts in costly food subsidies, were also pursued. However, bloody riots in the major urban centres and two military coup attempts in 1992 showed the political unpopularity of the adjustment programme, and the remainder of the fiscal and banking reforms were postponed. The election of a populist, anti-reform president in 1993 led to the suspension or reversal of some of the reform measures. The backlash against the IMF-type market reforms resulted in widespread government intervention in the economy at the time when almost all other Latin American countries were paring down their public sector involvement. As a result of a further sharp deterioration of economic conditions, however, a new recourse to the IMF credit and the acceptance of a new set of conditions were agreed to by the same administration.

Monetary and banking policies in Venezuela suffered from inefficient management and erratic direction. Under various administrations (except for a brief period in 1989), the government's monetary stance focused on holding down prices and raising government salaries, as a means of maintaining social welfare. During the first oil boom, monetary policy played a more active role. With the stagnation of domestic economic activity that coincided with the second oil boom in 1979, a policy of low domestic interest rates was initiated to reflate the economy. This easy credit policy, at a time when international rates were rising, led to substantial capital outflows. In subsequent years, monetary authorities made frequent use of interest rate ceilings and mortgage bonds as instruments of monetary policy. When interest rates became increasingly negative between 1986 and 1988 the pressure in the free market exchange rates intensified. One of the first reforms under the 1989 Structural Adjustment Programme was the removal of interest controls except in agriculture and housing. This liberalisation move was later ruled unconstitutional by the Supreme Court, and minimum and maximum interest rates were again established by the central bank. After 1989 some open-market operations in short-term, zero-coupon bonds and foreign debt conversions were added to reserve requirements, as a monetary policy instrument, to reduce private credit growth. With the fiscal position weakened in 1993, monetary policy was more widely used to maintain stability, support the bolivar and preserve international reserves. But the government forced the newly independent central bank to cut interest rates to stimulate the economy, and the rates became negative again in 1994.

Venezuela followed an elaborate system of price controls, in the name of consumer protection, education and welfare. In order to combat the rising price trend during the first oil boom, subsidy payments to farm products were raised; product prices of state-owned enterprises (particularly gasoline and other oil derivatives) were kept low and occasionally below cost and low-interest loans were offered to the agriculture and housing. In 1979, the onset of the second oil boom, price controls were relaxed and ceilings were removed. The subsequent expansion in aggregate domestic demand, the large wage awards and some price liberalisation measures, coupled with an upturn in external inflation, prompted the government soon to reverse its 1979 price liberalisation policy. After emergency measures were imposed during the 1983 exchange crisis, the government ordered a 60-day freeze on prices of all goods and services followed by a system of administered controls. A package of reforms worked out with the IMF in 1989 called for freeing of interest rates and eliminating price controls. In June 1994, price controls were reimposed, and by the end of 1994 the government controlled the prices of all 'essential' items, interest rates, rents and a variety of services, often in minute detail.

Venezuela's exchange regime was for the most part essentially liberal and free. Foreign exchange policies were established by the central bank in coordination with the ministry of finance. The two-tier exchange rate system that existed before 1974 (separating oil and iron from other goods and services) was unified in 1976 when the country formally accepted the obligations of Article VIII of the Fund Agreement. From mid-1976 until February 1983, the exchange rate of the bolivar was pegged to the US dollar at BS4.29=$1, and there were no restrictions on capital flows. Following some price deregulation in 1979 and the emerging domestic inflation, the real exchange rate began to rise. With domestic interest rates kept below global rates, large sums of foreign exchange began to move abroad, and private payment arrears began to accumulate. In late February 1983, the government suspended all sales of foreign exchange and moved toward exchange restrictions and multiple exchange rates for the first time in nearly two decades. As expected, a wide gap between the controlled and free market rates emerged, leading to distortions in cost/price relationships, production and trade. The bolivar rates were again unified under the IMF's 1989 reform package, and restrictions were removed. The banking crisis that erupted in January 1994 undermined public confidence in the currency, and the bolivar lost 50 percent of its value within a few months. By late June 1994, a new system of comprehensive exchange controls was again imposed. The bolivar was fixed at BS170=$1 (while the free market rate was BS197=$1). In June 1994, Venezuela froze all movement of foreign exchange in order to dampen speculation against its currency. Although exchange controls were later relaxed for foreign investors, the hope for new capital inflows, the prevailing environment of triple-digit inflation and a bolivar overvalued at 170 to the dollar did not prove particularly helpful.

Venezuela's trade policy followed a less tumultuous but still fluctuating pattern. During the first oil boom, Venezuela followed a relatively liberal policy. Non-oil exports were encouraged through low-cost financing and tax exemptions. In 1982, the 1979 liberalisation process was reversed and a number of imports were prohibited. Following the 1983–4 exchange measures, significant new restrictions were placed on foreign trade and licensing was again resumed. In later years, particularly after the trade liberalisation package of the 1989 IMF programme, most imports were free to come in against payments of *ad-valorem* custom tariffs. There were no controls or restrictions on exports. Export of agricultural commodities were encouraged by a fiscal bonus, that of manufactured goods by drawback of customs duties on inputs. Exports of hydrocarbon products and iron ore, however, were state monopolies. Venezuela became a member of the GATT (later, World Trade Organisation) in 1990. Along with the other members of the Andean Pact, it agreed to a reduction of external tariffs by 1 January 1994. Concurrently agreements were reached with Chile and the CARICOM countries for trade liberalisation and bilateral clearing of accounts.

Foreign direct investment was governed by the Andean Pact regulations whereby certain sectors were reserved exclusively for the government and national investors, and certain limitations were placed on equity participation and profit remittances. Investments in the financial sector were allowed, beginning in 1994. Investments in the petroleum and iron ore sectors were subject to specific regulations. Outlays in other sectors were freely permitted. Since 1992 certain external public debt could be converted into shares of equity in physical infrastructure, transportation, educational institutions, medical centres and other projects.

Of significant interest in the field of foreign direct investment was Venezuela's nearly total turnabout in its energy policy. Reversing its oil-nationalisation strategy adopted in 1976, Venezuela offered a direct stake in petroleum development to a number of American, Dutch and Japanese oil companies in 1992 – a move which was considered politically taboo for a long time. While the government maintained ownership of all new finds, foreign companies would receive part of the value of each barrel of crude produced as a return on their capital investment. Ironically, the turnaround in strategy occurred during President Carlos Andres Perez's second administration – the same president who booted the multi-national oil companies out of Venezuela in his first term in 1976. The crack in the old nationalisation policy, the first since 1976, was too obvious to overlook.

In sum, by 1994, the Venezuelan government exercised considerable control over the economy, including over wages, prices and exchange controls. Half of all banks were owned by the state. The privatisation programme was in limbo. Subsidies on public goods and services were considerable.[48] As a result of drastic and somewhat desperate government measures, Venezuela's credit-risk rating in mid-1994, published by the Economist Intelligence Unit, became South America's worst, and was at the same level as that of Algeria and Nigeria.

SUMMARY

The foregoing discussion showed that, while all members substantially increased their public expenditures as their public revenues rose in tandem with the oil booms (and uniformly failed to cut public spending when oil income dwindled in the prolonged 1983–90 bust), there were notable differences in the direction and focus of windfall allocation and shortfall disallocation. There were also significant variations among members with respect to saving and dissaving abroad (i.e. foreign lending vs foreign borrowing). The composition and mix of investment varied both in regard to sectors, size of projects and type of enterprise. Algeria emphasised the development of natural gas and exhibited the strongest preference for heavy industry at the expense of previously thriving agriculture. Nigeria and Venezuela concentrated on basic metals (e.g. aluminum and steel). In sectoral allocation, Indonesia paid due attention to agriculture while Nigeria and Algeria woefully neglected this traditional sector.

Infrastructural investments in both physical and human capital were almost unanimously emphasised, albeit with different results. For capital-deficit countries with a socialist bias (like Algeria and Ecuador), public policy heavily favoured capital accumulation over consumption and production of capital goods over consumer wares. In the capital-surplus countries of the Persian Gulf, the abundance of oil revenues allowed simultaneous gratification of consumption, investment and military build-up. Indonesia covered its flanks against an oil bust by pursuing a policy of relative austerity, while emphasising poverty reduction. In all countries, resource-based industrialisation and import-substitution became the norm.

As a percentage of government expenditure, Algeria, Gabon, Nigeria and Venezuela had the lowest share of defence outlays (i.e. less than 10 percent on average collectively) while non-combatant Libya, Qatar, Saudi Arabia and the UAE had the highest (i.e. as high as 42 percent). Iran, Iraq and Kuwait were special cases in this regard. Algeria, Indonesia and Venezuela borrowed heavily abroad even during their boom cycles, while Kuwait, Qatar and the UAE increased their foreign assets most of the time. Ecuador, Gabon and Nigeria went into debt when oil prices plummeted after 1983. Iraq accumulated a heavy debt burden during its eight year war with Iran, and could never recover from its ill-fated adventure into Kuwait.

All OPEC members made use of conventional macroeconomic policies that seemed to fit their needs. In all member countries, the public sector took the lead in capacity expansion and engaged in direct production and distribution, albeit in each country with different rapidity, scope and depth. Thanks to its enormous foreign revenues, Kuwait escaped the need for strict budgetary discipline. Indonesia exerted relatively better control over its fiscal policy (e.g. balance budgeting) and anti-poverty stance. Unique among members, Indonesia also developed a diversified tax base, and maintained a fairly effective fiscal discipline up to 1994. Algeria, Iran and Saudi Arabia, either by design or default, allowed deficit-financing by the banking system as part of their fiscal strategy. In monetary policy, Gabon and Indonesia adopted more effective anti-inflationary policies while Ecuador, Iran, Nigeria and Venezuela proved incapable of keeping prices in check. Timely exchange-rate adjustments were successfully made by Indonesia but were long neglected by Gabon, Iran, Iraq, Libya, Nigeria and Venezuela. The policy of windfall 'sterilisation' was hardly followed by members outside of the small capital-surplus countries, except Venezuela (in the early years) and Indonesia (through payment of *Pertamina's* debts). Small Persian Gulf members, plus Indonesia, adopted a fairly free-trade position and avoided quantitative controls over foreign exchange.

NOTES ON CHAPTER IV

1 Cf W.P. Webb, *The Great Frontier* (Boston: Houghton Mifflin, 1952).
2 For an extensive discussion of this point, see *Claiming the Future* (Washington: The World Bank, 1995).
3 Discussion on Algeria in this and the forthcoming chapters is based, *inter alia*, on: *Algeria: 1997 Statistical Appendix* (Washington: IMF Report no. 97021, 1997); *Algeria: 1996 Selected Economic Issues* (Washington: IMF Report no. 96071, 1996); *Algeria: 1995 Background Paper* (Washington: IMF Report no. 95054, 1995); *Algeria: A Country Study* (Washington: Library of Congress, 1994); *Algeria: The Transition to a Market Economy* (Washington: World Bank, 1994). For specific sectors of the Algerian economy, see essays in J.P. Entelis and P.C. Naylor (eds) *State and Society in Algeria* (Boulder: Westview Press, 1992).
4 See Jeal Leca, 'Algerian Socialism, Nationalism, Industrialisation and State Building', in Helen Desfosses (1975).
5 For a concise treatment of the Algerian response, see Patrick Conway, 'Algeria: Windfalls in a Socialist Economy', in Alan Gelb (1988), Chapter 10.
6 For insightful glances into the Algerian economy and polity, see Alan Richards and John Waterbury (1990), pp 18-19, 199–200, 252–5.
7 Discussion on Ecuador in this and the following chapter is based, *inter alia*, on *Ecuador: 1995 Recent Economic Developments* (Washington: IMF Report no. 95096, 1995); *Ecuador: 1997 Statistical Annex* (Washington: IMF Report no. 97100, 1997); *Trade and Growth in Ecuador* (Washington: The World Bank, 1994); *Ecuador: A Country Study* (Washington: Department of the Army, 1991).
8 For a fuller discussion of these points and other data on the Ecuadorian economy, see J. Marshall-Silva, 'Ecuador: Windfalls of a New Exporter' in Alan Gelb (1988), Chapter 11.
9 Discussion on Gabon in this and the following chapter is based, *inter alia*, on: *Gabon: 1995 Background Paper* (Washington: IMF Report no. 95129, 1995); *Gabon: 1997 Statistical Annex* (Washington: IMF Report no. 97054, 1997).

10 Discussion on Indonesia in this and the following chapter is based, *inter alia*, on: *Indonesia: Recent Economic Developments* (Washington: IMF Reports no. 97075, 1997, no. 96092, 1996, and no. 95083, 1995); *Indonesia: A Country Study* (Washington: Dept. of the Army, 1993); *Indonesia* (Washington: World Bank Report no. 7822, 1989, nos 8034 and 8455, 1990, nos 8317 and 9672, 1991, and nos 10333 and 10470, 1992); *Indonesia: Strategy for a Sustained Reduction in Poverty* (Washington: World Bank, 1990); *Bank Indonesia: Annual Report* (Jakarta: annual) and *Indonesian Financial Statistics* (Jakarta: monthly); *Bulletin of Indonesia Economic Studies* (Canberra: Australian National University: periodic).

11 Philip Shenon, 'Hidden Giant', *New York Times*, 27 August 1993, pp A1-6.

12 E.A. Gargan, 'Family Ties That Bind Growth', *New York Times*, 9 April 1996.

13 Discussion on Iran in this and the following chapter is based, *inter alia*, on: *Islamic Republic of Iran: Statistical Appendix* (Washington: IMF Report no. 96108, 1996); *Recent Economic Developments* (Washington: IMF Report no. 95121, 1995); Bank Markazi Iran: *Annual Report* (Tehran: annual); Plan and Budget Organisation: *Annual Report* (Tehran: annual); Statistical Center of Iran: *Statistical Yearbook* (Tehran: annual).

14 For details see M.G. Majd, 'The Oil Boom and Agricultural Development', *Journal of Energy and Development*, Vol. XIV, No. 1, 1990.

15 Discussion on Iraq in this and the follow chapter is based, *inter alia*, on: *Iraq: A Country Study* (Washington: Dept of the Army, 1990); Government of Iraq: *Annual Abstract of Statistics* (Baghdad: annual for some years).

16 For a concise discussion of the economic role of the state and the Baath Socialist party, see M.F. Sluglett (1990), Chapter 7.

17 For the breakdown of planned and realised expenditures, see Abbas Alnasrawi (1994), pp 142–3.

18 See Abbas Alnasrawi (1994), Chapter 7; F.W. Axelgard (1986); and *Iraq: Country Profile 1995/96* (London: The Economist Intelligence Unit, 1996).

19 H.C. Metz (ed.) *Iraq: A Country Study* (Washington: Library of Congress, 1990), Chapter 3.

20 Kamran Mofid (1990), Chapter 10.

21 Abbas Alnasrawi (1994), p 122.

22 Discussion on Kuwait in this and the next chapter is based, *inter alia*, on: Kuwait Central Bank, *Quarterly Statistical Bulletin* (Kuwait: quarterly); Kuwait National Bank, *Economic and Financial Bulletin* (Kuwait: quarterly); and Kuwait Ministry of Planning, *Annual Statistical Abstract* (Kuwait: annual).

23 See *Kuwait: A Privatisation Strategy* (Washington: The World Bank, 1993).

24 Discussion on Libya in this and the following chapter is based, *inter alia*, on: *Libya: A Country Study* (Washington: Dept of the Army, 1988); and Libyan Secretariat of Planning and Finance, *Developments in the National Economy, 1970-92* (Tripoly: 1993).

25 'Libya's Man-Made River', *The Middle East*, February 1996. See also 'Focus on Libya's Great Man-Made River', *OPEC Bulletin*, January 1997; and Raymond Bonner, 'Libya's Vast Desert Pipeline', *New York Times*, 2 December 1997.

26 The scheme was reported by intelligence sources to camouflage Libya's huge underground chemical arms plant. *New York Times*, 25 February 1996, p 8.

27 *Libya: Country Report* (Economist Intelligence Unit, 2nd Quarter 1995), p 20.

28 Discussion on Nigeria in this and the following chapter is based, *inter alia*, on: *Nigeria: 1997 Statistical Annex* (Washington: IMF, Report no. 97011, 1997); *Nigeria: Background Papers and Statistical Appendix* (Washington: IMF Report no. 95194, 1995); *Nigeria: A Country Study* (Washington: US GPO, 1992); Central Bank of Nigeria: *Annual Report* (Lagos: annual); *Nigeria: Structural Adjustment Program* (Washington: World Bank, 1994); and *Nigeria: Experience with Structural Adjustment* (Washington: IMF, 1997).

29 The discussion on Nigeria here draws on Henry Bienen, 'Nigeria: From Windfall Gains to Welfare Losses' in Alan Gelb (1988), Chapter 13.

30 S. Schats (1977), See also Ahmad Jazayeri (1988), Part III.

31 For details, see N.B. Tallroth, 'Structural Adjustment in Nigeria', *Finance & Development*, September 1987. See also *Nigeria: Experience with Structural Adjustment* (Washington: IMF, Occasional Paper No. 148, 1997).

32 Between the end of the second oil boom (1982) and early 1994, the naira was devalued from N0.68 to N22 per US dollar, thus raising the dollar value by 3235 percent.

33 See *Nigeria – 1996 Investment Climate Report* (Washington: US Department of Commerce, 1996).

34 The discussion on Qatar and UAE in this and the following chapter is based, *inter alia*, on *Persian Gulf States* (Washington: Library of Congress, 1994).

35 Information on Qatar's economy was hardest to obtain and least satisfactory of all members. Most standard statistical reference works do not even carry Qatar among countries covered. Data shown here have beencollected from various sources with various degrees of reliability.

36 Discussion on Saudi Arabia in this and the next chapter is based, *inter alia*, on Saudi Arabia Monetary Agency, *Annual Report* and *Statistical Summary* (Riyadh: annual); and *Saudi Arabia: Country Profile* (London: The Economist Intgelligence Unit, 1994–5).

37 See R.D. Crane (1978).

38 Although the guaranteed purchase price was gradually reduced from $1,000 a ton in the late 1970s to $550 a ton in the late 1980s, annual output grew steadily.

39 See *Saudi Arabia: Country Profile*, The Economist Intelligence Unit, 1994/5, p 9.

40 See 'Saudi Arabia', *Middle East Economic Digest*, 11 March 1994.

41 See 'Filling a Void', *The Economist*, 13 February 1982, p 33.

42 A Saudi bank manager, for example, boasted that by bypassing the chequebook stage and going straight to 'smart cards' he had reduced the staff by about 50 percent on average.

43 'Challenge for Saudi Arabia', *Financial Times*, 25 January 1984.

44 Discussion on The United Arab Emirates in this and the next chapter is based, *inter alia*, on *Persian Gulf States: Country Studies* (Washington: Library of Congress, 1994).

45 The Emirates are: Abu Dhabi, Dubai, Sharjah, Ras Al-Khaimah, Ajman, Fujairah and Umm al-Qaiwain.

46 Discussion on Venezuela in this and the following chapter is based, *inter alia*, on: *Venezuela: Recent Economic Developments* (Washington: IMF Report no. 96087, 1996); *Veneconomy* (Caracas: weekly and monthly); *Venezuela and the World Bank* (Washington: World Bank, 1992); T.L. Karl (1997), Part II; and *Venezuela: A Country Study* (Washington: Department of the Army, 1993).

47 Discussion in this section benefits from Francois Bourguignon, 'Venezuela: Absorption Without Growth' in Alan Gelb (1988), Chapter 15.

48 In 1994, regular gasoline cost 13 cents a gallon – one of the cheapest in the world. This price policy not only cost the government an estimated $500 million a year in subsidies but also resulted in an average annual gasoline consumption per registered vehicle of 1445 gallons – two-and-a-half times that of the US and seven times that of Japan.

V

Country Report Cards

Economic performance is measured by a number of different criteria. In the broadest terms, there are the overarching twin criteria of efficiency and equity – one comprising growth, full employment and rising productivity, and the other focusing on fairness in income distribution, equality of opportunity and social justice. Disaggregating these two criteria, there is a wide spectrum of success indicators centring on what is called net economic welfare, which takes into account not only the positive market value of goods and sources, but also the negative social cost of disamenities (e.g. pollution, environmental damage and perverse externalities). There are still other socio-economic indices that attempt to measure the 'quality of life' beyond a simple enjoyment of consumption.

Due to a lack of consensus on the selection of appropriate criteria, and the absence of comparable data for all members during the 20-year period, the analysis here will follow a double track approach. In one, each country's economic performance will be examined in terms of its own socio-economic agenda. Such a goal-oriented appraisal will show how each member has fared in achieving its own objectives by the choice of its allocation decisions and adjustment policies. In the other, each country's performance will be analyzed in terms of standard macro-economic variables, e.g. national and sectoral growth, employment, price stability, fiscal discipline and external balance. An additional criterion, i.e. a sustainable base for an eventually viable non-oil economy, is also added to this standard list.

116

ALGERIA

Algeria's initial 'socialist' development objectives emphasised (1) stimulating economic growth, increasing employment and raising the national standard of living (2) promoting income equity among both regions and groups and (3) increasing the overall productive potential of the economy. These objectives, in turn, were to be obtained by (a) full sovereignty over national resources (b) relying on domestic means and minimising long-term dependence on foreign financial and technical help and (c) state economic planning, assisted by workers' participation. Toward the end of the 1980s, widespread disappointment with this populist/statist strategy led to a tendency toward market-based policies, while fundamental national objectives of high-employment, growth, stability and diversification were also emphasised.

Algeria's total receipts from oil and gas exports during the 1974–94 period amounted to about $160 billion, with considerable fluctuations from year to year in both volume and value. By 1994, Algeria was also heavily in debt, to the tune of $30 billion. During the period, the country received grants from the French government and some multinational organisations. By the end of 1994, Algeria's economy was still on a life-support system financed by over $1 billion of French aid, World Bank loans and IMF standbys.

Algeria's economic performance during the first two oil booms (1974–82) was improbably weak.[1] The national growth rate was maintained at about 7 percent a year on average, thanks in part to the massive investment outlays. Yet, the annual growth rate had been even higher prior to the oil boom. Part of the problem was the rather high incremental capital output ratios (ICORS) of 7 to 10 in the mega projects (compared to 3 or 4 to 10 in developed economies). Low capital productivity was also frequently associated with the existence of unused excess capacity and supply-side constraints (i.e. bureaucratic inefficiencies, shortage of skills and the absence of free market pressures). Thus, while construction materials were still being imported, building materials industries operated at as low as 17 percent, and no more than 52 percent, of capacity. In some industries, like LNG (which operated at 40 percent of capacity during the early 1980s), low productivity was due to inadequate demands, i.e. faulty projections. Development plans 1974–84 fell short of their targets because of delays in project execution, cost overruns and deviations from planned investments. Due to a relative neglect of certain parts of the social infrastructure, severe shortages developed in housing, water supply and other social services. During the decade ending in 1994, GDP had 5 years of negative growth and an overall average of less than half of 1 percent a year, at best. GDP per capita and per capita private consumption also steadily declined.

Bound by its socialist principles, Algeria neglected agriculture and concentrated on massive industrialisation – wishing to become the 'Japan of Africa'. By the end of the second oil boom, while more than half of the labour force was still rural, farming accounted for only 7 percent of GDP. And the country, once self-sufficient in agriculture, imported 70 percent of its food consumption.[2] With high rates of population growth (initially encouraged by the government) emphasis on hydrocarbon and heavy industries, and rapid rural migration due to stagnation in the farming sector, the country was plagued with massive urban unemployment. The urban population of 20 percent at independence rose to 47 percent by 1977, putting an intolerable strain on housing and social services. Official unemployment statistics are not available for all years. Rapid increases in non-agricultural employment kept the rate moderate during the 1970s. After the second oil boom and reduced annual investments, some 16 percent of the economically active population were out of work. The employment situation worsened after 1988 under economic reforms, and out of the 6.5 million of economically active population in 1994, 27 percent (mostly youth and school graduates) were listed as unemployed, although the real (including the under-employed) rate was probably much higher.

Algeria's other goal of raising the national living standard and promoting equity in individual and regional incomes remained largely unfulfilled. Weak economic growth and rising unemployment translated into a substantial decline in average per capita income. There were no data on the incidence of absolute poverty. By World Bank estimates, the average annual real growth rate of GNP per capita between 1985 and 1994 was negative. As a result, people's annual livelihood deteriorated almost steadily during the second decade under review. Responsibility for the fall in the standard of living can be attributed to not only grave production inefficiencies, mis-management in public enterprises and the bloated bureaucracy but also child support benefits which encouraged a phenomenal growth in birth rates.

In other conventional national objectives, too, Algeria's record was unflattering. The objectives of internal and external stability were largely unrealised. Inflation, budget deficits and foreign exchange gaps plagued the country, particularly after 1985. Despite extensive price controls, Algeria's consumer price index was in or near the double-digit range for most of the period, with the 1994 figure reaching 39 percent. Actual inflation was probably higher because official statistics referred to officially set prices which were lower than those actually paid in private shops. Strict rationing, and the general shortage of consumer goods during the 1970s, helped to keep prices and wages in check. Between 1974 and 1994, the minimum monthly wage rates rose by more than 500 percent but up to the end of 1990 real wages were on the decline.

Algeria's commitment to sustaining a socialist economy and ensuring economic self-sufficiency gave the government an insatiable appetite for public expenditure. During most of the period, the central government budget ran a deficit each year, mostly on the accounts of advances and loans to money-losing public enterprises. The shortfalls were financed mainly by borrowing from the domestic banking system. After the second oil boom, budget deficits continued every year, averaging nearly 9 percent of GDP. Under the 1987 austerity programme, the fiscal deficit fell gradually through 1990. There was a small surplus in 1991 resulting from the third oil boom and the government's attempt at tighter demand management; but the budget came under pressure again in 1992 when a new social security system was instituted. Successive administrations' attempts to revive a destabilised economy through capital spending and the relaxation of earlier policy initiatives resulted in continued short-falls to the end. Only an IMF agreement in 1994 prevented the budget deficit from reaching an estimated 15 percent of GDP in that year.

External balance also eluded Algeria for most of the period. From the onset of the first oil boom until 1979, Algeria's current account was perennially in the red because of surging imports and payments for services. Thanks to capital inflows from foreign borrowing and workers' remittances, the overall balance was cumulatively in the black for the next six years. Between 1986 and 1990, rising outflows of interest, profit and dividends and rising farm imports put the current account under severe pressure so that the overall balance produced perennial deficits. In the 1991–4 period, the current and overall accounts showed small surpluses thanks to drawings from official sources. During the first oil boom, the Algerian dinar remained relatively stable. In the early 1980s, the dinar traded in the black market at two to three times the official rate. This ratio rose to six times in 1990. The Algerian currency gradually reduced its official rate in terms of the US dollar from AD 4.9=$1 in 1988 to AD 36.8=$1 in 1994.

Despite its national objective of relying on domestic means and avoiding foreign finance, Algeria engaged in heavy foreign borrowing to finance its hydrocarbons and industrial development. In some ways, however, Algeria made some slight progress in its diversification drive. The country was able to withstand the oil glut of the 1980s by exports of refined products, condensates, LPG and gas. While in 1994 the country was still dependent on oil and gas for over 95 percent of export revenues, the share of hydrocarbons in GDP had declined from 28 percent in 1974 to 22 percent. Despite its unusually heavy industrialisation drive, however, the share of industry (including construction and utilities) rose slightly from 18 percent of GDP at the start to 22 percent at the finish. And the share of the manufacturing sector, at nearly 11 percent, still remained much smaller than the 20 percent average for middle income countries; the sub-sector was also oriented toward import-substitution and heavily dependent on imported raw materials, spare parts and equipment.

Algeria also faced significant environmental challenges at the end including water shortages, soil erosion, land degradation and industrial pollution. Some 40 percent of the water supply was lost each year through poor distribution. Some 40,000 hectares of farmland were made idle each year due to soil erosion. Overgrazing, indiscriminate cutting and forest fires resulted in considerable deforestation. Industrial pollution (including hazardous waste) was threatening major urban centres.

ECUADOR

Ecuador, having previously experienced some harmful boom and bust cycles in cocoa and banana, chose to use its oil wealth to (1) integrate the various parts of the nation (2) improve living conditions, particularly of the poor and (3) build up its productive sectors. In its 1980–4 national plan, there were specific references to accelerating growth, high employment, increased labour productivity, greater price stability, better external balance and economic diversification.

Performance in all three areas showed a chequered record. During the first oil boom, 1974–9, the average annual growth rate neared 6 percent. Economic conditions began to weaken in the early 1980s, and the real growth rate of GDP averaged less than 2 percent a year between 1980 and 1985, trailing the estimated annual population increases of 2.8 percent. The sharp decline in oil export earnings, a major earthquake and the austerity measures adopted by a new left-of-centre administration after 1986 kept the 1986–90 rate of GDP growth to 2 percent – still not sufficient to cover population growth or to absorb the rising labour force. Under the third oil boom, a modest growth rate averaging about 3.5 percent a year was achieved due to a rise in crude oil export as well as in bananas and shrimp.

The behaviour of productive sectors was also not uniform. Agriculture was the laggard for most of the period, averaging about 4 percent annual growth. In addition to occasional flood and drought conditions, the sector suffered from such structural weaknesses as uncertainties associated with proclaimed agrarian reform, small sizes and poor quality of farms, small farms' inadequate access to credit, inadequate irrigation and extension services, high input costs and, due to price controls, insufficient price incentives. While agriculture fared better in Ecuador than in some other member countries, it was still disadvantaged compared to industry. It had fewer tariff exemptions on imported inputs; it suffered from subsidised imported food stuff (wheat and milk); it received only paltry extension services and credit from the government. The oil sector had its own ups and downs. Oil export receipts nearly doubled between 1974 and 1979. In the 1981–5 period, petroleum was still the star performer, rising at an annual average real rate of 10 percent due to larger crude production. Output and export were curtailed due to the 1987 earthquake and marketing problems but they both recovered later (particularly after Ecuador left OPEC), and the sector's growth in 1990–4 exceeded 8 percent a year on average. Under the first oil boom, manufacturing expansion was strong at an average annual rate of 11 percent, helped by industries that were energy-intensive. But, due to the limited size of the domestic market and the heavy reliance of this sector on basic metals and minerals, output growth in 1980–94 averaged about 1 percent a year.

High employment as another principal objective was not easy to achieve. Ecuador's active labour force increased faster than its population during the 20-year period – 2.1 percent *vs* 1.7 percent. While the unemployment rate remained in the low teens, underemployment was estimated to be as high as 50 percent. As expected, migration to the cities created labour shortages in rural areas, and unskilled urban workers constituted the bulk of unemployment in the cities. The structure of employment also changed considerably. While agriculture remained a large source of occupation in 1974 as well as in 1994, it became the only sector in the economy with a loss in total jobs. There was no detectable improvement in labour productivity.

The objective of internal and external balances was hardly reached. Budget deficits, internal inflation and payments imbalances plagued the country for nearly the entire period. Due partly to frequent changes in administration, Ecuador could not put its fiscal house in order. The annual budget deficit averaged about 5 percent of GDP between 1974 and 1979 and, with occasional improvements in some isolated years, continued to the end (reaching 7 percent of GDP in 1992). Deficits were partly financed by external borrowing. Expansionary policies, followed in the first half of the 1980s to offset the adverse effects of the depressed world oil market and increases in public sector wages and public development expenditures, necessitated a series of agreements with the IMF in 1983, 1986, 1989 and 1991, under which some external arrears were cleared and foreign debts were rescheduled under the Paris Club auspices. Despite a whole series of structural reform measures, the government failed to keep sufficiently tight control over public finances, and IMF disbursement was suspended in 1990 and 1992. The Duran Ballen's reform efforts resulted in a new stand-by facility in mid-1994, a Brady Plan deal with commercial banks and a rescheduling of Paris Club debt. The budget finally moved towards balance.

Inflation in Ecuador during the 20-year period was fuelled mainly by large public sector deficits, frequent currency devaluation, low agricultural and industrial productivity, high labour costs and the gradual phasing out of food and fuel subsidies. The consumer price index rose at an annual average rate of nearly 18 percent between 1973 and 1985. The weakening of economic conditions in the first half of the 1980s (including a bad harvest year), exacerbated by the March 1987 earthquake, had a continued strong adverse impact on prices through reduced government revenue, food shortages and currency devaluations. Between 1980 and 1994, consumer prices rose at an average annual rate of more than 37 percent (compared to the official target of 25 percent).

Ecuador's balance of payments showed relatively large, cumulative, current account deficits (including increased interest payments on debt) during the first two oil booms, although the overall balance was in surplus. Under the IMF stabilisation programmes launched in 1983, and a good export performance, both the current and overall accounts shifted to a small surplus in 1985. The overall balance was perennially in deficit during the eight years ending in 1994 despite the fact that the value of Ecuador's sucre fell from SU 33=$1 in 1982 to SU 1920=$1 in 1994. The deficits were marginally financed by the use of IMF credit and loans, as well as assistance from other multinational organisations, but they routinely resulted in the accumulation of arrears and the rise of external debt, which reached nearly $15 billion by 1994, despite having the receipt of some $3 billion in official loans and grants. Ecuador's per capita external debt and its ratio of debt to GDP were among the highest in Latin America.

The goal of diversification and reduced reliance on the oil sector was only modestly approached. Agriculture's share in GDP declined from around 22 percent to 12 percent 1974–94. The oil sector's contribution was marginally reduced downward from 13 percent to 12 percent. By contrast, the share of industry rose from 16 percent to 24 percent, as did the services' contribution from 49 to 52 percent. By 1994, oil and derivatives still dominated Ecuador's export trade with about a 40 percent share, down from 55 percent in 1974 and 47 percent in 1973. The country, however, was 80 percent dependent for its export earnings on oil and four traditional items (bananas, shrimp, coffee and cocoa). The pattern of government incentives to industry – low interest rates, low energy costs, various tax exemptions, duty-free imported inputs, accelerated capital depreciation and participation in the Andean (customs-union) group – was also significantly biased toward energy-intensive industries. As a result, the industrialisation strategy failed to develop strong, export-oriented, manufacturing industries.

Finally, the goal of improving living conditions as a primary objective was partially achieved. Before 1973, Ecuador was one of the poorest in Latin America. Per capita income in 1974 was estimated at $480; per capita GDP in 1994 was $1480. The average annual growth rate of real per capita GNP during the whole period was probably only about 1 percent. How the increased per capita income was shared out between the different social strata is not clear. Some observers believe that the oil bonanza actually increased income inequality and regional disparities. The position of middle class bureaucrats and professionals probably improved *vis-à-vis* the richest 10 percent and the poorest 20 percent of the population. By some estimates, 30–35 percent of Ecuador's 11 million population in 1994 were still below the poverty line.

GABON

Gabon's development priorities emphasised (1) economic growth (2) construction and maintenance of physical infrastructure (particularly roads and rails) (3) investment in agriculture and forestry (4) promotion of small and medium-size industrial enterprises and (5) education, training and health care.

Gabon's economy took off in 1974 with a boom in construction, trade and services, and the economic overheating caused by the sudden construction upsurge had to be stopped. After an extraordinary nominal growth averaging 40 percent a year in 1974–6, the economy declined by 26 percent in 1977–8, due to a combination of austerity measures taken to cool the economy and a decline in oil production and prices. In subsequent years, economic activity and development were led by the level of oil income and public expenditure. Between 1981 and 1985, real GDP behaved erratically, and, after 1986, followed a downward trend. Real non-oil GDP also declined between 1985 and 1994.

Different GDP components behaved in different manners. Agriculture lagged behind other sectors due largely to the massive rural exodus, transportation bottlenecks, soil erosion and declining profitability. The industrial sector expanded in response to the growth of oil and mineral extractions but concentrated on import substitution and processing (oil products, foodstuffs and timber) for a small market. It remained relatively undeveloped due to competition from neighboring countries, the high cost of local labour, and declining investment. The construction sector, which had a dominant position during the Trans-Gabon Railway construction, gradually lost ground after the latter's inauguration in 1985. The oil sector showed an erratic performance. Daily crude oil output increased by more than 10 percent between 1974 and 1977 but declined by more than 32 percent by 1981. Production was revived afterwards by 15 percent through 1985, reflecting the coming on stream of new oil fields. After another three-year slump, output was significantly raised each year, reaching nearly twice the 1981 level by 1994. Gabon's oil prices were also subject to great volatility.

Gabon's lacklustre overall growth, its primary objective, was due partly to shortcomings in its planning process. Public investment, 1976–80, exceeded the planned figures by 50 percent, while sectoral allocations also diverged considerably from their targets. Agriculture, forestry, mining, water and power experienced various production shortfalls, while tourism, hotels and administrative infrastructure absorbed three times as much as allocated. The three-year rollover planning process that was adopted after 1985 experienced a similar fate. Development planning was also ineffective in matching economic activity with the national workforce. During the first oil boom, Gabon experienced an acute shortage of manpower at all levels. Despite a planned emphasis on the Gabonisation of the workforce, more than 25 percent of wage earners in 1985 were expatriate workers (mostly in managerial and technical positions). By the end of the period, the employment situation had not changed markedly.

Planning deficiencies, aggravated by fluctuating oil revenues, resulted in serious internal and external imbalances. With both budgetary and extra-budgetary outlays rising by leaps and bounds, the total government budget gap rose from 5 percent of GDP in 1974 to 22 percent in 1977. Thanks to drastic measures adopted in 1978, the fiscal position improved for three years but deficits began to rise after 1983. Subsequent yearly fluctuations in public finance were due to the ups and downs in the oil price and the government's dependence on oil revenues. The third oil boom resulted in a small budget surplus in 1991 – the first in eight years and some relief was later provided by the currency devaluation in 1994; but the budget deficit was still equivalent to 2 percent of GDP. Persistent budget deficits resulted in the accumulation of domestic and foreign arrears.

Unlike most other group members, Gabon followed a tradition of low inflation. Both the rise and fall of CPI reflected changes in budgetary policy and economic activity. Between 1974 and 1984, the consumer price index rose at an annual average rate of about 13 percent. In later years, the rate fluctuated between 2 percent and 5 percent, except for a one-time rise in prices following the currency devaluation in 1994. In Gabon's relatively open economy, changes in aggregate demand spilled over into the balance of payments. Gabon's current and overall accounts surplus rose to a record level in 1974. But the positive balance was short-lived as the rapid growth of imports and payments for services – despite heavy external borrowing – threw the overall account into deficit by 1977. The sharp improvement in the terms of trade under the second oil boom helped produce a small surplus in the overall balance after three years of deficits. But while the trade balance was positive through 1985, the current account went into the red, reflecting the net outflow of private transfers (e.g. remittances of foreign workers in Gabon). Beginning in 1985 and continuing through 1994 (except for the 1990–1 third oil boom), Gabon experienced perennial current account deficits. A sharp deterioration in the current account in 1986 led to the suspension of debt service. Gabon was forced to conclude a series of stand-by agreements with the IMF in 1986, 1989 and 1991 in order to reschedule its commercial debt with Paris and London Club creditors. Performance under these agreements proved repeatedly disappointing. Due to significant shortfalls in programme implementation, only small tranches of the loans were disbursed, and domestic and external payments arrears were again accumulated. In January 1994, Gabon obtained a new 12-month stand-by arrangement with the IMF. Gabon's overall external debt in 1994 was estimated at $4 billion, (or the equivalent of nearly 80 percent of its GDP).

The objective of establishing and expanding physical infrastructure was partially achieved. Rail transport, communications and energy capacity were built and expanded, but the maintenance and upgrading remained inadequate. The Trans-Gabon

Railway was built with the oil money at an estimated capital-output ratio of 10, with hardly the possibility of ever becoming profitable. The inadequacies and poor state of trans-country roads, combined with high gasoline prices (a rarity among the group members), impeded the development of agriculture and forestry. In 1994, Gabon was the most urbanised country in Africa, with 73 percent of the population living in three major cities and nearly half of the entire inhabitants concentrated in the capital.

Gabon's objective of diversification did not progress well. Rising imports financed by oil revenues inhibited both agriculture and manufacturing. The joint ventures that were developed during the period were mostly public sector initiatives supported by heavy tariffs and quantitative restrictions. The modern sector was dominated by money-losing public enterprises. Public investment projects were also developed, at exceedingly high capital/output ratios naturally promising very poor returns. Industrial enterprises were created on a scale that exceeded domestic absorptive capacity, undermining competitiveness and diversification. In reality, the country's savings were largely remitted or invested abroad. By 1994, some 45 percent of the country still remained underdeveloped. And Gabon was still dependent on foreign labour, which constituted some 18 percent of the adult population.

Efforts made to reduce the dependence on the oil sector were not very effective. In 1994, oil still provided close to half of the public revenue and 50–60 percent of gross investment. As a percentage share of GDP, however, oil's contribution was reduced to about 32 percent in 1994 from 46 percent in 1974. Agriculture's contribution shifted from 12 percent to 8 percent. Manufacturing and mining's share changed marginally upward, but construction and public work which accounted for more than 14 percent of GDP in 1974–7, had their share reduced to around 3 percent by 1994. Services raised their contribution from 30 percent to 43 percent.

Finally, the ultimate goal of raising the people's standard of living remained unreached. A comparison of GDP per capita during the 20-year period is made difficult, inter alia, by disputes over population estimates. By all accounts, however, Gabon had one of the highest per capita GDPs in Sub-Sahara Africa in 1974 as well as 1994. Yet the magnitude of improvement in the living standard during the 20-year period was not clear. In nominal terms, according to the OPEC Secretariat, per capita GDP rose from about $2375 in 1974 to about $3865 in 1994. In real terms, however, according to World Bank estimates, Gabon's per capita income had an average annual decline of 0.6 percent, 1973–85, and an average yearly fall of 2.3 percent between 1985 and 1994. By 1994, the Gabon economy was in a dire situation. The overall fiscal deficit had widened further. The balance of payments remained under severe pressure. Large domestic and external payments arrears had accumulated. Private capital outflows were on the rise. Access to external financing was sharply limited. And due to the lack of competitiveness, non-oil private investment, both domestic and foreign was not generally promising.

INDONESIA

Indonesia's national socio-economic priorities included (1) expansion of agriculture, infrastructure, and industry (2) promotion of non-oil exports (3) import substitution to meet basic domestic needs and (4) mobilisation of domestic resources for development through increasing internal revenues. Included among top government priorities under the Suharto regime were 'food security', to the ultimate point of self-sufficiency, and rural reconstruction.

The objective of overall economic expansion was achieved reasonably, through realistic and suitable five-year planning. During the 20-year period, Indonesia showed an impressive record of sustained economic growth, especially after adopting market-oriented reforms in the mid-1980s. During this period, average annual growth exceeded 6 percent per annum.[3] Interestingly, even the profitability of state enterprises (with the exception of Pertamina, the state oil company, which was constrained in the raising of domestic oil prices) also improved after a series of corporate restructurings. As could be expected, the sectoral development was neither uniform nor flawless.

Altogether, Indonesia increased both its agricultural output and exports (particularly rubber and coffee). During the first two oil booms, annual agricultural growth was particularly noteworthy, and this sector (including forestry and fisheries) remained the most vibrant. The state-supported use of modern production methods and extension services, plus the appropriate maintenance of input and output prices at profitable levels, enabled Indonesia to substantially increase cash crops (corn, soybeans, palm oil, coffee and sugar) and to meet the goal of self-sufficiency in rice by the end of the second oil boom. The other goal of creating employment opportunities was also partially attained.

Non-hydrocarbon exports, both raw materials (timber, rubber, palm oil, coffee, tin) and manufactured products, more than doubled between 1974 and 1984. While infrastructural improvements contributed to this growth, the three timely rupiah devaluations of 1978, 1983 and 1986 proved highly stimulative. As a result, by 1987, the value of non-oil/gas exports exceeded hydrocarbons receipts for the first time in nearly a decade-and-a-half. Indonesia thus became the first and the only OPEC member to substantially reduce its export dependence on hydrocarbons. This trend was sustained through 1993 when a surplus was achieved on that sector's account for the first time ever. In 1994, the export of crude oil, petroleum products and natural gas made up only about 25 percent of total Indonesian exports, as against 75 percent in 1984. More significantly, the pattern of non-oil exports in the non-energy sector shifted toward manufactured products (garments, textiles, electrical machinery, cement, tyres and chemical products).

Import-substitution, as another major development objective, was given needed boosts. Manufacturing showed the second best performance in 1974–9, only slightly behind construction. In the subsequent five years, this sector grew vigorously, partly due to the improvement in competitiveness caused by the 1978 rupiah devaluation. The 1986 depreciation of the currency, and the liberalisation measures adopted in 1987, helped the continued expansion of manufacturing in the third plan. The value added in this sector between 1985 and 1994 showed about 10 percent annual real growth. The share of this sub-sector in GDP reached nearly 21 percent in 1994, accommodating about 11 percent of the labour force. Detailed data on employment were not officially published in Indonesia. Furthermore, since an individual working for only a few hours during the week was considered 'employed', the real amount of unemployment differed substantially from official estimates which, assuming full-employment in the rural sector, placed the rate of open unemployment at 2–3 percent for most of the period. Underemployment, however, remained widespread.

The goal of domestic resource mobilisation (e.g. reducing fiscal dependence on oil revenues) was assiduously followed. During the 20-year period every setback in oil export receipts was quickly made up by increasing domestic income and consumption taxes and raising import-tariff revenues. The inevitable gap in the fiscal balance was dealt with through foreign aid and loans. As a result, the share of oil revenues in the total government domestic revenues declined from about 55 percent in 1974–8 to about 35 percent on average, 1988–94. This cautious policy was of the utmost significance in reducing internal and external imbalances.

Inflation had been a major problem for Indonesia prior to the first oil boom, when it reached triple-digit figures in the mid-1960s. With increased government spending under the first oil boom, the CPI rose about 40 percent at first but, through a series of appropriate measures, the rate was brought down to less than 10 percent in 1978. The repeated rupiah devaluations were largely responsible for the pass-through effects of the terms of trade on subsequent inflation. Restrained financial policies, together with larger domestic supplies and lower import prices, helped reduce the CPI to an average annual rate of about 9 percent through 1994. This rate, which was still relatively low in comparison with that of some other members, was caused by periodic increases in administrative prices (petroleum, electricity, cement, transport) and adjustments in the minimum wages and salaries of civil servants. The rate was, in turn, kept relatively moderate by a lower rate of inflation in the price of tradables.

Indonesia's balance of payments, which showed marked fluctuations throughout the period, reflected large variations in export earnings (both oil and other raw materials). Except in 1974, 1979 and 1980, when Indonesia's current account showed a surplus, there were yearly deficits during the entire 1974–94 period. The overall balance of payments, however, was in surplus for three-fourths of the period, thanks to sustained increases in the inflow of foreign investments, foreign aid and long-term loans usually given on concessional terms. While these capital inflows helped to offset current account gaps, and kept international reserves at comfortable levels, they were also responsible for a large increase in official external debt, which gradually reached more than $96 billion (or 57 percent of GDP) in 1994. Although the highest debtor in the OPEC group during the entire period, Indonesia did not face difficulties in meeting its debt service obligations for a long time, for three main reasons: heavy reliance on long-term concessional loans; high non-hydrocarbon exports sustained through periodic exchange rate adjustments; and borrowing, mostly from official creditors rather than private sources. Yet, the speculative nature of some capital inflows, and the rising unrecorded borrowings by the private sector, increasingly intensified the economy's fragility and its unprotected exposure to eventual adverse developments.

In other unstated but implied common objectives, Indonesia's progress was also noteworthy. Sustained gains in growth and employment over the two decades resulted in dramatic improvements in most other social indicators of progress. Starting as a poorer neighbour to both India and Bangladesh in the late 1960s, Indonesia's steady economic growth not only lifted the per capita income from about $200 in 1974 to some $920 in 1994 (for a population 44 percent larger) but also reduced the poverty rate – making Indonesia one of the world's most successful countries in this regard. By World Bank estimates, while some 60 percent of the Indonesian people lived below the official poverty line in 1970, the percentage fell to about 17 percent in 1994.[4]

In the late 1980s and early 1990s the Indonesian government gave increasing importance to safeguarding the environment. Policies were developed to control water resource allocation, improve the sewage system, reduce urban pollution, and conserve land and timberland through local conservation, better management of forestry and public education. These measures had become necessary because, due to the heavy subsidisation of fertilisers and pesticides by the government, their application on farm lands had generally exceeded recommended levels by 50–75 percent. The result of the overuse was soil damage, the increased immunity of pests and cost/price distortions.

As has been frequently pointed out, Indonesia's economic achievements reflected smart macromanagement and an appropriate development strategy that focused on: infrastructure and rural reconstruction; diversification from energy into non-hydrocarbon exports (especially manufacturing); human resource development through basic education and better health care; and, last but not least, population control. Foreign financial and technical assistance on a grand scale had a large responsibility for the Indonesian success. But the country's increasing global integration, without instituting and observing the global management standards, did not rule out future troubles. In fact, Indonesia's so-called 'crony capitalism' (i.e. the prevalence of tax, customs and credit privileges for elite entrepreneurs, special trading monopolies and export cartels) was a clear source of worry for many.

IRAN

In Iran's pre-1979 regime, the focus of development was on (1) rapid growth, supplemented by (2) high employment and (3) supported by comprehensive public welfare services as a means of achieving distributional equity. Another goal was economic diversification through the establishment of new industries in metals, petrochemicals and nuclear energy to broaden the economy's revenue base, and to develop viable non-oil exports. After the revolution, the new regime initially tried to alter those priorities in favour of 'economic independence' (i.e. self-sufficiency), social justice and propagation of the Islamic culture. Later on, the promotion of economic growth and high employment, improvement in the welfare of the poor (*mostazafan*),and the state provision of citizens' minimum basic needs within a proper 'consumption model' were added to the pre-revolution list.[5]

The objective of rapid growth was faithfully pursued during the revised Fifth Five-Year Development Plan (1973–8). After adjustment for inflation, non-oil GDP grew by 13.3 percent a year on average, compared with 8.5 percent for total GDP. The oil sector had the smallest annual growth rate at less than one percent while industry and services achieved double-digit growth, and agriculture, 4.4 percent. Despite enormous increases in public and private outlays, however, the Plan's sectoral and overall targets both fell short of expectations. Part of the problem lay in the dismal behaviour of the oil sector. Responsible for the Plan's underachievement (i.e. average annual growth of 8.5 percent *vs* the 13.9 percent target) were, apart from the unexpected decline in oil revenues, (1) an undue haste in the Plan's 1974 revision, and the subsequent launching of new mammoth projects which lacked many of their essential ingredients (2) a lack of careful calculation of emerging demand for housing, electricity, water and transportation (3) too much confidence in the technocrats' ability to tamper with essential market forces and (4) the prevalence of incipient inflation, due partly to wage increases far beyond labour productivity.

In the post-1979 revolution period, coinciding with the second oil boom, GDP experienced a downward trend as a result of revolutionary disturbances, property damage, a fall in oil revenue, flight of capital and skilled labour, political infighting among successful revolutionaries, the foreign-assets freeze as a consequence of the American hostage crisis, the war with Iraq and a sharp overall decline in investment. During a 4-year recession (1979–82), agriculture was the only vibrant sector with an average annual growth of nearly 4.5 percent. The oil sector was the worst performer (except in 1981/2), due mainly to the decline in oil extraction and exports caused initially by a deliberate decision to 'save oil' and later by the war damage and Western economic sanctions. Industry was the second worst performing sector, mostly because of the mass exodus of private owner/managers and dependence on imported raw materials and intermediary goods which were drastically cut. Services also showed an average yearly decline. Altogether GDP fell by 33 percent in four years bringing the 1981/2 real GDP to the same level as 1973/4.

With the end of post-revolutionary upheaval and the recovery of oil production and exports, real output began to recover in the subsequent four years (1982–5) despite the global oil glut and worldwide recession. For the four-year period, the average annual real growth was a respectable 6.9 percent. Yet the real GDP in 1985/6 was still below that of 1977/8. The oil price crash in 1986 reversed the favourable growth trend. Intensified attacks by Iraq on oil facilities and soaring war damage brought about a new four-year recessionary condition, 1985–8, in which real output showed an average annual decline of 2.5 percent. With the ceasefire in the Iran/Iraq war, recovery of oil production and a new market-oriented strategy, the economy began to recover under a new Five-Year Development Plan (1989–94). Buoyed by the third oil boom, real imports soared to new records and real output increased by a notable 7 percent a year. The fastest post-revolution growth rate of GDP occurred in 1990–2 when imports reached historically high levels. The annual growth rate of 11 percent in 1990/1 subsequently dwindled to 2.3 percent in 1993/4 and 1.5 percent in 1994/5, as imports were severely compressed.

The objective of raising living standards was thus achieved during the first oil boom but living conditions severely deteriorated following the 1979 revolution. The average annual increase in real consumption after 1979 lagged far behind the yearly increase in population. Furthermore, due to inflation and food rationing, the consumption of red meat, rice, pulses, sugar and butter declined in favour of bread (which was available unrationed at a highly subsidised price).

High employment was easily maintained during the 1974–8 period. The buoyancy of economic activity, while accentuating an acute shortage of skilled and semi-skilled workers, was responsible for the very low rate of open unemployment – less than 1 percent of the labour force. The Islamic regime's encouragement of large families in the early years of the revolution resulted in a dramatic increase in the rate of population growth – estimated at 3.9 percent per year between 1979 and 1989. As a consequence, the workforce declined from 38 percent of the population to 26 percent and unemployment gradually rose to 14 percent. In the early 1990s, about one third of the labour force was employed in the public sector, half of them as regular government employees (or nearly double the pre-revolution number). Toward the end of the period, population growth was reportedly reduced to 2.5 - 2.7 percent a year, and unemployment was reduced to about 11 percent of the labour force. Private estimates put the number of unemployed or underemployed at close to 30 percent of the labour force, with the largest concentration in the 15–24 year old job-seekers.

The objective of diversification was partly achieved between 1974 and 1979 as the share of oil in GDP declined from 50 percent to 30 percent, while the share of industry increased from 13 percent to 19 percent. In this period, in contrast to the 'deindustrialisation' hypothesis of the Dutch-Disease model, a number of industries producing traded goods (motor vehicles, home appliances, carpets) showed healthy growth thanks to the relative abundance of imported components, subsidised inputs, low-cost credits and selective tariff protection. In agriculture, however, some tradable farm products (e.g. cotton, wheat, barley and beets) experienced stagnation or decline, and others (e.g. rice, oil seeds) expanded. In general, the agriculture sector suffered from the state's pro-industry bias, inadequate credit for mechanisation and modernisation, and increased fragmentation of land holdings. As a highly labour-intensive sector, agriculture was also adversely affected by rising labour cost (both farm wages and labour's opportunity costs, due to increasing non-farm, rural employment and rural/urban migration). On the whole, however, the expected adverse impact on agriculture during the first oil boom was quite negligible because the government's agricultural policy (land reform, public investment, subsidies, protection and capital assistance) provided the necessary counterweights.

Changes in the composition of national product in the post-revolution period exhibited a mixed picture. Agriculture's share rose from about 10 percent of nominal GDP in 1977/8 to about 21 percent in 1994. Industry's share declined only marginally. Services, on the other hand, went up from 40 percent to 42 percent (after reaching as high as 54 percent in 1988/9). The oil sector's share declined almost steadily from near 30 percent just before the revolution to about 18 percent in 1994. While oil still constituted the largest source of the government budget to the very end – effectively over 70 percent of the total – its share in total export receipts fell from upward of 95 percent in 1974 to about 80 percent in 1994.

In social welfare, the 1974–8 period showed a notable increase in GNP per capita from $870 to $2115. During the post-revolution period, however, despite the Islamic Republic's clearly populist stance and a strong penchant in favour of the lower classes, the lot of the average Iranian deteriorated badly, reflecting declines in national output, the ravages of the war with Iraq and the depressed oil market. According to official data, per capita income in 1994, at constant prices, was about 60 percent of the figure in 1978. In the absence of reliable data on internal income distribution, the indications are that in the first two years after the revolution wholesale takeover of the properties of the old regime's officials by the 'poor' and the not-so-poor helped in the redistribution of existing wealth. More directly, such policies as wage increases, food subsidies, tax exemptions, land distribution and cash payments to 'martyr' families reduced income inequalities. In subsequent years, however, the data published by government agencies on household expenditures adjusted for inflation indicated no improvement in alleviating poverty in either the urban or the rural sector.

By all indications, the revolution's major beneficiaries included some mid-size, near-city, farm operators, a number of strategically placed bazaar merchants, certain privileged importer/distributors, the well-connected deal-makers and speculators and, above all, those at the centre of rent-seeking activities. Worst off were white-collar workers, fixed-income recipients, retirees and those on the fringe of the formal economy who paid their taxes. By some private estimates, often quoted by the opposition, some 60 percent of the people in 1994 lived below the official poverty line. In sum, the consensus was that the distribution of income deteriorated after 1981. However, the rural and provincial origin of the ruling clergy, and the regime's ideological bias in favour of the 'depressed regions', induced them to allocate sizeable resources (both private and public) toward rural reconstruction. As a result, both physical and social structure – roads, water supply, health clinics, telephone lines and electric power – improved considerably in most rural regions.

With respect to other conventional performance criteria, Iran's record was not a successful one. Internal price stability proved difficult for both the pre- and post-revolutionary governments to maintain. Between 1974 and 1978, physical and management bottlenecks (i.e. ports and road congestion, insufficient supply of construction materials, power shortages and inadequate coordination) caused the inflation rate to rise much faster than anticipated. Expansionary monetary and fiscal policies during the first oil boom resulted in an average annual growth in inflation of 15.5 percent, which, although still below Third World averages at the time, compared unfavourably with less than 4 percent a year between 1968 and 1973. During the second oil boom, 1979–82, prices, as measured by the cost-of-living index, rose sharply by 19 percent a year on average, reflecting the effects of revolutionary economic dislocations, shortages of domestic and imported commodities caused by labour strikes, production stoppages, the Iraqi invasion in September 1980, import compression and the government's expansion of liquidity to rescue the shaky banking system. Larger oil revenues, availability of imports and output recovery after 1982 reduced the index temporarily but the intensification of the war and tightening of Western economic sanctions, combined with the rapidly rising money supply, reignited inflationary pressures, and the average yearly price rise registered nearly 25.5 percent, 1986–9. Despite specific requirements in the 1989–94 Five-Year Plan to keep private liquidity in check, inflation was officially put at 23 percent a year during the period. Private estimates indicated twice that figure. Responsible for the upsurge in inflation were such additional factors as reduction in consumer subsidies, price decontrol, wage increases and, not least, sharp, official and *de facto* devaluations of the rial.

Government finances were perennially under pressure due to a number of factors, partly beyond the government's control but largely due to wrong policies. Despite smaller government financial involvement in the economy after the revolution, fiscal stability was not easy to achieve, mainly because of the government's heavy reliance on (fluctuating) oil revenues and the inflexible nature of non-oil income (i.e. taxes and revenues from government business operations). Lower oil revenues after the revolution, combined with scant contribution from non-oil taxes and the maintenance of a highly over-valued exchange rate, reduced the government's annual oil income and resulted in a budget deficit year after year, with the deficit ranging between 4 to 8 percent of GDP after 1983 up to the early 1990s. Development expenditures were the major casualty of the reduced incomes, when public investment fell to only about 4 percent of GDP compared to 14 percent before. Mainly responsible for the rising budget deficits was the fact that some 60 percent of the Iranian economy escaped taxation of any kind, due to exemptions granted to various groups and activities, and some 20–25 percent was considered the 'informal economy', where tax evasion was widespread and tax collection conspicuously weak.

Iran's external balance was characterised by significant volatility. The overall balance registered sizeable surpluses throughout the first oil boom (except in 1975). During the period, Iran significantly enlarged its foreign loans, investments and grant programme by more than $5 billion. Total foreign exchange assets in 1979 amounted to $15 billion, against an estimated, long-term, external debt of some $7.5 billion. For most of the years after 1979, Iran experienced a deficit in external accounts, mostly related to changes in oil export receipts. Additional import needs caused by the war with Iraq, Western trade sanctions, flight of capital and the government policy of paying back its external debt aggravated the situation. The trade balance was positive in two thirds of the period but the services account was perennially in deficit, pushing the current account also into the red in most of the years. The principal activities in the capital account occurred largely between 1990 and 1993, when surging imports were almost totally financed by short-term credits and resulted in the accumulation of unpaid arrears for the first time in the country's post-World War I history. The current account between 1990 and 1993 was in deficit, as much as 10.5 percent of GDP, before a drastic compression of imports in 1994 to 50 percent of the 1991–2 level helped improve the situation. Iran's external debt, which stood at a negligible amount in 1989 (despite the long and costly war with Iraq), ballooned to $23 billion by March 1994 (including more than $11 billion in arrears). Despite strong Washington pressure on Japan and the European Community, Iran was successful in avoiding the Paris Club and IMF rules. Its official creditors reluctantly agreed to bilateral debt reschedulings up to the year 2000. Some $14 billion of the country's foreign debt was rescheduled by the end of 1994.

The environment was another casualty of the revolution, the war with Iraq, diplomatic isolation and public mismanagement. Despite a constitutional mandate and some faint government efforts, the damage to the environment in the post-revolution period was extensive. Published private reports refer to considerable deforestation, systematic misuse of pasture land, excessive soil erosion, a reduction in flora and fauna species and high urban air pollution. Responsible for this deterioration was, *inter alia*, a neglect of ecological issues in development planning, excessive population growth, uneven distribution of inhabitants and economic activity around major cities, and lax enforcement of environmental regulations.

In sum, by 1994, Iran had a much larger population (60 million) than at the beginning (35 million), more extensive physical infrastructure, more and better roads, railways, sea and airports, greater electric power-generating capacity, vaster telephone and telecommunication lines (including state-of-the-art faxes, e-mail and mobile phones), more multi-purpose dams and modern irrigation networks, more numerous manufacturing, petrochemical plants, and oil refinery plants, considerable weapons production capability and much larger education and health facilities. Yet the country had a much smaller per capita income and private per capita consumption, a larger per capita domestic and external debt, higher unemployment, bigger budget deficit, more inflation, a wider balance-of-payments gap, a more overvalued currency, more numerous money-losing public enterprises and a shorter life-expectancy for its oil deposit at the on-going extraction rate. Self-sufficiency in food products, which was supposed to be reached by 1990, did not materialise. Iran was more than ever dependent on imports, and imports still dependent largely on oil exports.

Any objective evaluation of Iran's overall performance, however, ought to allow for the enormous cost of the war with Iraq, and the massive damage to the country's productive sectors between 1980 and 1988. The Iraq/Iran war of 1980–8 imposed a calamitous burden on the economy. The damage consisted of physical destruction of infrastructure, farm land, industrial equipment and oil installations, and the draining of labour resources from productive employment. While a good part of the military cost was covered by the war materials accumulated during the Shah's regime and inherited by the Islamic Republic, the consequences of the war on the economy were indeed devastating. A detailed study of the 8-year war put the total direct and indirect costs at a grand total of $644 billion.[6] President Ali Akbar Rafsanjani later repeatedly referred to a figure of over $1 trillion without elaborating on individual items. A United Nations team subsequently estimated the direct cost (mostly damage to the infrastructure) at $97 billion (albeit at an overvalued exchange rate).[7] Regardless of which figure was to be taken seriously, the war was a major catastrophe for the economy.

IRAQ

Iraq's development objectives since the mid-1970s aimed at (1) infrastructural expansion (2) industrial diversification (3) agricultural self-sufficiency (4) improvements in social services and education and (5) reduced dependence on petroleum.

An evaluation of Iraq's efforts to achieve these objectives is made uncommonly difficult not only because of the paucity of data and near total lack of transparency on the part of Iraqi government agencies for most of the period under review but also because of significant distortions in the economy caused by the two wars with Iran and Kuwait. From shortly after the end of the second oil boom until the end of 1994 the Iraqi economy was a casualty of those two devastating wars, comprehensive trade sanctions and internal insurgencies.

By all indications, Iraq's goal of rapid economic growth was reached only under the first two oil booms. The decade of the 1970s, especially the second half, witnessed an unprecedented expansion of the Iraqi economy in both oil and non-oil sectors. The four-and-a-half times increase in oil revenue between 1974 and 1980 enabled Iraq to increase public spending on infrastructure, agriculture, industry, social services, and arms through imports. Iraqi imports increased from $2.6 billion in 1974 to $21 billion in 1982. During this period, public investment rose in real terms. Manufacturing, transport and communications, and construction all went up. Only agriculture had poor growth.

The bold rush into huge projects undertaken in the second oil boom under the banner of 'development despite the war' was rather imprudent and short-lived. Early in 1983, 'austerity' became the slogan. Except for defence-related outlays which were largely financed by aid from friendly Arab neighbours (e.g. Saudi Arabia and Kuwait), all other spendings were curtailed. From then on, Iraq carried on a war or war-like economy, with all its attending restrictions deprivations, destruction and disruptions.[8] The invasion of Kuwait in August 1990, and the ensuing UN embargo and sanctions, resulted in a catastrophic fall in output and income unlike that of any other group member. The greatest impact of the 1980–8 war and its destruction was reduced domestic capital formation which, as a percentage of GDP, dropped to 17 percent in 1988 from 45 percent in 1981–2. The sharp decline reflected the plunge in oil revenues and drastic cuts in imports. Private consumption was also particularly hard hit by war conditions, as the nearly 14 percent positive annual growth in the 1970s turned into a negative growth average of 6 percent in the 1980s.

All economic sectors were affected by the wartime economy. Agriculture performance proved disappointing due to ineffective planning, poor administration, uncoordinated investments, rural migration in search of higher city wages, and soil salinity. Once regarded as the Middle East granary, food imports became a perennial feature of Iraq's total imports. In 1974, agriculture contributed about 8 percent of GDP and employed half of the total labour force. By 1986, despite an estimated interim investment of $4 billion, the sector's contribution remained the same. By 1990, two thirds of food needs had to be met by imports (including 80 percent of wheat and 75 percent of rice) compared with only one third in the mid-1970s.

In the oil sector, crude output rose from less than 2mb/d in 1973 to 3.5 mb/d in 1979. Oil export proceeds increased from less than $2 billion to more than $21 billion due to both rising volume and rising prices. Crude oil output reached its peak of 3.7 mb/d during the first half of 1980, benefiting from a substantial drop in Iran's post-revolution extraction. After the outbreak of hostilities with Iran, Iraq's access to its deep-water terminals in the Persian Gulf were lost, and the 2.5 mb/d oil exports were transported by two pipelines to Mediterranean terminals through Turkey and Syria. In April 1982, Syria closed the line through its territory. Production then plunged to less than 0.98 mb/d in 1982, partly due to considerable damage to the oil facilities at Kirkuk and Basra. Crude oil output picked up after 1983 and reached 2.8 mb/d in 1989, because of increased access to new oil export terminals and a new oil pipeline spur linking Iraq's southern oil fields with Saudi Arabia's Red Sea port of Yanbu. After the Persian Gulf War, all export pipelines were shut down.

Industry (comprising mining and manufacturing, construction, water and power) constituted a small 15 percent of Iraq's GDP in 1974. Manufacturing expanded rapidly under the first two oil booms. Among industrial activities, the iron and steel industry was the forerunner, importing iron ore to produce steel products for the domestic market. Fertilisers and petrochemicals were the other two industries receiving special attention. Construction and water and power were the subsectors with highest sustained annual growth. However, locating most of the major industrial projects (steel and petrochemicals) in the south, near Iraq's only outlet to the Persian Gulf, as well as in proximity to energy and feedstocks, made them easy targets for Iranian attacks during the war and ensured that they were put out of action. Also, inadequate planning and the absence of linkage with forward activities made excess capacity and low return a perennial industrial problem. As could be expected, the construction sub-sector experienced the largest boom due to the availability of domestic resources, ease of entry, unrestrained imports, and eager local and transnational enterprises. Nevertheless, the rapid growth of urban population resulted in the worsening of the existing housing shortage, thriving land and real estate speculation, city overcrowding and the rise of suburban slums. Furthermore, the steady growth of construction was halted after 1982 when Iraq developed payment difficulties, and construction activities lost their momentum toward the end with annual declines of as much as 16 percent between 1985 and 1990.

A corollary of rapid growth in the 1970s was the provision of employment for all job-seekers. In fact, the rising demand for labour led to an inflow of some two million labourers from North Africa and Southeast Asia. Not surprisingly, the largest number of new entrants was employed in the public sector. Public sector employment went up from about 400,000 in 1974 to more than 630,000 in 1980, the overwhelming majority in administrative and non-productive tasks. The labour shortage, particularly in the technical and skilled categories, was vastly intensified as the war with Iran wore on, resulting in an unprecedented entry of women into the labour force, especially in administrative and managerial ranks. Between 1980 and 1990, the labour force expanded from 3.2 million to 4.7 million, and the men under arms increased from less than 450,000 to 1.2 million. The ratio of armed forces to the national work force increased from less than 3 percent in 1974 to more than 21 percent in 1990. The increase in the ranks of men under arms throughout the 1980s was largely achieved at the expense of employment in agriculture and industry which lost some of their skilled and semi-skilled workers.

With oil revenue providing some 85 percent of government revenues on average between 1974 and 1980, government finances were easily managed. As oil revenues declined, and war costs rose rapidly after 1981, the budget came under increasing pressure each year to the very end. No budgetary figures were published between 1981 and 1989. The published budget for 1990 was superseded by events and become irrelevant under the UN trade embargo. The treasury is reported to have floated domestic dinar bonds to finance part of the state expenditures.

During the first two oil booms, the official consumer price index rose between 9 and 15 percent a year on average, reflecting primarily foods and housing and related services. The inflationary effect of wartime expenditure, in the face of a diminishing productive base and difficulties involved in importing goods over land (due to the closure of Persian Gulf terminals), became endemic after 1983. In the absence of official data, annual inflation was privately reported to have run as high as 45 percent up to 1990,[9] and as high as 500 percent thereafter – placing the country among the ranks of hyperinflationists. Foodstuff prices, according to an FAO report, sky-rocketed by 4000 times after 1990, as the state printed money to pay for its soaring expenditures by as much as one billion dinars every day.[10] Punishments of violators of state price guidelines (through amputation of hand and foot, and execution) did not deter perpetrators.[11]

Iraq's external balances on current account between 1974 and 1980 showed sub-stantial yearly surpluses, offset in part by deficits on services and transfers. By 1980, Iraq accumulated $40 billion of reserves. The 1981–5 development plan began with awarding contracts totaling $20 billion for new projects, at just about the time that oil revenues headed for a dramatic slide. In the early 1980s, an estimated $4 billion a year also left the country remitted by foreign workers. Some $3 billion of trade arrears were thus accumulated between 1982 and 1984 and had to be rescheduled. Under a series of drastic adjustment measures adopted after 1983, foreign contractors were told to finance the foreign currency portion of their projects themselves; new loans were obtained from the Euromarket and the Arab financial institutions; and bilateral credits were received from a number of countries. Imports were cut to $8-10 billion a year, and Iraq's customers were provided with some 300,000 barrels of crude oil a day by Saudi Arabia and Kuwait on Iraq's account. By 1986 Iraq was unable to service its estimated $45 billion debt to the West, even those debts rescheduled from 1982. There were also unspecified debts to Eastern bloc countries and others.

A brief respite was achieved between 1987 and 1990 as a result of some timely reforms. But, after the invasion of Kuwait and the loss of oil export revenue, annual deficits in the external account appeared again, despite the fact that annual imports in 1990-4 were down, on average, to only about one tenth of the average yearly figure in 1987-90. The Iraqi dinar which before the 1991 Persian Gulf war was officially worth $3.2 plunged so deeply in the black market by the end of 1994 that $1 was equal to more than 2000 dinars. Post-1991 hyperinflation in Iraq rendered the official rate (still at 1D = $3.2169) totally irrelevant as a unit of account in 1994 free market transactions.

135

According to official data, Iraq's outstanding non-military debt in 1982 stood at $1 billion. By 1983, the four OPEC Arab members had reportedly loaned Iraq up to $25 billion, and obligations to other countries were estimated at about another $25 billion. In 1989, due to the earlier collapse of oil prices, the total debt reportedly rose to $80 billion (with $30 billion owed to Arab neighbors). By 1994, the total was unofficially put at $92 billion. With compensation due to Kuwait, Iran and others under two different UN ceasefire resolutions – estimated to approach several hundred billion dollars – Baghdad was destined to become a permanent debtor, even if the UN sanctions were lifted.

The objective of diversification also remained mostly unrealised. The share of various sectors in GDP underwent considerable changes over the period, mostly on account of changes in the oil sector's fortune. Oil's contribution to nominal GDP was about 50 percent in 1974; it rose to 60 percent in 1980 but followed a downward slide thereafter, falling to 18 percent in 1989. After the invasion of Kuwait, and the UN embargo, this sector's share hit the record low of less than one percent.[12] Other sectors increased their relative shares as oil's was reduced. Agriculture's share of about 8 percent in 1974 went up to about 16 percent in 1989 and 30 percent in 1991. Industry's share of about 15 percent at the outset fell to 10 percent in 1991 as a result of war damage and lack of imported inputs. The share of services went up from 28 percent in the beginning of the period to about 49 percent in 1989. The extraordinary plight of the economy in 1991–4 made any further comparison meaningless.

Diversification was to be achieved, in part, through a reduced share of the state in the economy. But the 1987 privatisation programme was another casualty of inadequate preparation and subsequent adverse events. Legal and financial frameworks for the transfer of state enterprises were not attractive enough for quick sales. The inefficient and money-losing status of state enterprises in the face of high returns from rent-seeking activities outside the productive sector were additional disincentives. As in other countries in similar situations, privatisation favoured those with close political or other connections and involved widespread corruption and abuses.[13]

The objective of improved social welfare was approached at first. During the first two oil booms, both public and private consumption rose faster than GDP. The average Iraqi did not experience any real hardship in daily life until mid-1982 when the oil price plunge, the rise in the value of the US dollar and the escalating cost of the war with Iran brought the 'guns and butter' strategy to an end. The austerity programme announced in 1982 was tightened nearly month by month until the war's end. The stringent trade sanctions imposed on Iraq by the United Nations in 1991 pushed the country to near economic ruin, social disintegration, breakdown of law and order, rampant crime, corruption and overall chaos. All basic commodities were sparingly rationed. Drug shortages resulted in the closure of healthcare facilities. Child malnourishment increased at an alarming rate.

A discussion of Iraq's economic performance will not be complete without a brief glance at its two devastating wars. The first casualties of the war with Iran were the extensive oil installations (including the loading terminal at the Persian Gulf, pumping stations, refineries and pipelines). The war also made heavy demands on manpower, as 40–45 percent of the regular male workforce was sent to the front and only partially replaced by women. The military draft was an immediate drain on employment in the goods-producing sectors. Financial aid from other Arab states somewhat eased the burden of the war and enabled Iraq to continue with some of its major development projects (particularly in transport, electric power and agriculture). The invasion of Kuwait, in August 1990 under the pretence of border violations and oil price manipulation, ended up with further damage to civilian and military infrastructure which was still severely damaged from the Iran/Iraq War.

A reflection of this calamitous outcome was the change in Iraq's per capita real GDP, which having risen from $1745 in 1970 to $4100 in 1980 declined to $630 by 1991.[14] The 1994 level was hardly greater than that of the 1940s. Mounting economic hardship and a disastrous fall in living standards made the Iraq of 1994 a tragic shadow of its 1974 self. The oil bounty brought Iraq a stagnant agriculture, high-cost manufactures and rising dependence on foreign food, foreign labour and foreign skills. Above all, the authoritarian regime, backed and strengthened by ample oil revenues, was emboldened in its nefarious and ill-fated military adventurism.

KUWAIT

The aims of the Kuwait economic development, as stated in its First Five-Year Plan, 1968/1972, included, *inter alia*, (1) an increase in per capita income (2) a more equitable income distribution (3) greater diversification of the economy (4) training indigenous skills and (5) greater Arab economic integration. To these ends, Kuwait adopted five main areas for its development focus: expansion of various social services, creation of a reserve fund for future generations, overseas investment, industrial diversification, and aid to other countries through the Kuwait Fund for Arab Economic Development.[15]

Kuwait's first national objective, i.e. a better and fuller life for Kuwaiti citizens, was easy to achieve. Thanks to the rising oil export revenues and returns from foreign direct and portfolio investments, Kuwait's per capita income rose during the 1970s and 1980s. Although population increased steadily throughout the period, per capita GDP in 1994, at about $15,000, was still much higher than the 1974 level (Table 8). Due to significant differences in eligibility for comprehensive welfare benefits between Kuwaiti citizens and expatriate workers, however, the per capita income figures did not reflect the standard of living among all inhabitants.[16]

Social welfare provisions made Kuwait the world's archetypical cradle-to-grave welfare state, where the bulk of the government budget was spent on salaries, health, education and subsidies. Some 90 percent of the economy was controlled by the state. Within a short period of time, Kuwait thus raised its standard of living from a poor pearling and trading level to one of the highest in the world. The lavish welfare programme included education that was free from preschool to university and compulsory to age 14, free healthcare at home and treatment abroad at the state's expense, and the right to free housing for low-income families. As much as 90 percent of electricity costs and 75 percent of water costs were paid for by the government. Expenditures on health ranked third in the national budget. Every citizen was guaranteed state employment, and the state employed directly or indirectly 97 percent of Kuwait's citizens. The array of public services offered by the state included various subsidies for land purchases, food, gasoline, shelter and utilities. There were no direct taxes on income or wealth. There was, instead, direct financial support for needy families. The social security system included generous retirement pensions. In short, prior to the Iraqi invasion in 1990, Kuwait's welfare regime probably had no parallel in the world. While data on domestic income distribution are not available, the scope and intensity of social welfare provisions may be regarded as having satisfied, at least partly, the objective of greater income equality among Kuwaiti citizens.

The goal of industrial diversification was much harder to reach. Unlike most other group members, Kuwait remained essentially a 'one-pillar edifice' throughout the period due to its small size and extremely limited absorptive capacity in both agriculture and industry. Nevertheless, during the first two oil booms, a small industrial base in the non-energy sector was gradually built. About 60 percent of manufacturing output was generated by the public sector, and the rest was divided equally between the private and joint sectors. The private sector share consisted of several thousand small industrial units – mostly operated by skilled foreign workers. As in many other member countries, however, the criteria for selecting industries were not subjected to close scrutiny. Consequently, a good deal of inefficiency and waste of resources ensued. During the 1985/6 economic slow down, a large number of hard-pressed industrial companies had to be rescued by government buyouts.

Kuwait's agricultural potential was extremely limited, due to harsh climatic conditions, scarcity of arable land, scarcity of fresh surface water and inadequacy of trained manpower. Thus, despite strenuous government efforts throughout the period, agriculture's contribution to GDP remained less than one percent. Local production met only a fraction of domestic needs for most foodstuffs. Altogether, however, some transformation occurred in the economy's real productive sector. In 1974, oil and gas constituted 72 percent of the total GDP, industry 6 percent, and services 22 percent. In 1994, the oil and gas sector (including refining) stood at about 42 percent, industry had climbed to 11 percent and services (including defence) to 47 percent.

The objective of providing economic security for future generations was served in earnest at first. Under the first two oil booms Kuwait's oil revenues exceeded its total public expenditure, and the large surplus was mostly placed in foreign assets accruing to the Reserve Fund for Future Generations. Kuwait's reversal of fiscal fortune began after the second oil boom. In summer 1982, the country's unofficial market, the Souk-al Manakh collapsed under a $90 billion mountain of bad cheques written by influential speculators.[17] The bail-out of the Souk-al-Manakh and the emergence of the depressed oil market drastically reduced the fiscal surplus in a rapidly sliding fashion.[18] Fiscal developments in 1991 and 1992 were critically impacted by the cost of the war and subsequent reconstruction operations. As a result, a deficit equivalent to 70 percent of 1989 GDP was recorded in 1990/1, financed mainly by drawing on official foreign assets. During reconstruction years, in 1992–4, the annual deficit was reduced to 22 percent of GDP but by 1994, the gap was still about 12 percent. The devastation caused by the 7-month Iraqi occupation raised Kuwait's public expenditures of $6–8 billion a year in the 1980s to more than $26 billion in 1991. This huge rebuilding cost, combined with a temporary stoppage of oil exports, was partly responsible for increasing budget deficits. The Kuwait Investment Authority's income, which averaged more than $10 billion a year in the late 1980s, was also reportedly reduced to only $3 billion in 1993. Partly responsible for the sizeable budget deficit was also renewed heavy defence spending.[19] Kuwait's budget deficit for the 1994–5 year was expected to reach nearly $6 billion, and a balanced budget was not anticipated until the year 2000. Lax supervision of domestic and foreign investments badly damaged the Kuwait Investment Authority's good reputation. The unexplained disappearance of some $5 billion from official state investments in Spain added to the country's fiscal problem.

While the objective of creating a reserve fund for future generations was pursued satisfactorily for years, and was resumed after Iraq's expulsion from Kuwait, the goal of the Kuwaitisation of the labour force was not eminently successful. There were, to be sure, significant changes in the composition of population. The 1975 census estimated Kuwait's total population at less than one million of which 47 percent were Kuwaiti nationals. The labour force comprised 30 percent of population, 70 percent of which were non-Kuwaiti. During the Iraqi occupation an estimated 1.4 million expatriates left Kuwait and the population fell to about 400,000. By 1994, the population was back to 1.6 million of which some 800,000 were foreigners (mostly Egyptians). Foreign workers accounted for nearly 85 percent of the labour force. The Kuwaitisation objective remained to be met.

Unemployment was not a problem for much of the period, as the government continued to provide guaranteed employment to Kuwaiti university graduates, and foreign workers had to be sponsored by a Kuwaiti employer before entering the country. Wages in Kuwaiti industries were below corresponding levels in industrialised countries but considerably above both developing and newly industrialising nations. Furthermore, Kuwait's labour market was highly segmented, with the public sector employing more than 90 percent of the Kuwaiti labour force at salaries and fringe benefits well above the private sector's. By 1994, however, employment opportunities for the growing number of nationals joining the labour force were becoming a problem.

With respect to conventional economic objectives, Kuwait was only partially successful. GDP growth essentially followed the oil fortune. Thus, from 1974 to 1981, aggregate real GDP expansion averaged 7 percent a year, with non-oil GDP between 7 and 11 percent. Continued weaknesses in the oil sector caused non-oil GDP to grow about 4 percent in real terms between 1981 and 1985. In the second half of the 1980s, GDP continued upward until the Iraqi invasion. During 1990, GDP fell nearly 50 percent and, in 1991, output was at its lowest level since 1974. The economy rebounded to more than its 1988 level by 1994 mostly due to a remarkable recovery in the oil sector.

Due to its near total integration in the global economy, Kuwait successfully escaped severe domestic inflationary pressures despite rapid increases in domestic public expenditures during most of the period. Until the Iraqi invasion in 1990, inflation was not much of a problem for Kuwait, for three basic reasons: the high savings/GDP ratio, reflecting the low absorptive domestic capacity; the high share of US goods and services in Kuwait's large total imports (around 45 percent of GDP);

and the close link between the Kuwait dinar and the US dollar. The cost of living rose by about 8 percent a year in 1974–9 – one of the lowest rates among members – reflecting the high share of the oil revenue 'saved abroad'. During the first half of the 1980s, CPI decelerated from 7 percent in 1982 to 1.4 percent in 1985. Responsible for this lower inflationary rate were mainly falling import prices (due to the continued strength of the US dollar), weak domestic demand and reduced government spending. In the 1985–90 period, CPI rose by less than 3 percent a year on average. The Iraqi invasion and occupation resulted in rising inflationary pressures because of the shortage of goods and the rapid expansion of domestic liquidity as the Kuwaiti dinar was linked to the Iraqi dinar at par. After the liberation, monetary stock was expanded to meet the demand for reconstruction and rehabilitation. Consumer Cooperative Societies which dominated the foodstuff sales in Kuwait were largely helpful in holding down the percentage increase of consumer goods prices in the low double-digit range.

During the first two oil booms, Kuwait experienced large merchandise trade surpluses and a smaller but still a significant positive balance on the services account (due in large part to income from foreign investments). Between 1974 and 1978, the surplus on current account amounted to $32 billion. Another $43 billion was obtained between 1979 and 1981. But the escalation of the war between Iran and Iraq in the mid-1980s cut sharply into Kuwait's thriving pre-war re-export trade with both countries. For the period 1982–90, however, the balance showed a surplus of merely $51 billion. Throughout this period, there was a considerable volume of private foreign investments by Kuwaiti citizens in the form of foreign CDs and purchases of other foreign assets. Kuwait's investment abroad insulated the country against volatile oil prices, fluctuating domestic oil output and gyrating foreign demand for oil better than any other OPEC member.

For the most part of the 1980s, up to 1989, the overall balance was perennially positive. Between 1990 and 1992, owing to physical damage, rising imports needed for reconstruction and dwindling investment income (due to heavy drawdowns to pay for the cost of liberation) low oil exports resulted in an unprecedented current account deficit of $26 billion, followed by two subsequent years of negative balances totaling $8 billion. Oil output recovery helped the external account to return to a surplus of $4.6 billion in 1994.[20] After the 1991 Persian Gulf war, Kuwait took out an unprecedented $5.5 billion of international loans.

For a good twenty-year period, Kuwaiti citizens enjoyed some of the highest living standards in the world, as though their resources were infinite. Priorities seldom figured high in annual resource allocation. The Iraqi invasion and its aftermath was not only a politico-military shock but a socio-cultural awakening. The cradle-to-grave welfare system was neither sustainable nor even workable anymore. The public sector which provided ready employment for almost every job-seeker was no longer capable of absorbing the rising reservoir of Kuwaiti labour. And if the private sector were to employ only Kuwaiti nationals, and pay them the wages they demanded, it could not compete anywhere in the world.

In 1994 Kuwait, in addition to the task of economic and environmental recovery from the devastating and ruinous invasion by Iraq, faced certain basic shortcomings in its economic structure, namely imbalances between aggregate production and consumption, and the predominance of public sector activities over private sector, which impeded free-market reforms. At this crossroads, the government's overseas cash reserves were reportedly down to one-third of their 1989 level; half of its annual budget was devoted to salaries and another third to the purchase of advanced weapons. With oil still accounting for 90 percent of all revenues, and population growing at 4 percent a year, neither high economic growth nor the sumptuous welfare state was easy to underwrite. As a critical observer noted, self-reliance was still an elusive goal for Kuwaitis: they had to live on oil exports, be taught by Egyptians and Palestinians and defended by Americans. That was the way of the past, and looked as though it was that of the future too.[21] Kuwaiti leaders, needless to say, differed from this pessimistic evaluation, and saw a new period of peace and prosperity for their country.

LIBYA

In Libya, Muammar Qadhafi's *Green Book,* which served as the country's effective post-1969 constitution, described basic national socio-economic objectives as: (1) the expansion of the production base and the creation of new export sources to reduce independence on oil (2) self-sufficiency in food (3) strengthening national defence (4) provision of essential services (housing, education and healthcare) for the Libyan people (5) the development of physical and human infrastructures and (6) a more equitable distribution of income and wealth. The periodic three- and five-year development plans routinely described main priorities in specific details.

The objective of broadening output capacity was served under the first two oil booms but, after 1981, the development process faced severe setbacks. Libya was one of the world's poorest countries before the advent of oil. The country's dearth of basic natural resources such as cultivable land and water allowed for only a rudimentary and fragile socio-economic infrastructure. Economic activity was limited to subsistence agriculture and livestock herding. The economy was also partly sustained by modest foreign aid.

The 1973/4 oil price surge gave the growth target the needed financial boost. The 1973–5 plan was generally successful as all expenditure targets were nearly reached and the plan's allocations well spent. The 1976–80 plan also achieved an annual growth rate of nearly 10 percent, although sectoral performance deviated from planned targets. Due to wide variations in the oil sector contribution, there were major deviations from the 1981–5 plan targets. Non-oil sectors averaged less than 3 percent annual growth compared with the planned target of 10 percent. Total real GDP experienced a 4 percent annual decline compared with the 1 percent growth rate projected in the plan. Benefiting from the third oil boom, aggregate output rebounded with vigour, producing a real GDP increase of about 4.5 percent on average in 1990 and 1991 despite the contraction in the non-oil sectors. Subsequent developments were overshadowed by Libya's deteriorating diplomatic status in the world community. With the reversal in oil income and a continued decline in the non-oil sectors under the 1992 UN sanctions, GDP faced new contractions, and total real GDP was estimated to have fallen by an average rate of 4.5 percent a year in 1992 to 1994.

The objective of reducing dependence on oil was only partly achieved. The oil sector, during the first two oil booms, accounted for 50–60 percent of real GDP, 97 percent of total exports and more than 85 percent of public revenues. Between 1981 and 1986, the oil sector accounted on average for almost half of nominal GDP, 70 percent of public revenues and virtually all exports. Between 1987 and 1994, the hydrocarbons sector ended up accounting for about 21 percent of GDP, still more than 50 percent of total government revenue, and about 93 percent of total exports. The increase in non-oil exports from about 3 percent of total at the beginning to about 8 percent at the end was also just in line with the basic objective.

Despite strenuous efforts toward 'food security' as an overriding national goal, the results were disappointing. Agriculture, constrained by the dearth of cultivable land area (less than 2 percent of total territory), fragmented holdings, the vagaries of the weather and water shortage, was adversely affected by the growth of economic activity in urban construction and trade which drained both labour and capital from farming. It also suffered from the spread of desertification, chronic manpower shortages, and increasing water salinity in the fast depleting aquifers. Between 1974 and 1979, the agriculture sector (including forestry and fishing) accounted for about 2 percent of GDP. In 1993–4 Libya had to import some $1.2 billion of farm products, of which some $500 million was spent on cereals alone. The sector's share of GDP reached nearly 8 percent in 1994 but it continued to be heavily subsidised and was hardly sustainable on its own – a fact that made the Libyans decline joining GATT.[22] By UNDP estimates, per capita food production in Libya in the 1990s was 20 percent below the 1979–81 level.

Libyan industry, (including mining, manufacturing and construction) had high annual growth rates in the second half of the 1970s thanks mostly to construction activities. Manufacturing's performance decelerated in later years due to shortages of skilled and managerial manpower, the smallness of the domestic market, relatively high costs and interrupted flows of inputs as a result of lower (and erratic) oil export proceeds. Labour problems such as absenteeism, frequent turnover and military callups were added factors. Capacity utilisation was additionally hampered by insufficient local resources, interrupted availability of water and power, and shortages of spare parts. The absence of internal planning and long-term marketing resulted in periodic overstocking and output reduction. Construction's flourishing trend during the 1970s was due largely to increasing state investments in housing, schools, hospitals and other social infrastructure. During the first half of the 1980s, building activity was on the decline because of the completion of previous projects (only partially revived by the new Great Man-Made River project) and sharply falling development expenditures resulting from budget constraints. Between 1987 and 1994, industrial performance was particularly uneven. Manufacturing thrived for a while but was adversely affected later under the impact of UN sanctions and foreign exchange restrictions. Construction was boosted by allowing families to have more than one dwelling if the addition was for their children. Altogether, both manufacturing and construction considerably increased their shares of GDP at the expense of the oil sector.

The services sector (including finance, trade, education and health) increased their value-added steadily throughout the period albeit at uneven rates. This sector's growth was accelerated toward the end of the period due to the measures taken to encourage private sector participation in trade and transport, and to reduce public services other than health and education. The services sector accounted for about 60 percent of non-oil GDP in 1994.

Finally, the stated goal of a more equitable distribution of wealth and income showed no clear evidence of being reached or even approached. Anecdotal indications point to a similarly observed trend in other OPEC members where the technocratic and military elites improved their income status at the expense of other social strata. The attractiveness of rent-seeking activities and the prevalence of corruptive practices also underscored the possibility of a growing unfairness in income distribution.[23] The secondary objectives enunciated in the *Green Book* – the elimination of the wage, the transformation of workers into partners in small industry, trade and services, and the replacement of money with a barter system – never materialised as the people remained aloof and unenthusiastic about them.

With respect to common objectives – internal and external balance, and economic stability – the Libyan record was mixed. The domestic budget deficit (the difference between total expenditure and non-oil revenue) was estimated to have been around 50 percent of GDP during the first oil boom. During the first half of the 1980s, when oil receipts were in continued decline, the overall fiscal position deteriorated, and the total 1986 deficit was equivalent to more than 25 percent of GDP. The brunt of the reduction in annual expenditure fell upon the development budget which was cut by nearly two-thirds. After 1987, both government revenues and expenditures declined as percentages of GDP, reflecting reduced oil income and restrained spending. Public revenue in 1994 was only 20 percent of GDP compared to about 50 percent in 1974; public expenditure was 27 percent compared to 55 percent. The overall deficit was almost entirely financed by domestic Bank borrowing (including direct advances from the Central Bank).

Libya's inflation record was difficult to evaluate due to: administrative controls and commodity subsidies; the limited coverage of CPI (Tripoli only); and the fact that the CPI's weights were based on a 1969 budget survey of the poorest families. Within these constraints, the price index showed moderate single digit rises between 1974 and 1979. During the first half of the 1980s the index rose by more than 10 percent a year on average. Between 1985 and 1989, official CPI rose in single digits, but the pace accelerated in 1990–1. Official statistics were not available for subsequent years. Indications are, however, that prices further increased in 1992–4 culminating in an estimated 50 percent by the end of 1994, reflecting UN sanctions and the sharp depreciation of the Libyan dinar on the black market.

Libya's balance of payments behaviour in the early years reflected principally the developments in hydrocarbon export receipts and the public sector's external expenditure policy. The current account was officially in surplus every year between 1974 and 1980 – despite considerable remittances by the foreign workers in the country.

By contrast the capital account was perennially in deficit due mainly to official transfers and investments. The overall balance, however, was mainly affected by the negative 'errors and omissions' item which ranged in equivalence between 27 and 83 percent of annual net merchandise trade. Official reserves peaked at $14 billion at the end of 1980. Between 1981 and 1986, BOP reflected largely the reduced level of oil export receipts, offset partly by the government's cautious fiscal stance. The current account was in deficit in every year except for 1985. The same was true with the overall account, except in 1986. With the completion of a number of projects, petrochemical exports became an important foreign exchange earner. Since the deficits were financed almost exclusively through a rundown of reserves, the latter declined to $6 billion at the end of 1986. Part of these reserves was frozen in January 1986 by the United States under economic sanctions. BOP developments after 1986 were affected mainly by the conditions in the global oil market, the government's domestic fiscal position and the extent of the private participation allowed in imports (10 percent of total in 1991). Current account surpluses and reserves accumulated between 1990 and 1992 were drawn upon to pay for subsequent deficits. Libya's trade and payments were severely hampered by UN trade sanctions after 1992, and by the further overseas assets freeze in 1993. The country's credit risk was lowered and foreign borrowing and investment were effectively ended. A noteworthy development was the increase in non-oil exports which accounted for about 8 percent of total exports in 1994 (composed largely of chemicals, food, live animals and re-exports). Since BOP deficits were routinely financed by a drawdown on official reserves, government-to-government barter deals or the temporary postponement of payment debts, the country's stock of foreign debt in 1994 was privately estimated to consist of only $3 billion in short-term debt and $2 billion in medium-term obligation – representing a debt/GDP ratio of less than 25 percent.

Full employment of labour was not a problem in Libya until the second half of the 1980s. Given the relatively small population, the traditionally low level of participation of women in the work force and the country's young median age, economic development was in fact perennially constrained by labour shortages. A large number of workers, both skilled and unskilled, had to be brought in from other countries. Some 35 percent of the labour force was estimated to be non-Libyan. During the 1990s, unemployment among nationals gradually became a cause of discontent as Libyans, not used to low-paid menial jobs, found difficulty in finding desirable employment. With an estimated population growth of 3.6 percent a year, the full employment of the native labour force became increasingly dependent on the expulsion of expatriate workers.

Within the two decades, the golden period for Libya was the 1974–80 years when GDP nearly tripled. During the 1980s, total output even in nominal terms gradually declined to less than the 1980 level. Even with the Persian Gulf war's windfalls, output in 1994 was less than a decade earlier. In the meantime, population grew from 2.3 million to 5.2 million. Per capita GDP in 1994 was thus actually far less than the $11,700 reached in 1981 and barely higher than the figure in 1974.[24]

NIGERIA

In Nigeria, national goals followed successive political upheavals, military coups and occasional experiments with civilian rule. But three basic objectives were consistently embraced by all political and military leaders. They included (1) economic growth and modernisation (2) Nigerisation of the economy in both the oil and non-oil sectors and (3) improvement in the distribution of income among ethnic groups and regions. The main components of the growth objective were increased production of food for both domestic consumption and exports, rapid industrialisation, a drastic reduction in unemployment and the creation of 'a just and egalitarian society'. Nigerisation called for reducing the role of foreigners in the economy, through increasing Nigerian ownership and employment in transnational corporations and through strengthening the private sector. Income redistribution meant reducing the uneven share of different regions of the country in physical infrastructure.[25]

The most populous country in Africa with great wealth of natural and human resources (including oil and gas), a vibrant private sector and an active labour force, Nigeria was a leading candidate for major, global power rank. Yet, during the years of on-again off-again military rule, the country fell into the grip of ethnic and tribal conflicts, a politically volatile atmosphere and endless tales of mismanagement and malfeasance almost unique among the group members. The result, for the most part, was political stalemate and economic paralysis.

Nigeria was, less than any other group member, successful in approaching – much less reaching – its lofty objectives in any of the major areas of interest. The objective of growth was only partially achieved, and only for part of the period. During the first oil boom, 1974–8, real GDP grew by 6 percent a year on average, and non-oil GDP by 9.5 percent. Agriculture, however, had an annual growth rate of only 2 percent and farm exports were halved as farmers rushed to the cities to earn more money in urban activities. The oil sector had negative growth in two of the five years, due to adverse shifts in demand for Nigeria's crude. The second oil boom raised Nigeria's oil exports in both volume and value to new record highs, and helped GDP's continued growth. But the new affluence was short-lived. The crisis began in 1981 when Nigerian crude production plummeted to 1.4 mb/d from 2.3 mb/d in 1979. The value of Nigerian oil exports declined from a peak of over $25 billion in 1980 to about $6.5 billion in 1986, and real GDP declined on average by 2–3 percent a year in 1981–6. In addition to the four-year decline in world demand for oil, Nigeria's oil exports were further cut by the 50 percent increase in domestic oil consumption. A decline in the Nigerian oil price from about $37 to $30 was partly responsible for lower oil revenues, and domestic oil price subsidies were partly to blame for the increase in domestic consumption during the period.

After 1986, real GDP growth recovered somewhat on account of the liberalisation measures under a structural adjustment programme (SAP) but by 1987 the real GDP was still only 7.5 percent higher than it had been in 1973, while population had increased from less than 70 million to probably near 100 million. Spurred by the initial results of the SAP, Nigeria's real GDP grew nearly 5 percent a year between 1987 and 1992, primarily reflecting a recovery in both agriculture and manufacturing. The deterioration in economic and financial conditions after 1992 resulted in an economic slow-down, and GDP growth was 1.5 percent in 1993 and only 0.3 percent in 1994.

The goal of adequate farm output for both domestic consumption and exports was never achieved. Despite successive efforts to raise food production under the 1970s 'Operation Feed the Nation', and the early 1980s 'Green Revolution', the results were not successful. Once a modest exporter of farm products (cocoa, ground-nuts, palm kernels and rubber), Nigeria became a large net importer of foodstuffs (rice, maize, wheat and sugar) by the early 1980s. After the 1986 adjustment pro-gramme, a series of policies, particularly price decontrol, trade liberalisation and market incentives, combined with the large depreciation of the real effective exchange rate, helped to increase agricultural growth. By conservative estimates, the sector grew at a yearly average of 3.5 percent between 1987 and 1994, and food imports fell to one-fifth of the value in 1986. Some of these enlightened policies were reversed in the meantime, and total agricultural exports in 1994 were less than 2 per-cent of total exports, down from 8 percent in 1988. The relatively poor performance of Nigeria's agriculture up to the mid-1980s was due to such factors as the ambiguities of land ownership rights, poor infrastructural facilities, massive food imports at the appreciated exchange rate, loss of skilled workforce to urban construction, fewer family child labourers due to the spread of primary education, inefficiency in the dis-tribution of domestic inputs and technical support, inadequate producer prices and high production costs. Rising rural wages, combined with the replacement of family labour by hired workers, and increases in farm-gate prices due to an appreciated exchange rate, drove Nigerian agriculture out of competition with rival countries. By 1994, palm oil, groundnuts and cotton – all formerly exported – were consumed at home and had to be supplemented by imports.

Rapid industrialisation also remained an elusive goal. The industrial sector – con-sisting of food and beverages, fuel products, chemicals, metals, cement, wood and assembled motor vehicles – was essentially geared to the domestic market. Heavily import-dependent for up to 70 percent of input, hamstrung by a vast array of con-trols and stifled by other bottlenecks (e.g. water and power shortages), domestic manufacturing was relatively high-cost and inefficient. The protective shield around import-substitution industries, through tariffs, import licences, excise taxes and exchange rate policies, helped to ensure their survival, albeit at the cost of significant price distortions.

Many of the public industrial projects (including the inefficient iron and steel) were chosen without necessary feasibility studies as to their economic viability and without adequate attention to domestic administrative and managerial capacity. Furthermore, the relatively low-interest rate policy, combined with rising wages and the appreciating exchange rate, encouraged the expansion of other capital-intensive, assembly-type industries based on imported inputs such as large-scale, mechanised state agribusinesses designed to produce grain, root crops and livestock. By 1994, industry accounted for about 10 percent of GDP, down from 19 percent in 1974.

Responsible for the poor overall economic performance in both agriculture and industry were the high share of infrastructural outlays in total allocation, very high capital output ratios in these investments, laxity in back-up requirements, the stop-go nature of planned capital formation and the inefficient and *ad hoc* character of foreign exchange allocation to projects. Overall capacity utilisation in manufacturing in the early 1990s reportedly hovered around only 40 percent.

During the first two oil booms, significant strides were made in expanding and improving roads, rails, seaports and airports. This sector grew fastest between 1974 and 1979. For the rest of the period, improvements in physical infrastructure were particularly noticeable in the trunk roads in urban areas. Nevertheless, considerable bottlenecks emerged in port facilities, means of transportation and road mainte-nance. The impact of the expanded infrastructure on bridging the gap between dif-ferent regions remained unclear. Urbanisation projects (including a new capital city in Abuja) were haphazard and inordinately costly.

The goal of high employment seemed beyond the planners' reach. Although there were no reliable data (due to the seasonal nature of activity in rural areas and the large size of the services sector composed of petty trade), available indications point out considerable urban unemployment. This was true particularly among the youth migrating from the villages in search of the high urban wages associated with the oil booms. Official estimates in 1994 put the national unemployment rate at 3 percent (5 percent urban and 3 percent rural); the peak was officially 7 percent in December 1987 (10 percent urban and 6 percent rural). Yet private estimates indicate much higher rates of nearly 30 percent, with a large number of working days lost through strikes in some years.

While there were no reliable statistics on domestic income distribution, World Bank analysts believe that while overall poverty might have declined between 1987 and 1994 income distribution actually worsened, and the poorest of the poor were disadvantaged. In all probability, the civil servants and industrial proletariat – bore the brunt of the oil fortune reversal after 1981, and the fall in foreign exchange earnings. By contrast, small farmers and farm labourers were the primary beneficiaries of the post-1986 adjustment programme.

There was scant progress toward economic diversification. In particular, the goal of industrialisation and modernisation was far from even partial realisation. Rudimentary data seem to indicate that mining and manufacturing accounted for about 7.5 percent of nominal 1994 GDP, not much better than the 7 percent of 1973/4. Construction activity, which accounted for 15 percent of GDP in 1979, became the smallest component of GDP (2.5 percent) in 1994. Agriculture, with considerable ups and downs throughout the two decades, ended up at about 38 per-cent on average for 1993–5. The services sector had its share fluctuating between 30 and 40 percent. But the oil sector's ranking, as originally intended, was noticeably changed downward.

In the unstated but universal national socio-economic goals, e.g. price stability, fiscal solvency, external balance and environmental protection, Nigeria encountered serious setbacks. Increased public spending by the Nigerian federal and state govern-ments and the resulting budget deficits were financed internally and mostly through Central Bank credit. Under the first two oil booms, the budgetary operations of the federal government showed rising deficits averaging about 7 percent of GDP. The budgetary gap followed primarily the fluctuations in petroleum receipts and changes in current expenditure ranging between 4 percent of GDP in 1984–5 and 17 percent in 1993, with an 8 percent a year average. The average deficit would have been much larger without the third oil boom in 1990. Yet the government ended up the year 1994 still with a budget deficit equal to 10 percent of GDP.

Despite the statutory annual price guidelines issued by a government board, the Nigerian consumer price index during the first two oil booms rose by about 16 percent a year on average. During the first half of the 1980s, the official CPI fluctuated widely with a high of 40 percent in 1984 and a low of 5.5 percent in 1985, reflecting variations in fiscal, monetary and exchange-rate policies. Expansionary fiscal policy, rising money supply, a decline in real imports, distribution bottlenecks and significant monopolistic practices in domestic production and distribution were mainly responsible for the double-digit inflation. In the second half of the 1980s and beyond, the rate of inflation again fluctuated widely due to the government's stop-go measures. The rate ranged from 55 percent in 1988 to 7.5 percent in 1990 (the third oil boom). Subsequent expansionary policies and a substantial devaluation of the naira pushed the index up again towards 50 percent in 1992 and more than 100 percent in 1994.

Nigeria faced external imbalances during most of the period. Up to 1980, the balance-of-payments on current and overall accounts showed an erratic trend, following daily oil output and prices. However, the overall balance registered persistent deficits after 1981 and up to the end of the period. In October 1986, the Nigerian government failed to make the first repayment on its rescheduled, uninsured trade debt, and in January 1987 it defaulted for the first time on its interest payments. After the rescheduling of arrears under the Paris and London Clubs sponsorship, imports were drastically cut, but external deficits continued every year until the end of the period due largely to domestic cost/price increases, the decline of competitiveness in non-oil exports and the burden of external debt service. The stock of public foreign debt increased from about $4 billion in 1980 to nearly $34 billion by 1994. Part of this increase was due to cross-currency revaluations (i.e. a rise in the value of non-dollar denominated debt) and the conversion of the 1982–3 trade arrears into public obligations.

Nigeria's major environmental problems were reflected in soil erosion and degradation, water contamination, deforestation and rapid exhaustion of natural resources (including the flaring of natural gas, in which the country ranked as the world's worst culprit). Many of these problems were rooted in rapid population growth, which itself was related to rural poverty as farmers desired more children to help the family survive on the land – a vicious circle.

In short, Nigeria turned out to be unsuccessful in achieving most of its objectives. By 1994, the economy's structural weaknesses included an agricultural sector without its export base, a manufacturing sector highly dependent on imported inputs and widely uncompetitive, and a bureaucracy plagued with a cumbersome regulatory framework and legendary corruption. High tariffs and other effective protection given to import-intensive, domestic, consumer goods industries kept them from improving productivity. By contrast, low to negative rates of protection given to other industries using local resources placed them at a competitive disadvantage with imports.

As a result, GDP per capita, which peaked at nearly $1,200 in 1980, dwindled to less than $400 in 1994, with a third of the population considered poor, and one-tenth extremely poor. In real terms, per capita income and private consumption in 1994 were only slightly higher than they were in the early 1970s. By World Bank estimates, with GDP growing at 5 percent a year, and population at 3 percent – the rates experienced in 1987–92 – per capita income could rise by only 2 percent annually. At that rate, it would take about 30 years for Nigeria to recover its peak living standard achieved in 1980.[26]

QATAR

Qatar's threefold national development agenda included 'prosperity, justice, and stability' which translated into (1) sustained economic growth coupled with a high standard of living (2) improvements in income distribution (3) internal and external balance and (4) political peace guaranteed by strong defence.

The evaluation of Qatar's economic performance during the 1970s is rendered difficult by the lack of data on national accounts as well as on employment, wages and prices. Most standard statistical reference sources do not cover Qatar, and those which do provide only rudimentary information. Based on available statistical information, the objective of economic growth was reached during the first two oil booms but faltered afterwards. In the absence of official national income accounts, the indications were that the implementation of major capital projects by the government in 1974–81 (mostly schools, hospitals, public buildings and roads), combined with a sharp rise in private residential housing, boosted real GDP at an average annual rate of about 4 percent. With the emergence of the depressed oil market after 1982, GDP began to decline each year, reaching less than 60 percent of the 1981 level in 1986. Total GDP peaked at about $9 billion in 1981 before declining, with marginal fluctuations from year to year, to about $7.5 billion in 1994. Real GDP growth was negative in three of the five years 1990–4.

Fluctuations in the rate of growth of GDP components were responsible for the overall performance. Oil production in Qatar peaked in 1974 with an average daily production of 518,000 barrels, and its lowest level of 269,000 b/d was reached in 1983. Reliance, therefore, was increasingly shifted to the massive North Field natural gas reservoirs for the production and export of LNG through three joint-venture projects with foreign partners. During the 1970s, Qatar's economy was almost totally dependent on crude oil production and exports. Oil provided almost all of the country's export earnings, over 85 percent of budgetary revenues and some two-thirds of the GDP. In the post-1982 depressed oil market, oil's share in GDP declined to 20 percent in 1986. During the 1987–94 period, the oil and gas sector accounted for about one-third of GDP, 70 percent of total exports and 65 percent of budgetary revenues.

Agriculture and fishing contributed a very small percentage of Qatar's non-oil GDP. In 1974, of Qatar's 35,000 hectares of arable land (less than 3 percent of land area) only about 6,000 hectares were cultivated. This figure rose to 13,000 hectares in 1984, along with notable increases in cereals, fruits and vegetables, and forage. While the country became largely self-sufficient in summer and winter vegetables, reliance on food imports remained high, as fishing also declined in the 1990s. Altogether, agriculture and fishing accounted for only 1 percent of GDP and 1.6 percent of non-oil GDP in 1994.

Industrial production in 1974–9 showed marked increases in refined petroleum, water desalination, and electricity. Output also expanded in fertiliser, steel and cement, 1980–5. By 1994, the industry share rose to 19 percent of total GDP and 30 percent of non-oil GDP. Except in the hydrocarbon sector, in which Qatar had a relatively recognisable advantage, the balance sheet of large industrial undertakings did not indicate high profitability. For example, for ten years after its inauguration, Qatar Steel Company operated at very large losses and ran into debt.

Construction also played a significant role in Qatar's growth up to 1982. By 1983, Qatar's massive construction boom was ended as most of the country's infrastructure (schools, roads, communications, hospitals and housing) was completed. Increased demand for housing and a few major projects kept the sub-sector active until the early 1990s when leisure projects (an international tennis stadium and a 27-hole golf complex) became attractive. The construction industry was mostly sustained by an expatriate labour force. The sector's share, which peaked at 7 percent of total GDP in 1989, declined to 5 percent in 1994.

The services sector provided the largest share of Qatar's GDP after 1982. By the end of the 1970s, this share had reached 22 percent of total GDP; by the late 1980s, the figure was 34 percent and, by 1994, nearly 49 percent. The sector included finance, insurance, real estate and household services, but the share of public services outweighed all others with a more than 50 percent contribution.

In terms of its declared agenda of prosperity and high living standards, progress was evident. Per capital GDP, according to World Bank figures, rose from about $7,240 in 1974 to over $12,800 in 1994 (in nominal terms). Given Qatar's relative price stability after the initial stormy period (1974–7), Qatari residents' purchasing power also probably improved slightly – or at least did not deteriorate. However, in relation to the record GDP per capita of about $36,000 in 1981 (OPEC estimate), the 1994 situation was decisively worse in every sense.

In the absence of data on domestic income distribution, the goal of 'justice' is difficult to evaluate. Yet, the lot of all Quatari citizens was undoubtedly improved, and they obviously fared better than their expatriate co-workers. Within the citizenry itself, the oil windfall was, however, not evenly enjoyed, as some immense and publicised private fortunes were reported in the press. The poorest of the 'poor' in Qatar were still much better off than most middle income recipients in most other group members.

The objective of economic stability was reasonably achieved in part of the period through periodic and timely adjustments to the fluctuations in the world oil market. The exchange rate of the Qatari riyal was carefully maintained, and price rises were kept in check. The overall record, however, was mixed. Excluding capital expenditure and net lending, the budget was in surplus from 1974 until the end of 1980. Between 1980 and 1984, the budgetary account alone showed persistent surpluses

while the overall balance fluctuated due to government lending and equity participation. In the post-1986 oil price market, public finance came under further pressure,
and after 1988 the budget account was in total deficit every year until the end of the
period, when the gap amounted to 11 percent of GDP. Budget deficits were largely
financed by a drawdown of both government domestic deposits and foreign assets,
plus borrowing from domestic sources. Later on, foreign borrowing for budgetary
purposes also increased significantly.

Statistics on price movements remained rather weak. Nevertheless, rapid increases
of an average 23 percent a year were estimated for 1974–8. Price increases were
brought down to an average of 2 percent in 1982–4, reflecting mainly lower demand
pressures along with falling domestic liquidity. Between 1985 and 1989, the consumer
price index was also subdued due to weakened economic conditions and a significant
fall in development expenditures. Between 1990 and 1994, the CPI remained low, at
about a 2 percent annual average, and the non-oil GDP deflator was broadly
unchanged due to virtual GDP stagnation.

In the absence of an official balance of payments compilation by Qatari authorities
during the 1970s, the external developments were estimated with large margins of error.
The 1974–9 period showed large overall surpluses as oil receipts exceeded all payments
including capital and net official transfers, and there were substantial inflows of funds
to finance Qatar's major industrial projects. The current account surplus peaked at
$2.7 billion in 1980 and began to decline thereafter due to rising levels of imports as
well as official foreign grants. The overall payments surplus, which had reached more
than $900 million in 1981, the highest level since 1974, was subsequently reversed in
1982. Beginning in 1986 and continuing through 1994, the current account was in
yearly deficit except for 1990 (the third oil boom). Thus, despite substantial capital
inflows, the overall deficit rose to an average of about 7.5 percent of GDP in 1994.

Due to the narrow base of domestic non-energy production, Qatar depended
almost entirely on imports for its domestic consumption and industrial output. More
than half of annual imports typically consisted of machinery, transport equipment
and manufactured goods. Foodstuffs normally amounted to about 15 percent. To the
very end, exports consisted mostly of crude oil and natural gas, and their related
products (e.g. fertilisers and petrochemicals). There was also a small export of steel.
Foreign borrowing for development and current budget financing continued from
year to year. In 1994, Qatar's total foreign debt was estimated at about $2.5 billion,
or nearly one third of GDP. A large part of the foreign public debt was related to the
hydrocarbon projects.

Finally, the goal of 'political peace' was reasonably served despite protracted
regional crises, and a subsequent internal palace coup. Qatar skillfully kept itself away
from serious involvement in the Persian Gulf conflicts. Nevertheless, by IMF's
estimation, Qatar, in mid-1994, stood at a 'critical juncture'. GDP in 1994 fell 4.1
percent, and further falls were expected in subsequent years. The country's bond
rating was given a subinvestment grade. The situation was not expected to substantially improve before the end of the decade. The country's hopes were almost entirely
placed on the development of its enormous gas reserves – second only to those of Iran
among OPEC members.

SAUDI ARABIA

The national development agenda in Saudi Arabia, detailed more fully in the second plan (1975–80) included (1) defending the country and maintaining internal economic and social stability; (2) raising the people's living standard and welfare through a high rate of balanced economic growth; (3) reducing dependence on oil as a source of national income; (4) developing indigenous human resources through education and healthcare; and (5) building of the physical infrastructure required for the attainment of these goals. In practice, the establishment of a viable modern economy at almost any cost was an overriding objective.

In no other group member was the link between state oil revenues, government expenditure and economic change so virtually 'umbilical'. The pattern of gross domestic output was determined principally by government spending, and the latter by budgetary developments. Changes in state revenues were, in turn, mainly affected by oil export earnings, reflecting daily crude oil output and world petroleum prices. This extraordinary dependence on oil was the cause of most subsequent national economic difficulties as the kingdom underwent some extraordinary cycles of boom and bust.[27] Saudi Arabia had a larger annual oil export income than any other OPEC member in every single year during the 20-year period. But, with the exception of Iraq and Kuwait which were mutually involved in the 1990/1 Persian Gulf war, no other OPEC member experienced such devastating volatility in its yearly petroleum revenue, as shall be detailed below.

The Kingdom managed to attain its goal of rapid economic growth in the first half of the 20-year period. During the first oil boom, all major economic sectors grew rapidly, with the construction sector showing the largest expansion and the farm sector the smallest. Real GDP growth was remarkable at 9 percent a year on average. The unexpected near doubling of Saudi oil income 1980–1 following the Iranian 1979 revolution produced an unplanned increase in development spending on the already saturated infrastructure. During the second oil boom, the Kingdom's nominal GNP reached nearly six times that of 1974's, reflecting, in part, considerable increases in domestic prices. Between 1979 and 1981, the non-oil sector grew at an average annual real rate of over 11 percent and total GDP by about 5 percent on average.

Saudi Arabia's reversal of fortune began after 1982 when the oil sector experienced a 9 percent decline in real terms, followed by a precipitous five-year annual average decline of 18 percent. A period of recovery began in 1988 and was strengthened by the Persian Gulf war, with total GDP averaging about 7 percent real growth a year in the four years ending in 1992. And there was a final period of economic stagnation between 1992 and 1994 when real GDP virtually stood still (and per capita GDP declined). Altogether, during the period under review, the Kingdom implemented four five-year development plans, with the first two achieving their growth targets but the last two badly missing their objectives.

The primary goal of improving the people's lot was achieved in terms of increased income per head, particularly in the first few years. Per capita GNP reached nearly $17,000 in 1981 from $1200 in 1972. Reflecting the decline in oil revenues in the 1980s, however, GNP per head fell back to $5,500 before the Iraq/Kuwait war.

Improvement in oil production and exports during the 1990s helped increase the figure to nearly $7,000 by 1994[28], or about 45 percent of the 1981 peak. Thus while Saudi citizens had considerably larger per capita incomes in 1994 than in 1974 in nominal terms, the rise in their real incomes was not certain. Saudi GNP in 1994, at constant 1970 prices, was estimated at about $90 billion or about $5,325 per head (based on a 16.9 million population). Real GNP in 1974 was estimated at $39 billion, giving a per capita figure of $5,580 (based on a population estimate of 7 million). Admittedly, the reliability and comparability of these figures are subject to serious question. But their main significance lies in casting doubt on the plausible but facile conclusion that with nearly $1 trillion of oil income between 1974 and 1994, the average per capita real income must have become considerably larger.

Data on internal income distribution are not available. Yet there are strong indications that the lot of even the poorest Saudi nationals improved during the early years. The comprehensive system of subsidies – free land for building a house, interest-free loans for home construction, free education and health care, job opportunities and guaranteed employment for most citizens – enabled the population at large to enjoy some of the fruits of this oil wealth. But with rising population, reduced subsidies, emerging unemployment and environmental deteriorations, the all-embracing national welfare went into a downward slide. And, more significantly, the glaring income disparity between the ruling elites (the royal family, well-placed technocrats and other fortunate beneficiaries of the oil rent) and ordinary citizens palpably widened, and was widening at the end. The gap was also notably wide between the income of Saudi citizens and that of the so-called guest workers.

The objectives of internal and external stability were only achieved in part. Inflation was kept in check during nearly the entire period after the first oil boom – a feat unparalleled among most other group members. But the national budget and the balance of payments registered significant deficits after the second oil boom, and only marginally improved under the third. Saudi Arabia was uniquely successful in holding down price rises for most of the period through a judicious control of domestic liquidity. Domestic prices accelerated after 1973 in response to increased economic activity, the bottleneck on supplies and materials, port-congestion and transport jams. The official CPI initially rose 35 percent in 1974/5 but it was brought down to 4 percent in 1977/8 and 1 percent in 1982/3. The steady decline in the official price indices reflected the restraint on public spending, continued high growth of real output, the elimination of bottlenecks on ports and roads, a decline in import prices due to the riyal's appreciation and state subsidies. The CPI and GDP deflator experienced virtual stability up to 1989 and produced only a temporary spike of 3.5 percent a year following the Persian Gulf war. Subsequent price declines were due largely to the sharp administrative cuts in the prices of gasoline, electricity, telephone services and port duties in early 1992. The 1993–4 CPI hovered around 1 percent.

In other macroeconomic areas, the authorities were not so successful. Saudi Arabia's largely open economy made it more than almost all other group members susceptible to global economic activity, changes in the volume of world trade, foreign price movements, international exchange rates, and interest-rate trends. The value of the Saudi riyal, riyal interest rates and the Kingdom's terms of trade were all exogenously determined. Partly for this reason, and largely because of the government's far-flung domestic commitments and the negligible source of domestic non-oil revenues, a long-term balance in the fiscal and external payments accounts proved elusive for much of the period. Fiscal developments showed a particularly volatile and chequered record. Between 1974 and 1978 total budgetary expenditures increased almost eight fold, producing a deficit as early as 1977/8. As oil revenue began to rise sharply under the second oil boom, a budget surplus equal to 22 percent of GDP materialised in 1981 but by 1983 the budget came into balance as oil revenues declined. A steady fall in government revenues, due mainly to a drastic fall in oil export receipts, afterwards led to the emergence of budgetary deficits. Although government expenditures were cut on development projects, and non-oil revenues raised, the budget deficit rose to a massive, non-sustainable 25 percent of GDP in 1987. Despite weak oil prices, however, the Kingdom made good progress in reducing the budgetary deficits to less than 10 percent of GDP in 1989. This successful adjustment process was subsequently halted by the Persian Gulf war in 1990, and the gap mounted to more than 17 percent in 1990–1 before it was brought down to 8 percent in 1994. For the entire decade, therefore, the Kingdom's budget was in the red by a yearly average of $10 billion. Meanwhile, foreign investment income also fell sharply because of a fall in government foreign assets and lower international interest rates. The latter's share in total revenue declined from 12 percent in 1989 to 5 percent in 1992. Iraq's invasion of Kuwait reportedly cost Saudi Arabia some $55 billion, mostly in direct payments to the US ($12.8 billion) and other allies. Since higher oil prices following the invasion brought in fresh revenues, the true cost is estimated at about $30 billion. The rise of Saudi oil output from 5.5mb/d before the Iraqi invasion to 8mb/d afterwards also somewhat compensated for the war's cost. Saudi's budget deficits were financed by drawing on foreign assets and through domestic and foreign loans.[29] In addition to losing foreign assets, the Kingdom was also drawn into debt. Saudi Arabia had no public debt as late as the early 1980s. After 1983, when the government budget became perennially unbalanced, the treasury began to borrow from domestic sources in the form of Treasury bills to avoid drawing on foreign reserve assets. By the end of 1994, the total figure was estimated at the equivalent of $100 billion (or 77 percent of GDP).

The Kingdom's growing tendency to spend beyond current means was finally halted by the advent of the Persian Gulf war. Riyadh's generous welfare system, its unstoppable consumer and producer subsidies, gigantic weapons purchases and foreign assistance pushed the Treasury to break with past practice, and in 1992 the Kingdom borrowed $4.5 billion from international banks for the first time ever. More than half of that amount was repaid in 1994, and the Kingdom's official external debt became less than $2 billion by the start of 1995. The medium and long-term external debt of state enterprises, however, totaled around $3 billion.

Saudi Arabia's external balance also came under heavy pressure after the second oil boom. Due to the openness of the Saudi economy and the relatively small domestic non-oil sector, a major part of government spending (and private sector income) was routinely translated into demand for foreign goods and services. Between 1974 and 1978, the sizeable surpluses in the merchandise account were offset by services and private transfers. Thanks to a steady rise in foreign investment income, the overall account was still in surplus up until 1982. But a drastic decline in oil revenue (due to both reduced exports and lower prices) turned the 1983 account into an $18 billion deficit. The current account moved into deficit in every year up to 1994. The overall balance was also in deficit, averaging $2.4 billion annually, 1986–94. Saudi Arabia's persistent current account gaps were due largely to the shortfalls on services and private transfers.[30] Responsible directly for continued deficits – and, indirectly, for the budget gap – was also the government's determination to maintain the riyal's value *vis-à-vis* the US dollar at $1=SR 3.75 over nearly a decade from 1985. Despite declining oil revenues and their effects on the balance-of-payments and central bank reserves, there was a solid commitment not to devalue the riyal.

The objective of diversification had a rosier outcome than internal and external stability. In the early 1990s, the Saudis were proudly claiming that the desert Kingdom was no longer a one-product economy. Only one percent of the labour force was engaged in the production and export of crude oil. Agriculture, manufacturing, construction, and public utilities had all raised their shares in GDP. Indeed, during the 20-year period, the composition of GDP underwent considerable changes. In 1974, agriculture accounted for about 1 percent of GDP, industry 5 percent, private and government services 17 percent, and oil 77 percent. During the first oil boom, the oil sector contributed almost three quarters of GDP, about 90 percent of government revenue and virtually all export receipts. In 1994, agriculture accounted for 7 percent of GDP at current prices, the oil and gas sector 32 percent, industry 20 percent, government services (including defence) 18 percent, and other services 23 percent.

From the standpoint of sectoral contribution to GDP, the diversification strategy seemed successful in developing other non-oil related domestic productive sectors. But the feat was achieved at an exorbitant cost. The prime example was agriculture, and the creation of an artificial farm sector kept alive by an expensive irrigation system, extravagant public subsidies and temporary foreign farm workers in a desert area where water was the key constraint on economic development. Helped by free land from the state, interest-free loans, subsidised seed, fertilisers and equipment, and guaranteed purchase prices, the Kingdom's farm output, increasingly in the hands of large commercial enterprises, rose to over 10 percent of GDP in 1992. Wheat production rose from just 300,000 tons in 1975 to more than 4 million tons in 1992, and Saudi Arabia became the world's sixth largest wheat exporter. Agriculture's achievements were thus obtained at enormous financial and environmental costs. In financial terms, the farming sector was an unbearable burden on the budget.[31] Excessive utilisation of scarce underground water and sub-optional cropping patterns in favour of wheat (instead of mere suitable crops like barley and sorghum) were not cost-effective either.

From an environmental view, too, the success made little sense. Some 87 percent of total water consumption each year was drawn from non-renewable water sources of which some 80–90 percent was used in agriculture. As a result, the country was expected to start running out of water in 25 years – four times faster than its oil depletion. As they said, the Saudis used the proceeds from the sale of oil at highly profitable prices to 'export water' at heavily subsidised rates.[32]

In 1994, Saudi Arabia boasted about being the Arab world's leading industrial producer with a manufacturing sector larger than those of Greece, Portugal, Singapore or Hong Kong. Although still relatively small in terms of its contribution to GDP and employment, the Kingdom's industrial sector developed rapidly under the government's aegis. During the first two oil booms, non-oil manufacturing activity rose at an average annual rate of over 12 percent in real terms in such industries as food and beverages, household needs, building materials, petrochemicals and metal products. However, chemicals and plastics accounted for two thirds of total investment in the sector. With the notable exception of some giant enterprises producing petrochemicals, basic metals, pharmaceuticals and pipe and cable, the majority of Saudi factories remained small, with limited local value added. In 1994, according to a Saudi publicity document, there were more than 2,200 manufacturing units in the Kingdom with an invested capital of more than $40 billion and an annual output of more than $15 billion (including $3 billion of non-oil industrial exports). However, Saudi industrialisation was admittedly not yet a self-sustaining process without extensive public incentives. The venture was also costly in environmental terms. The Kingdom's environment was damaged by the rapidly expanding population's use of non-renewable national resources (subsidised by the government). Expansion in heavy industry and other development outlays produced air pollution (caused by cars and the petrochemical industry), a shortage of municipal waste facilities and land contamination from the improper disposal of hazardous industrial and military wastes.[33] Saudi Arabia was the second largest flarer of natural gas after Nigeria in 1994.

The construction sector played a significant role in the rapid growth of the Saudi economy. By the end of the first oil boom, the sector's share in real non-oil GDP reached nearly 18 percent, reflecting the bulk of gross fixed capital formation. Physical infrastructure, utilities and residential housing by the private sector were responsible. In a sense, Saudi's economic growth in the first two oil booms was oil-financed and construction-based. The construction boom, however, ended after 1982 due to a decline in public projects and infrastructural saturation. The sector's share in the economy gradually declined to 1 percent in 1994. The decline had a contractionary impact on the whole economy, and was accompanied by lower profit margins and lower wages.

Water and power was another sub-sector that expanded rapidly during the first two oil booms. In addition to increased urban consumer demand for electricity, gas and water, these utilities were increasingly needed as essential inputs in agricultural and industrial development. Furthermore, for both diversification and welfare considerations, these items were provided by the state at concessional rates, which, in turn, spurred their use and abuse. The share of this sub-sector rose from less than 3 percent to 6 percent in the first ten years. Despite all this, however, by the mid-1990s, Saudi industry was hobbled by power shortages. New office towers and homes stood empty for months waiting for power hookups as the demand for electricity was growing three times as fast as the economy.[34]

Due to limitations in domestic resource endowments for extensive agriculture or industrial development, Saudi Arabia's services sector occupied a strategic significance in the economy. Trade, finance, real estate and government accounted for the largest share of non-oil GDP. In the mid 1980s, the sector provided 45 percent of non-oil output and 40 percent of civilian employment. During the first half of the 1980s, roughly one-fifth of private consumption was spent on imported services. The sector maintained to the end its number one ranking in GDP share at about 41 percent.

Finally, the objective of developing Saudi human resources was attended to in earnest, but costs were high and the results were mixed. The Saudi economy enjoyed high employment in the earlier years of the oil boom.[35] In addition to fully using the domestic workforce, the Kingdom had to import foreign workers. Official reports put the workforce in 1980 at about 2.4 million – 1.4 million Saudis and 1 million foreigners compared to a total of 1.75 million in 1974. While 79 percent of the workforce in 1974 was employed in the productive sectors, the figure for 1994 was only 62 percent. More than 74 percent of the labour force were foreign workers. As much as 90 percent of private-sector jobs were filled by foreigners since they were both cheaper and more industrious and more willing than the Saudis to get their hands dirty. Due to steady growth for nearly a decade, unemployment was not a problem for the Kingdom. In the late 1980s, however, some unemployment emerged among Saudi nationals. Rising unemployment was one of the casualties of heavy investments in human resource development in the face of the oil shortfalls. In 1994, only one third of the graduates from Saudi universities could find jobs in the public sector. Others had a hard time finding employment in the private sector due to their lack of skills, expectation of high salaries and unfamiliarity with a business environment. At the peak of the oil booms, Saudi citizens were all but guaranteed high-paying jobs in public agencies or large state-owned enterprises. With oil prices falling in real terms, the state was no longer able to absorb young job-seekers. And the latter lacked the incentive, enthusiasm, hard-working habits, business loyalty, punctuality and other work-ethic qualities required in the burgeoning private sector.

Altogether, at the threshold of 1995, the Saudi economy was more than that of most members dependent on the outside world for its livelihood and prosperity. The Kingdom was still dependent on the oil sector for about 90 percent of total merchandise exports and 70 percent of government revenues. Moreover, the non-oil sector's contributions to the national product were not self-sustained without public expenditures and imports. Imports of goods and services amounted to more than 40 percent of GDP. And the urban private sector was almost totally reliant on expatriate labour.

THE UNITED ARAB EMIRATES

While the federation of seven separate emirates forming the UAE since 1971 had no formal, unified, long-term, coordinated development planning, four main objectives were pursued by the federal government (1) strengthening the country's physical infrastructure (2) expanding the social services network (3) diversifying the economic superstructure and (4) expanding the territory for *entrepot* commerce.

The primary objective of building up the country's physical infrastructure was the easiest to achieve thanks to the inflow of petrodollars and the availability of foreign contractors and modern construction technology. With a land area of about 78,000 square kilometres in 1994 the UAE had about 2,000 km of paved roads, fifteen modern seaports, special oil loading facilities, six international airports, two national airlines, satellite telecommunications, high-capacity radio relay, mobile telephones, an elaborate power grid, 22 water desalination plants, sophisticated broadcast services, and a data transformation network. In short, during the 20-year period, the UAE's physical landscape was transformed beyond recognition and expectation. By 1994, the UAE boasted of having some of the world's best golf courses, football stadiums and luxury resorts, calling itself the 'sporting capital of the Middle East'.

By the government's own accounts, the country crossed over from the middle ages to the modern world in two decades or so. Small fishing villages and trading posts were transformed into ultra modern cities; tents, huts and shacks were replaced by modern housing with up-to-date amenities. Luxury apartments and villas provided a dream world for modern architects; roads and highways were lined with trees, shrubs, plants and flowers. Abu Dhabi considered itself the 'garden city' of the Persian Gulf.

Another declared goal, expanding social services, was also clearly reached. All through the years, public services at an average yearly share of 11.5 percent of GDP were the largest among all services (e.g. trade, transport, storage, finance and real estate). The objective of diversifying the economy's superstructure was met in a partial manner. In 1974, nearly 78 percent of the UAE's GDP was made up by the oil sector while agriculture had less than a 1 percent share; industry (including construction) had 5 percent; and services 16 percent (including 3 percent in government services). Starting with a nearly four-fifths contribution to the national product, the oil sector almost steadily ceded part of its share, and ended up with only 34 percent of total GDP in 1994. Despite the enhanced significance of non-oil activity, however, the energy sector still accounted for some 80 percent of government revenues and roughly two-thirds of indigenous exports.

The industrial sector (including non-oil manufacturing, water, power and particularly construction) had a sharp rise over the 1974–8 period, with its value-added increasing about fivefold. The Jebel Ali industrial complex, built by Dubai, became a magnet for industrial investment. Under the second oil boom, manufacturing continued its rapid growth through 1982; and even during the 1983–6 recession this sub-sector was not unfavourably impacted. The water and power component of the industry sector actually expanded during this slack period; construction was the only activity which lost ground. Between 1987 and 1994, water, power and construction all experienced fast growth. In 1994, the manufacturing sector alone employed some 9.5 percent of the labour force and comprised about 8 percent of GDP. The industrial sector as a whole accounted for more than 20 percent of GDP in 1994.

Agriculture's contribution to the UAE's economy, while remaining low during the first oil boom, at about 1.5 percent of the non-oil GDP, was the main source of livelihood for the four oil-less emirates. With 80 percent of the country covered by sandy desert, and much of the rest by uncultivable salt marshes, farming took place on about 50,000 hectares of land regularly part of the oases scattered over the country. Fishing and animal husbandry were also carried out by small-scale operators. Under a host of government incentives, the area devoted to farming more than doubled, mostly in northern and eastern coastal plains. In the first decade under review, agricultural value rose almost twice as fast as total GDP. During the second decade, 1985–94, several farm products increased significantly and made the emirates self-sufficient in certain food items. The sector's share rose to 2 percent of GDP in 1994. The objective of diversification was partially achieved, if measured in terms of non-oil's share in the economy. But the economy's dependence on the energy earnings remained high, and much of the UAE's industrial diversification was also based on hydrocarbons.

The stated goal of becoming a regional trade *entrepot* met with probably the biggest success. The country aspired to be the region's major trading *entrepot* and the leading transport and communications centre. And in this endeavour, the government achieved notable success. By 1994, the UAE was a major hub of regional and international air and sea traffic. Dubai, for example, tried to become the principal entry for business and commerce to the Persian Gulf region. Port Rashid, opened in 1970, ranked as the region's leading import-transit port in 1994, and among the world's 15 leading ports for container traffic. The Jebel Ali Free Trade Zone also boasted of housing several hundred companies from around the world. The free-trade zones in all the emirates offered foreign manufacturers and distributors inexpensive storage and re-export sites, adequate infrastructure for trade and industry, duty-free benefits and extensive port facilities. Thanks to those provisions, the emirates turned out to be the least vulnerable to retrenchments as oil prices tumbled.

In terms of other conventional national criteria, the UAE's record was markedly better than most. Growth was moderate but inflation was low; external balances were favourable, and budget deficits were easily financed out of substantial returns on foreign investments. In the first two oil booms, real GDP rose at an estimated average rate of about 5.5 percent, mostly due to soaring construction activities. With a decline in oil export receipts after 1982, all the major components of domestic expenditure also declined, and output fell by an average yearly rate of about 4 percent in 1982–6. Recovery began modestly in the late 1980s mainly due to expanding regional trade following the Iran/Iraq ceasefire. Real non-oil GDP continued its steady growth of about 4 percent a year through 1994, mostly because of increased activities in banking, port operations, construction and services. Total GDP had a slower growth rate. The volatility in output almost directly reflected the ups and downs in the oil sector's fortune.

The behaviour of prices in the UAE was probably better than in almost all other members, except in the early years. As a result of a more than fourfold increase in domestic expenditure in five years, the first oil boom witnessed high rates of inflation due to port congestion, supply shortages and other infrastructural bottlenecks. Some estimates suggested a rise in consumer prices of 30 percent a year between 1974 and 1977, slowing down later to an 8 percent annual rate in 1978–82 as supply constraints were gradually removed. Between 1983 and 1985, domestic prices were believed to have actually declined by 2–4 percent annually as the local currency prices of imports were reduced because of the dirham's appreciation *vis-à-vis* the major European currencies and the Japanese yen (reflecting US dollar appreciation). Between 1986 (and the US dollar depreciation) and 1990, prices began to rise again at an estimated average annual rate of about 3.6 percent. CPI ranged between 4 and 5.5 percent in 1991–4.

High employment was easily achieved for UAE nationals. In 1974, the UAE's labour force of some 300,000 was employed 21 percent in agriculture, 32 percent in industry (including construction), 22 percent in government services and 25 percent in other services. To cope with the sharply higher level of economic activity during the first oil boom, a liberal immigration policy was adopted. There was virtually no open unemployment in the economy as citizens seeking jobs were guaranteed employment primarily in the government sector, and expatriate workers were required to secure work certificates before entering the country. Some 80 percent of the total 1994 labour force were expatriate workers.

Fiscal developments were not remarkably better than most. The rapidly rising oil income during the first oil boom unmatched by rising expenditures resulted in considerable annual budget surpluses giving rise to net foreign assets. With the start of the oil bust in 1982, the budget showed a persistent deficit each year, reaching 16 percent of GDP in 1986. In response, domestic subsidies, foreign grants, domestic loans and investments, and development outlays were halved and other expenditures curtailed. The subsequent annual deficits between 1986 and 1994 reflected sharp fluctuations in oil earnings while outlays remained largely unchanged, and yearly deficits reached 3.5–11 percent of GDP. These deficits were financed from the income on foreign assets, sales of these assets or domestic borrowing. Altogether, the UAE ran a budget gap perennially after 1982, and even during the third oil boom. In the absence of consumption taxes and with inflated public sector employment and extensive subsidies, the budget deficit became endemic.

Developments in the UAE's external payments were difficult to trace due to inadequate recording and reporting of routine transactions. Data provided by the Central Bank were only indicative of the magnitude of principal items. According to these figures, the UAE's balance-of-payments during the 1974–8 period was in surplus in both the current and overall accounts, while official transfers, including grants and loans to LDCs, doubled during the period. Official foreign assets were accumulated during the first oil boom of about $10 billion. The surplus in external accounts continued under the second oil boom, and peaked in 1980–1. Beginning in 1982, the depressed world oil market caused oil earnings to enter into a declining trend, affecting the volume of imports and payments for services. Developments between 1986 and 1994 reflected changes in oil export earnings and showed wide swings. Trade and current accounts showed persistent surpluses during the entire 1974–94 period, albeit with marked variations of 50 to 100 percent from year to year. Private capital outflows were a substantial part of the UAE's payments balance. The entry for 'net errors and omissions' was believed to mainly reflect such outflows. The capital account – comprising official loans, equity participation and other public and private capital flows – was in deficit every year of the period after 1986. The overall account, however, was in surplus for most of the years. Despite a steady increase in non-oil exports of goods and services, the non-oil trade gap widened steadily over the 20-year period due to the growth in imports. Among the entire OPEC membership, the

UAE had the enviable distinction of having no official foreign debt, and considerable net foreign assets privately estimated to run as high as $100 billion in 1994. There was some private external debt.

In terms of overall economic development, the UAE managed to increase its per capita GDP over the two decades, in nominal terms. Thus, while population rose from an estimated half a million in 1974 to an estimated 1.8 million in 1994, per capita GDP (according to OPEC figures) expanded from an estimated $18,265 to $19,473. The 1994 per capita GDP, however, was only about two thirds of the peak level reached in 1981. Equity in income distribution was not a formally declared objective but budgetary transfers from wealthier emirates to the poorer ones, and a very elaborate welfare regime within each, were significant measures in that direction. However, while GDP per capita in the UAE as a whole was near $20,000 in 1994, the figure for the rich Emirate of Abu Dhabi (with the largest per capita hydrocarbon wealth in the world) was $32,000, and that of the poorest northern Emirate of Ajman only $6,300.

The UAE's impressive achievements were, however, obtained at a high cost. The significant independence of individual emirates and rivalry among them resulted in a good deal of waste and duplication in the development projects. In the absence of a coordinated federal development plan, separate industrial projects, without sufficient infrastructural support, created massive unusable capacity in many areas. The country, for example, built six international airports, many within half-an-hour's drive from each other. The dry-dock constructed in Dubai for 500,000-ton oil tankers remained idle as the optimum tanker size was only half that much. Altogether, there were too many banks, too many ports and too many plants dealing with limited demand. The UAE was labelled one of the most over-banked countries in the world, with lax supervision reflected in the collapse of the Bank of Credit and Commerce International (BCCI) in 1991.

Environmental costs were also not negligible. As in other Persian Gulf sheikhdoms, water was both scarce and not optimally used. Averaging 60–100 mm a year of rainfall, the UAE had an acute water shortage. And yet about 70 percent of the total water consumption was absorbed in subsidised agriculture, even though a number of modern water-conservation techniques were also used. As a result of the expansion of irrigation using underground water, the water table in many parts of the UAE fell seriously and soil salinity increased.

In sum, the UAE was unique among group members for having achieved most of its national objectives, i.e. expansion of infrastructure, improvement of social services and promotion of its entrepot status in regional trade. But the country's goal of developing its non-oil economy was realised at an evidently very high cost. And much of the non-oil economy – construction, real estate, power, transport and even trade within the region – depended on public sector contracts and the government's oil revenue. Also, the autonomy enjoyed by the Emirates, and rivalry among their numerous and separate official bodies, resulted in a lack of overall policy coordination, a good deal of project duplication and excessive importation of labour.

VENEZUELA

Venezuela's socio-economic objectives were initially wrapped up in the political slogans of 'sowing oil' to reap other benefits and 'managing abundance with a mentality of scarcity'. These banners in practice called, *inter alia*, for the reform of agricultural institutions in order to increase farm production, improvements in social welfare benefits, expansion and diversification of industries to reduce dependence on oil and an increase in domestic employment. On the eve of the first oil boom, underlying economic conditions in Venezuela seemed all too conducive to the implementation of this socio-economic agenda. The country's per capita income, currency value, and future prospects were among the top-ranked in the Third World, and highest in Latin America. There was an oil-driven, thriving economy, exceptionally low inflation, less than $1 billion foreign debt, high rate of investment, and a consumption sector relatively self-sufficient except for food.

The challenges, however, were equally formidable. The public sector's share of the economy was large, and state-owned enterprises (SOE) included such fields as electricity, transport, communications, metals (iron ore, aluminum, steel) and, of course, petroleum. However, in addition to the small and inefficient agriculture (with 5 percent share in GDP but 30 percent share of the labour force), the economy suffered from high unemployment, considerable income inequality, poor public services; capital-intensive, highly-protected and non-competitive industries, and virtually no non-oil exports.

In its primary attempt to invigorate its agriculture sector, Venezuela was anything but successful. Despite repeated campaign promises by both major parties to revive the farm sector, agriculture remained Venezuela's most neglected economic activity. Though the country had more than enough land to feed its population – and did so in its pre-oil past – there was little appreciable improvement in that regard during the 20-year period. Extensive efforts were made by the government to encourage agricultural development through investments in irrigation and other infrastructure, as well as various input subsidies and price support. Nevertheless, output grew at a moderate pace in good years (1974–8 and 1985–8) and stagnated in the others (1989–90 and 1993–4). This checkered output record was a reflection of misdirected public policies and inefficient private management. Import controls, guaranteed purchase prices, input subsidies and other government incentives encouraged the production of some crops for which Venezuela lacked comparative advantage. Outmoded techniques, small-scale operation and insufficient investment reduced land and labour productivity. After the 1989 reform which called for the reduction of input subsidies, increased import competition and elimination of a preferential exchange rate, a number of other institutional deficiencies (e.g. lack of adequate property and water rights) worked against sufficient investment. In 1994, Venezuela still imported 70 percent of its food. It was the only Andean nation not self-sufficient in coffee. Agriculture still accounted for about 5 percent of GDP, employing some 13 percent of the labour force.

The aim of improving social welfare – the second primary objective – also proved hard to reach. In early years, increased household consumption as a share of GDP, and the overall increase in the share of labour in the total product, indicated some general improvement in domestic welfare. Between 1973 and 1981, households also benefitted from the subsidised prices of major consumer goods. But while subsidies on food and fuels continued during the 20-year period, and wages and salaries were repeatedly adjusted upward periodically through to the end, persistent high inflation eroded the real value of household earnings.

The cost of living index (Caracas) was kept in a moderate range of 7–9 percent a year during the first oil boom, as rising oil revenues allowed larger imports to mitigate inflationary pressures and certain price categories were under various controls. A wide range of agricultural commodities that were subsidised also helped keep inflation down. Sales of petroleum products at prices below production costs and subsidised interest rates were additional factors. During the 1979–80 price liberalisation drive, inflation rose to 21 percent a year due to imported inflation and general wage increases. The 1983 price freeze and the re-imposition of price controls resulted in the 1981–6 inflation coming down to an annual average of about 11 percent, despite upward price adjustments in public utilities, higher oil product prices and the bolivar depreciation. After another currency depreciation in 1986, expansionary demand policies and uncertainties regarding the bolivar's exchange rate helped inflation to surge to a record level of 81 percent in 1989 and to average about 36 percent a year in 1990–3. Economic turmoil in 1994, and new price decontrols, pushed prices up nearly 70 percent in that year as government spending was increased and massive outlays made in the oil sector.

Social welfare was also the victim of considerable monetary and fiscal mismanagement, Venezuela engineered arguably one of the world's worst inequities in the distribution of income and wealth. While reliable data were not always available, there were more than mere anecdotal indications of income gaps. Thus while per capita income in 1981, estimated at $4285, was the highest in Latin America, there was abject poverty too, particularly among slum-dwellers around Caracas, who included some 2–4 million illegal immigrants from neighboring countries in search of jobs and better wages. Nominal GDP per head, estimated at $2250 in 1974, nearly doubled by 1981 but afterwards almost steadily declined to an estimated $2830 in 1994, or much lower than the 1974 level in real terms. By official reckoning, years of low oil prices, political instability, waste, mismanagement and capital flight combined to double the percentage of Venezuela's people living below the poverty line from less than 15 percent to more than 31 percent in 20 years.

The objective of diversification was vigorously pursued but the outcome was highly disappointing. Several state-owned industrial conglomerates in steel, electricity and aluminum were set up with the promise of replacing crude oil as the country's main export by the end of the century. Manufacturing activity increased during the first oil boom, and the sector grew at about 8.5 percent a year on average in real terms. There was an anaemic growth of only 1.6 percent a year during the 1979–83 stagflation period. Manufacturing growth between 1984 and 1994 – modest and uneven as it was – still concealed basic structural deficiencies. Protected by high tariff and import quotas for most of the period under review, this sub-sector increased its contribution to GDP from 10 percent in 1974 to only about 14.5 percent in 1994 despite colossal investment of as much as $100 billion throughout those years. A very high capital/output ratio (attributed to the long-gestation nature of industrial enterprises, poor management and marketing, cost overruns and unused capacity) symbolised the sector's low efficiency. Oil money financed the importation of expensive modern machinery and equipment but the technical and managerial skills needed to keep them functioning optimally were inadequate. A strong focus on metals and petrochemicals as areas where Venezuela was presumed to have comparative advantage proved to be a costly mistake. The state-owned Guyana complex in steel and aluminum was a perennial fiscal haemorrhage.

With much higher output capacity than the domestic market could absorb, much more sophisticated technology than could be locally accommodated and weak management and marketing talent, these resource-based industries proved internationally non-competitive and were particularly hard-hit by the depressed global markets in the early 1980s. The net value-added by the public manufacturing sector did not even cover wage payments in some years. A study devoted to the impact of resource-based industrialisation in some oil-exporting countries has found the Venezuelan case one of the riskiest because of poor timing, bad sectoral composition, little foreign participation to assure foreign market access and an inordinately large scale for the domestic market.[36] Other studies have added the emphasis on short-term financing as another high risk.[37] By press accounts, Venezuela's publicly-owned industrial complexes could not be sold to the private sector under the 1994 privatisation programme for even a small fraction of their total investments.

The objective of increasing domestic employment was modestly achieved during the early years, but the relatively high growth of population and Venezuela's capital-intensive industrialisation strategy, made this goal difficult to reach during most of the 1980s and 1990s. The level of employment increased steadily during the first oil boom, reflecting heightened economic activity and higher demand for labour, particularly in construction and services. Due to economic stagnation and subsequent recession after 1979, unemployment steadily increased to a peak of 8 percent in 1985, largely concentrated in the manufacturing and services sectors. The rise in total employment in the face of output decline pointed to a large reduction in labour productivity, particularly in the services sector. In the second decade, unemployment reached 10 percent. During the entire 20-year period, some 80 percent of the workforce was employed in the private sector, and the informal economy was estimated to account for some 40–50 percent of total employment.

In terms of other conventional objectives, i.e. growth and stability, Venezuela's record was marred by many setbacks. Economic growth was notable in the early years, but became anaemic and haphazard during the 1980s and 1990s. As a result of an ambitious oil-financed investment programme between 1974 and 1978, GDP expanded by 5 percent a year in real terms despite substantial declines in oil output (due to a new conservation policy). However, as annual oil income stagnated, the massive investment outlays were financed increasingly by resort to foreign (short-term) credits. Between 1979 and 1985, the combination of a sharp decline in private investment, a massive outflow of private capital and sizeable losses by non-oil state enterprises, caused the economy to stagnate. In late 1985, a reflationary domestic programme got underway and was maintained in subsequent years through 1988, stimulating an average real growth of 4 percent a year. Following the adoption of IMF's structural adjustment measures, the economy fell into another severe recession in 1989. Helped by the injection of new credits from the IMF and the World Bank, and partly due to the third oil boom, real GDP expanded by nearly 8 percent a year on average in 1991 and 1992. In 1993, a political crisis involving the removal of President Perez on charges of misuse of funds introduced further uncertainties regarding the continuation of IMF-imposed reforms. Since the Venezuelan Congress was unwilling to adopt unpopular measures to harness domestic expenditures, the Central Bank resorted to tightened credit, and the economy contracted by 1.8 percent a year between 1993 and 1994. On the whole, during the 20-year period, Venezuela's real annual GDP growth lagged behind both population growth and expansion of the labour force.

Public sector finances in Venezuela were almost perennially under pressure from start to finish. Public sector revenues rose substantially in 1974–5 but only moderately between 1976–81. Total public sector expenditures, meanwhile, expanded appreciably each year. As a result, the early surpluses disappeared soon after 1976 and the overall deficit reached 9 percent of GDP by the end of the first oil boom. The gap was financed by resort to domestic and foreign financing. The second oil boom temporarily eased the fiscal burden but, by the end of the boom, expenditures exceeded receipts by 3 percent of GDP. In 1982 expansionary fiscal policy designed to revive the economy led to further sharp increases in public outlays and a total deficit of nearly 12 percent of GDP. A reduction in the operating losses of public enterprises, cuts in development expenditures and the devaluation of the bolivar (raising the local currency value of oil income) helped improve government finances dramatically, and the deficit was wiped out in 1984–5. Following the collapse of oil prices in 1986, the situation deteriorated again, particularly because the government embarked on a massive investment programme to reactivate the economy. By 1988, total deficits again reached nearly 9 percent of GDP, financed by running down reserves and periodic currency devaluations. Following a wide-ranging structural programme under IMF tutelage, cutting subsidies, lifting price controls and floating the bolivar, the deficit was brought down slightly. The third oil boom reduced the urgency of fiscal action temporarily, and the proceeds from the sale of the state telephone company and the national airline turned the balance into a small surplus in 1990. In 1991, the balance reverted to a deficit as the government was unable to compensate for a fall in oil income through domestic sources. The gap widened again in 1992 as a result mainly of wage increases, the suspension of earlier price rises of public goods and higher expenditures by public enterprises. The actual deficit in 1993, plus non-payment of government bills, kept the budget shortfall at 5 percent of GDP in 1993. A package of fiscal measures under the so-called Sosa Plan was approved by the parliament in early 1994 but, because of the government bail-out of the banking system, the fiscal deficit again eclipsed all previous records.

Venezuela's balance-of-payments position fluctuated widely between 1974 and 1994. During the first two oil booms, high oil income financed rising imports and net capital inflows kept the overall balance in the black. But after the 1982 Mexican debt crisis, it became clear that Venezuela also was unable to service its large short-term debts obtained between 1974 and 1979 and that the prevailing exchange rate was not sustainable. Thus, the overall balance registered a deficit as foreign banks curtailed their lending to the public sector and cut export credit lines. Private capital continued to leave the country in early 1983. External debt at the end of 1983 was estimated at $35–38 billion (or 60 percent of GDP) compared to about $7 billion in 1978. Exchange controls and a moratorium on public debt payments in 1983 reduced imports and slowed down capital flights. The oil price collapse in 1986 led to the intensification of exchange restrictions rather than the more appropriate currency devaluation or a reduction in spending. Instead, the government tried to boost the economy through increased public outlays. Capital flights were resumed and BOP deteriorated. With the floating of the bolivar in 1989, trade deficits turned into a surplus once again and capital flights were halted for two years. Trouble emerged in 1992 as the oil price continued its slide and imports rose. The three-year surplus turned into a deficit. Private capital inflows were dramatically reduced after the February attempted coup. Exchange controls were reimposed in 1993 and carried on into 1994, reducing the BOP deficit.

Finally, the objective of effective diversification was far from achieved. The composition of GDP during the 20-year period showed notable changes in some sectors, but these changes did not appreciably reduce the country's reliance on petroleum. In 1974, the oil sector accounted for 40 percent of GDP, industry (including mining, manufacturing, construction, water and power) for 23 percent, agriculture 5 percent, and services 32 percent. As the cornerstone of the Venezuelan economy, petroleum exports accounted for 97 percent of total exports and two thirds of the central government revenues. In 1994, petroleum constituted about 22 percent of GDP, nearly 83 percent of total exports and 64 percent of public sector revenues. Industry was responsible for 23 percent, agriculture 5 percent and services 50 percent. Some of these changes, notable as they were, somehow concealed the absence of self-sustained diversification. In the oil sector, for example, daily output had been on the decline during the 1970s, due to the exhaustion of the old oil wells and the government's policy of oil conservation. By 1983, the initial policy of reducing reliance on oil was clearly reversed, as the authorities declared their intention to expand crude oil production towards 3.6 mb/d by the year 2002. The construction industry, which doubled its contribution to GDP from 4 percent in 1974 to 8 percent in 1994, was, except for public infrastructural outlays, engaged mostly in the development of unsustainable private office buildings, commercial shopping centres and luxury residential housing. The rise in the share of services at the expense of other productive sectors was also not necessarily an indication of improved sectoral transformation. With respect to the larger contribution of non-oil exports, too, the figures were not indicative of a much brighter picture as the bulk of non-oil exports (aluminum, steel and iron ore) enjoyed 30 percent export subsidies, and the public enterprises which exported them were themselves money-losers.

NOTES ON CHAPTER V

1   This section on Algeria, for the early years, draws on Patrick Conway's 'Algeria' in Alan Gelb (1988). A thorough examination of Algeria's policies and performance can also be found in *Algeria: Country Economic Memorandum* (Washington: The World Bank, 1994).
2   J.C. Randal, 'Algeria ...', *Washington Post*, 9 March 1983.
3   For a further detailed discussion of Indonesia's achievements, see Philip Barnes (1995).
4   For a more detailed discussion of the early period, see Bruce Glassburner's 'Indonesia' in Alan Gelb (1988), pp 222–3.
5   For details of Iran's objectives and performance before and after the 1979 revolution, see Jahangir Amuzegar (1977) and (1997).
6   See Kamran Mofid (1990).
7   UN Security Council, S/23322, dated 24 December 1991.
8   Cf CARDRI (1989); and *Economic Report: Iraq* (London: Lloyds Bank, 1983).
9   Some private Arab sources claim that true inflation amounted to 95 percent in 1980, 140 percent in 1981 and to as much as 370 percent in 1988. See Alnasrawi (1994), p 104.
10 *The Middle East*, February 1996, p 15.
11 *Middle East Economic Digest*, 24 June 1994.
12 The figures for post-Iran/Iraq war are based on UN *National Accounts Statistics 1992* (New York, 1993).
13 K.A. Chaudry, 'On the Way to Market', *Middle East Report* (May-June 1991).
14 Abbas Alnasrawi (1994), p 122.
15 See M.W. Khouja and P.G. Sadler (1979), Chapter 8.
16 Due also to the large volume of returns on foreign investments, Kuwait's GDP and GNP exhibited much greater differences from one another than was the case with most other OPEC members, as did differences between respective per capita. Thus, while GDP peaked in 1980, GNP's peak was reached in 1989.
17 The government later agreed to bail out thousands of 'small' investors to prevent widespread bankruptcies.
18 Annual surpluses include foreign investment incomes that are excluded in Kuwait's official annual budgets. Without these incomes, the formal budget showed perennial deficits from 1982 on. After 1987, these deficits were financed by the issue of treasury bonds and bills.
19 Part also was due to a shrinkage of capital reserves since the war which, according to private estimates, had dwindled from $100 billion in 1990 to about $30-$50 billion in 1994, and not all liquid.
20 Due to a number of political and economic factors (including early years' low official ceiling imposed on interest rates) there was a significant volume of annual, informal, private capital outflows (as much as KD1 billion/$3.5 billion in 1981). These transactions were routinely reflected in large items in the balance of payments account as 'errors and omissions'.
21 See Douglas Jehl, 'In Kuwait, Work is a Concept that Is Foreign', *New York Times*, 24 September 1996.
22 By some private estimates, the actual price of a ton of wheat produced in Libya in 1980 was about $1350 compared with the European market price of about $240.
23 *Libya: Country Report* (London: The Economist Intelligence Unit, 3rd Q. 1995). In 1994, the General People's Congress established 'clean-up committees' to look into the origin of wealth and incomes of Libyan citizens suspected of accepting bribes or of misuse of public funds. As in similar edicts in other member countries, the work of the committees produced little, as some of their own members close to the ruling elite qualified for scrutiny.

24 By OPEC Secretariat's calculation, Libya's per capita income in 1994 was $4,500 compared with $5,630 in 1974 in current dollar terms. Other organisations give different 1994 figures.

25 This section draws for the early period on Henry Bienen, 'Nigeria', in Alan Gelb (1988) and for the subsequent years on Rashid Faruqee, 'Nigeria', in *Adjustment in Africa* (Washington: The World Bank, 1994).

26 Cf Brian Pinto, 'Nigeria during and after the Oil Boom: A Policy Comparison with Indonesia', *World Bank Economic Review*, No. 3, 1987.

27 For details, see M.G. Nehme, 'Saudi Development Plans...', *Middle Eastern Studies*, July 1994; and *OPEC Bulletin*, May 1987, p 3.

28 The figure for 1994 is based on the 1992 official census which indicates a population of 16.9 million of which 12.3 million were Saudi nationals. The census figure, however, has been greeted with great scepticism by independent analysts who put the figures close to 12 million and 8 million respectively.

29 Due to an antiquated legal framework, considerable lack of transparency of ownership within the Kingdom and the highly secretive operation of the Saudi Monetary Agency, the government's foreign assets and exchange reserves were always subject to guesswork and informal estimates. By the end of the second oil boom, the Kingdom's total foreign assets were estimated by Western banks at between $160 billion and $180 billion. Before the Kuwait war, the Agency reportedly could count on assets of around $120 billion as Riyadh had paid for the Iran/Iraq war's cost. At the same time, wealthy Saudi nationals reportedly had stashed unrevealed billions of dollars in private overseas accounts. Total foreign assets reportedly stood at about $70 billion.

30 Of the Kingdom's estimated $120 billion GDP in 1994, an estimated $16 billion consisted of remittances by some 4 million foreign workers.

31 At the peak of the wheat programme in 1992 the Saudi Grain Silos and Flour Mills Organisation reportedly paid producers $1000 a ton for a crop that was valued at $150 in the United States.

32 For a fascinating account of Saudi Arabia's water resources and their use, see Caroline Montagu (1994).

33 *Saudi Arabia: Country Profile 1995-96* (The Economist Intelligence Unit, 1996).

34 See Daniel Pearl, 'Short Circuits', *Wall Street Journal*, 20 August 1996.

35 Little information was published on population and employment in Saudi Arabia. Periodic population censuses were believed to be tainted with various political considerations. The figures here should be treated with caution.

36 Richard Auty, 'The Internal Determinants of Eight Oil Exporting Countries' Resource-Based Industry Performance', *Journal of Development Studies*, April 1989.

37 François Bourguignon (1988), p 322.

VI

Cross-Country Performance

As Chapter V showed in detail, OPEC members' successes and setbacks in achieving their national socio-economic objectives differed widely. In certain identifiable areas such as physical and human capital formation, output growth, social welfare, price stability, fiscal management and external equilibrium, the differences were highly revealing and easily comparable. In less tangible categories of equity, social cohesion, political development and prospects for future development, the results were equally varied but less clear and quantifiable. The variety of experiences among members reflected not only their initial levels of economic development, their different resource endowments and their exogenous circumstances but also the different mixes of their chosen growth strategies and economic policies.

Based on the country report cards detailed in Chapter V, the first part of this chapter attempts to make a cross-country appraisal of the members' performances with respect to the common national objectives. Attention will be focused on infra-structural improvement, GDP and per capita growth rate, equitable distribution of income, internal and external stability, non-oil export promotion, national security, assistance to other developing countries and care of the environment. On the strength of these findings, the latter part of the chapter will endeavour to identify cases of relative success or setback among members. At the end of the chapter, the members' ranks in an overall development index suggested by the United Nations Development Programme will be compared.

PHYSICAL AND SOCIAL INFRASTRUCTURE

Investments in infrastructure (roads, railroad tracks and power-generating capacity) were massive in all group members, but the relative magnitude of improvement produced was markedly different and the quality of services created was not uniform by international standards. In the two decades between 1970 and 1990[1], paved roads increased nearly six times in Gabon, Indonesia and Iran, five times in Iraq, two times in Ecuador and Nigeria, but far less in Venezuela and Algeria (Table 1). In railroad tracks, Gabon was first in terms of extension, followed by Saudi Arabia, Venezuela, Algeria, Iran and Indonesia in descending order. Power generating capacity was enlarged by a spectacular 58 times in Saudi Arabia, 13 times in Iraq, 12 times in Indonesia, eight times in Iran and nearly six times in Gabon, Algeria, Ecuador, Nigeria and Venezuela. As a result, energy consumption per capita surged nearly five times in Indonesia and Saudi Arabia, three times or more in Algeria and the UAE, more than double in Ecuador and Iran. Telecommunication facilities were also expanded markedly, although data for comparison over the 20-year period are not available. Except for Nigeria, Indonesia and Gabon, which lagged behind, main telephone lines per capita among group members exceeded the average for all LDCs by a factor of two to eight. But, among those countries which compared well with the developing world average, waiting time to obtain a telephone connection was often long – as long, for example, as 10 years in Algeria. Television sets per 1,000 people in 1992 among group members were also noticeably more numerous than the average for all developing countries, particularly in GCC countries where ratios were between two and three to one. Only four members fell behind the average (Table 8). Similar advances could be observed with regard to the number of radio sets per capita.

Group members allocated a greater share of their national income to education and health than any other developing region. Improvements in social indicators were consequently dramatic. Large portions of GNP were devoted to education among nearly all group members, particularly Algeria, Saudi Arabia, Kuwait, Iran and Venezuela – much more than the developing world average. Education spending per student as a percentage of GNP per capita was highest in Kuwait, followed by Saudi Arabia, Qatar, Iran, Algeria, Venezuela, Ecuador, Indonesia and the UAE in descending order. Within the 20-year period, average adult literacy for the group improved from 47 percent in 1974/5 to 72 percent in 1993/4. School enrolments at all levels of education went up remarkably, and at certain levels by a factor of four and five in certain group members (Table 7).

Nevertheless, the results were not commensurate with big spending. On the whole, educational systems did not prepare graduates with the necessary skills, sufficient discipline or the right attitude for the available jobs. The poor returns on human capital investments were evidenced by high school dropout, high unemployment and low labour productivity at all levels and all round. The concentration of unemployment among school graduates and inadequate international competitiveness in non-oil exports were the outcome. This was particularly true in Algeria, Ecuador, Iran and Venezuela, but not limited to them. Making huge investments in human knowledge and skills unrelated to domestic market needs and linking years spent in school to guaranteed high-wage jobs in the public sector were the cause of both low overall productivity and unemployment among first-time job seekers, particularly in Algeria, Iran and the Persian Gulf members.

High investments in health also produced remarkable results. A child born in 1994 in Indonesia could expect to live 15 years longer than his or her parents; in Saudi Arabia the figure was 25 years. Infant mortality declined more than 50 percent in Indonesia, Kuwait and Libya and more than two-thirds in Iran, Qatar and the UAE. Iraq was the only exception, due to the imposition of UN sanctions. Other health-related indicators of development also showed substantial improvement for all group members for which statistics are available. People's access to safe water was particularly enhanced in countries with limited initial access, e.g. Indonesia, Ecuador, Iran and Saudi Arabia. Population per physicians and nurses was reduced remarkably in all group members, particularly in Saudi Arabia, Algeria, Nigeria, Gabon and Kuwait, in that order. Public health expenditures as a percentage of GDP also went up, particularly in Saudi Arabia, Algeria, Indonesia, Nigeria, and Iran (Table 6).

REAL OUTPUT AND PER CAPITA GROWTH

In economic growth, OPEC members as a whole had perhaps the least expected, and most ironic, performance. As Table 10 shows, despite enormous and unprecedented revenues from oil exports, the estimated average annual real growth of GDP in all member economies (except for Indonesia) between 1973 and 1974 was lower than their annual growth rate during the 1960–73 period, before the first oil price explosion. During the first oil boom, GDP growth rate in the capital-surplus oil exporters (due partly to rising receipts from overseas portfolios) was quite notable with Saudi Arabia occupying the first rank, followed by Kuwait, Libya and the UAE. Among the high-absorbing, capital-short members, Algeria and Iran were the star performers, followed by Indonesia, Iraq and Nigeria. Their performance compared with those of Ecuador and Venezuela. Gabon was the only laggard.

During the second 10-year period following the second oil boom, and coinciding with the depressed oil market of the 1980s, Indonesia was the only member showing a solid performance, almost matching its record of the earlier periods; its yearly growth rate was double and triple the rates of its closest competitors – Algeria and Ecuador. Between 1990 and 1994, coinciding with the third oil boom, Indonesia again was the best performer. The cessation of war with Iraq and some needed reforms in Iran also helped improve the Islamic Republic's economic achievement. A series of structural adjustment programmes in Ecuador and Nigeria were instrumental in raising their annual growth rates substantially, as was the case in Venezuela. In stark contrast, Gabon slipped farther down the growth ladder. Iraq uniquely suffered from its misadventures and subsequent UN sanctions. The Persian Gulf war also badly hurt Saudi Arabia as reduced earnings from overseas investments affected its growth. Libya suffered from UN sanctions, and Kuwait rapidly rose from the ashes of war.

During the 20-year period, Indonesia was the only member with both high and sustained annual growth, exceeding all high-absorbing members by substantial margins and even surpassing the best performance by the capital-surplus countries. Venezuela and Gabon were among the growth laggards, while Iraq was virtually devastated under international embargoes. Kuwait's performance was exemplary among the low-absorbers, while Libya and Qatar were the least successful in achieving even minimal growth. All capital-surplus, low-absorbing countries showed unmistakably that the key to economic growth and prosperity was not in the availability of capital but in its productive use.

Detracting further from the unflattering annual GDP growth rates in almost all group members were the relatively high rates of population growth. Between 1974 and 1994, the average annual growth rates of population in the world's low and middle-income countries ranged from a low of 1.7 percent a year to a high of 2.5 percent, and those in high-income countries from 0.6 percent to 0.8 percent. In the same period, the corresponding rates for eight OPEC members in the low and middle-income category stood between 2.5 percent and 4.8 percent. In another two countries in the same low and middle-income group, the figures were between 1.5 percent and 2.3 percent. The three high-income economies had a growth rate as high as 10.2 percent a year largely due to the inflow of guest workers. Between 1974 and 1994, OPEC's population grew nearly 58 percent, or at an average annual rate of 2.9 percent, well exceeding the 1.8 percent for all developing countries. At the same time, the workforce rose less rapidly (Table 2), raising the national welfare bills.

These relatively high population growth rates had different root causes. In some countries such as Algeria, Libya and post-1979 Iran, a larger population was encouraged as a matter of socio-political ideology. Iran also had its disproportionate influx of war refugees from Iraq and Afghanistan. The high-income, labour-short countries in the Persian Gulf followed an extremely liberal immigration policy, and imported large amounts of foreign labour. In all countries, significant improvements in nutrition and health care helped increase fertility, reduce infant and adult mortality rates.

WEALTH, POVERTY AND INEQUALITY

A rapidly rising population in almost all group members, combined with a relatively modest GDP increase in most, naturally resulted in a slow increase, or an actual decline, in per capita real income. Ten members experienced a fall in their living standards in at least half of the 20-year period. Only Indonesia and Ecuador managed to avoid this setback. Real per capita incomes in Iraq, Kuwait and Venezuela during the 1990s were equal to the levels prevailing in or before 1960. Libya and Saudi Arabia had reached their highest real per capita incomes in the 1960s, Algeria, Gabon, Iran, Nigeria and the UAE in the 1970s and Ecuador in the 1980s. Indonesia was the only group member with a per capita real income in the 1990s higher than ever before.[2] Per capita income in Iran from 1980 and in GCC countries after 1985 declined or remained stagnant up to 1994. As a result of falling growth rates and rising population, many group members failed to eradicate absolute poverty – defined internationally as an average consumption of $1 per person per day in terms of the

1985 US dollar. Even Indonesia was still among the ten developing nations with the highest number of poor people in its midst, i.e. 14.5 percent of its population in 1990–3. Ecuador and Nigeria shared a similar dubious distinction with Indonesia (having nearly 30 percent of their populations below the international poverty line). Nigeria was also among heavily indebted poor countries in 1994.[3]

Poverty in all group members reflected unemployment or under-development, insufficient or unusable education and poor health conditions. Jobless rates for most group members were unavailable or unreliable for most years. With the growth of population, the active labour force increased but, as a percentage of total population, the workforce in all members (except Qatar and the UAE, who relied on expatriate labour) was still considerably lower than the average for all developing countries (Table 2). Part of the reason was that composition of the population was skewed toward youth due to a very rapid rate of population growth during the 1970s and 1980s. Despite the relatively low ratio of the labour force to total population, the official unemployment rate for Algeria, Ecuador and Iran was in double-digits, and for some others rather high too. Furthermore, up to 40 percent of the labour force in Algeria, 36 percent in Venezuela and 30 percent in Iran was considered underemployed. In Saudi Arabia, some 30 percent of the nation's high school and university graduates were unemployed, even though Saudi citizens held fewer than 9 percent of private sector jobs. Altogether, according to World Bank estimates, the unemployment rates in the Middle East and North Africa region (excluding the GCC countries) during the early 1990s were the highest in the world.[4]

Income maldistribution and inequality, poverty depth and poverty severity differed among group members but affected most. As Table 11 indicates, the Gini index in countries for which data were available showed Venezuela as the country with least income equity, followed by Ecuador, Algeria, Nigeria and Indonesia in that order. Furthermore, the ratio of income share of the highest 10 percent of households to the lowest 10 percent in the latest years for which data were published ranged from 30.5 percent in Venezuela, signifying the largest income inequality, to 6.6 in Indonesia, reflecting the smallest. Income distribution also became more unequal in Venezuela between 1970 and 1990 while it improved considerably in Indonesia between 1976 and 1993. Countries for which data are not available did not necessarily fare better. While absolute poverty must have been eliminated in Kuwait, Qatar, Saudi Arabia and the UAE due to extensive state subsidies (particularly to citizens), income inequality might have increased. Among other members, the picture was mixed. An internal World Bank report, for example, shows that while only 6.5 percent of Iran's population lived in absolute poverty in 1985, the figure rose to nearly 9 percent in 1990 before falling to about 7 percent in 1994.[5] People below the 'poverty line' as defined by the country itself accounted for a much larger percentage.

Another phenomenon generally observed but inadequately quantified was the widening gap between rural and urban wages and between the wages of urban employees in the formal and those in the informal economy. Minimum wage legislation, labour-union pressure and other political factors helped to raise the take-home pay of workers in the modern urban sector, while the majority of the workforce engaged in peasant agriculture or in the informal urban services sector was left behind.

This was particularly true in Indonesia, Iraq, Nigeria and Venezuela. Chronic structural unemployment and under-employment were again the root causes of income inequality.

INTERNAL BALANCE: PRICE STABILITY AND FISCAL DISCIPLINE

The maintenance of internal balance – price stability and budgetary discipline – varied considerably among group members and over time. As a rule, inflation was subdued in the small Persian Gulf members which pursued relatively stable currency and liberal trade regimes. In contrast, countries with multiple currency rates and trade restrictions experienced high domestic price inflation. Not only Iraq, which created hyper-inflation in the aftermath of post-1990 UN sanctions, but Ecuador, Venezuela, Nigeria, Iran, Algeria and Libya all underwent annual double-digit hikes in domestic consumer prices for almost the entire period (Table 12). For all of these countries, also, inflation accelerated in the ten years after 1985 compared to the previous ten years.

Almost the entire membership also incurred budgetary deficits year after year. Algeria topped the list with a fiscal gap for every year under review, followed in terms of frequency by Iran, Indonesia, Nigeria, Saudi Arabia, Ecuador, Libya and Qatar all of whom had budgetary shortfalls in more than two-thirds of the period. As can be seen from Table 13, even the rich Persian Gulf sheikdoms could not escape the red ink. Kuwait could claim to be the least affected, thanks to its prudent provision of income from foreign assets. A combination of rising social-welfare expenditures, bloated bureaucracy, a limited tax base, project cost-overruns and large military outlays was responsible for the deficits.

The budget revenue's dependence on oil and gas income also continued to loom large, exposing government expenditure and public investments to the vagaries of the global oil market. This dependence in the case of Kuwait, Nigeria, Qatar and the UAE exceeded 80 percent on average. For Iran, Saudi Arabia and Venezuela, it was about 70 percent or more. Libya relied on oil income for nearly 60 percent of its total fiscal receipts. Algeria, Ecuador and Gabon had nearly 50 percent of their fiscal revenues from oil and gas. Only Indonesia managed to keep its need for oil income below 40 percent.[6]

EXTERNAL BALANCE AND FOREIGN DEBT

With respect to external balance, group members experienced various degrees of setback. While OPEC as a group had once been the only capital exporter among all developing regions, its resource deficit on goods and services became one of the largest of all such regions by 1994. Even in the GCC countries, where 'saving abroad' was one of the most significant balance-of-payments features, overseas assets began to decline, particularly after the Persian Gulf war. Net interest income from foreign deposits and investments fell from more than 10 percent of GDP in 1985–9 to less than 5 percent in 1990–4. With net national savings rates (defined as gross national savings rate minus capital depreciation minus depletion of non-renewable resources) perennially negative, many group members had to resort to foreign credits for domestic investments. And with real domestic interest rates generally kept low, and often negative (i.e. below the inflation rate), a good deal of capital left the countries for more remunerative opportunities abroad. By World Bank estimates, the stock of group members' capital abroad was larger than that of any other region as a share of GDP. The countries with the least hospitable environments for private investment, as measured by capital flight, were Iran, Iraq and Venezuela all in contrast with Indonesia up to 1994. Political stability, the rule of law and policy predictability were often cited by private investors as their major concerns.[7]

As Table 14 indicates, all group members experienced an overall balance-of-payments deficit some of the years. All but Kuwait and the UAE had a deficit on their annual current accounts in most of the period. Some members even ran into balance-of-trade gaps in several years. Ecuador, Nigeria, and Gabon were deficient in their overall payments balance two-thirds of the time. Algeria, Iran, Iraq, Libya, Saudi Arabia and Venezuela had such shortfalls in half of the period. Only Qatar and the UAE managed to avoid external gaps for most of the period (Table 14).

As a result of consecutive deficits in the external balances over the years, the central government or the public sector as a whole in all group members went into external debt. While foreign debt in 1970 was negligible among such members as Algeria, Indonesia, Iran, Iraq, Nigeria and Venezuela, these countries all joined the ranks of the heavily-indebted by 1994. Except for Iraq, whose external debt was mainly due to its military adventurism, the figures for most others represented foreign borrowing for project financing, imports of industrial inputs or military purchases. In almost all group members, the size of the external national debt, and therefore the country's overall indebtedness, was exaggerated by not taking into account sizeable *private* assets held abroad.

SECTORAL CHANGE AND NON-OIL EXPORT PROMOTION

In the area of diversification, which was invariably and by far every member's primary goal, the experience was vastly different and, by and large, not up to aspirations. The share of oil sector in GDP was reduced in all group members. Best performers in reducing their reliance on the oil sector were the oil exporting countries in the Persian Gulf plus Libya, where such dependence was cut by more than half, and Venezuela which reduced it by 45 percent (Table 4). In all these countries, however, the reduction in the share of petroleum was made up by sharp rises in the share of the services sector, e.g. more than 600 percent in Qatar and more than 100 percent in Saudi Arabia, Kuwait and the UAE. The increase was about 50 percent in Gabon, Libya and Venezuela. Agriculture lost its share in Ecuador, Gabon and Indonesia but retained it in Algeria and Venezuela, and gained substantial ground (albeit at very high costs) in Saudi Arabia, Libya, the UAE, Qatar and Kuwait. The industrial sector, which was the target of diversification efforts in all group members, had its share raised in all countries except Nigeria and Venezuela. But the largest share of this sector in GDP, belonging to Indonesia at 30 percent, was still less than one-third of the total; for most, it was 24 percent or less, and in some only about 11 percent. In the major oil exporters of the Persian Gulf, the industrial sector included petroleum refining and petrochemicals which were basically dependent on crude oil and gas (Table 4).

Changes in the allocation of the workforce were also highly noteworthy in all group members. The share of the labour force in agriculture fell dramatically in Algeria, Iraq, Qatar and the UAE; the decline was 50 percent or more in Kuwait, Libya and Venezuela. The share of industry in labour employment went up markedly in Algeria, Gabón, Indonesia and Libya, but it actually declined in Kuwait and Nigeria; and it remained virtually the same in Ecuador, Iran and Venezuela. The net gain in employment occurred in the services sector, ranging from highs of 150 percent in Nigeria and 94 percent in Gabon to lows of 20 percent or less in Kuwait, Qatar and the UAE (Table 2).

The least satisfactory achievement in the diversification efforts was in reducing dependence on oil exports as the main source of foreign exchange receipts, a mainstay of government finance and a share of GDP. Improvements in the shares of agriculture and industry were translated in import substitution but not much in industrial or commercial farm exports. Depending still on oil and gas exports for about 90 percent or more of total foreign exchange earnings in 1994 were Gabon, Kuwait, Libya, Nigeria and Saudi Arabia. Dependence for more than 80 percent of foreign currency income continued in Iran, Iraq and Qatar. Algeria and Venezuela received more than 70 percent of their foreign exchange income from oil exports. Ecuador and the UAE were dependent on less than 50 percent. Only Indonesia managed to reduce such reliance to about 15 percent (Table 5).

At the same time, the fact that non-oil exports were lagging in the face of still significant dependence on imports did not augur well for economic viability once oil exports were gone. With the exception of Indonesia and the UAE, which could pay for all or a substantial portion of their import needs out of the proceeds of their non-oil exports or re-exports, no country could afford to pay even half. In 1994, merchandise imports in Algeria, Gabon, Iran, Qatar and Venezuela were about three to four times that of non-oil exports, in Saudi Arabia and Kuwait five to six times, in Iraq, Libya, and Nigeria seven times or more (Table 5).

The diversification process among the group members was held back by no other factors as much as a poor human-resource base, lack of indigenous technology, mismanagement of the export proceeds and the pursuit of wrong macroeconomic policies. Export diversification required appropriate technology and skilled labour so as to enhance domestic production capacity. To achieve this, technology transfer from abroad played a supplemental role but could not substitute for domestic research and development efforts.[8]

NATIONAL SECURITY AND MILITARY BUILD-UP

None of the group members publishes data on its military outlays and arms trade. Data on military expenditures from multicountry statistical sources are also limited in transparency and comprehensiveness. Information from available sources, however, shows that military spending as a share of GNP in the Persian Gulf segment of group members was the highest in the world.[9] Part of the large expenditure was related to the Arab-Israeli conflicts but a large portion of the arms build-up was related to internal civil strife or disputes among members themselves (Table 15). During the two decades under review, Iraq, Saudi Arabia, Kuwait, Libya, Qatar and the UAE had the largest shares of GNP devoted to military spending. In contrast, defence spending as a share of national income was the lowest in Venezuela, followed by Nigeria, Indonesia, Ecuador, Gabon and Algeria in ascending order. Thus, while the share of the group members among the GCC countries in the early 1990s was two to ten times that of the average for all developing countries, the ratios for Indonesia and Venezuela were half as high, and that for Nigeria one fourth. Iran had a relatively large share during the war with Iraq but Iraq had persistent, high military expenditure per capita relative to all members. By US official estimates, OPEC members as a whole spent about 10.4 percent of their GNP on defence in each year between 1984 and 1994, as compared with 3.3 percent in the OECD countries. OPEC members' annual share of military expenditure in total central government expenditures was also three times as much.

According to a UN report, nearly 60 percent of weapons importers worldwide do not participate in the UN Register of Conventional Armaments. Among the reluctant nations, Iran, Kuwait and Saudi Arabia are cited by name. Among countries cited for having purchased expensive arms despite having more desperate needs, Iran is reported to have bought two submarines from Russia 'at a cost that could have provided essential medicines to the whole country many times over'. Nigeria is also cited for having purchased 80 battle tanks at a cost that could have immunised all of its two million unimmunised children and provided family planning services to nearly 17 million of the more than 20 million couples lacking such services.[10]

ASSISTANCE TO POORER COUNTRIES

While all OPEC members insisted in calling themselves oil-exporting developing countries faced with the usual trials and tribulations of economic development, a decision was made by all, shortly after the 1973/4 oil price rise, to reduce the increased oil imports burden of the Third World's oil-less nations through an increase in compensatory aid. The oil-importing developing countries wanted to buy oil at a discount but OPEC members preferred to charge everyone a uniform price and somehow subsidise those who could not afford to pay the full cost. Financial assistance to the needy countries was carried out through (1) direct bilateral channels (2) each individual country's national aid organisation and (3) multi-national financial institutions such as the World Bank, the International Monetary Fund and various

UN agencies. Kuwait was the first OPEC country to set up its own aid agency, the Kuwait Fund for Arab Economic Development. Other national units established later include the Abu Dhabi Fund for Arab Economic Development, the Venezuelan Investment Fund, the Organisation for Investment, Economic and Technical Assistance of Iran and the Libyan Arab Foreign Investment Company. Eight other development aid institutions, financed wholly or largely by OPEC members, include: The Arab Fund for Economic and Social Development, the Arab Bank for Economic Development in Africa, the Islamic Solidarity Fund, the Arab Fund for Technical Assistance to African and Arab Countries, the Islamic Development Bank, the OPEC Fund for International Development and the Arab Gulf Programme for the UN Development Organisations. The OPEC Fund for International Development is the only aid agency representing all OPEC members.[11]

Between 1974 and 1994, total commitments by OPEC aid institutions reportedly amounted to $58 billion, out of which some $41 billion was disbursed to nearly 130 countries in Africa, Asia, Latin America and the Caribbean countries. Members' total official development aid is shown in Table 16. As that table shows, Saudi Arabia was by far the largest donor among aid-giving members, followed by Kuwait and the UAE. Six other countries' total contributions amounted to less than $2.5 billion by comparison. Three countries who were recipients of aid from various sources were not among the aid donors. As a percentage of GNP, Saudi Arabia was also number one, followed by Kuwait throughout the period. But, these countries' foreign assistance, while it was not always officially acknowledged, also included non-economic aid (e.g. military assistance to Iraq in its war with Iran, and similar assistance elsewhere).

CARE OF THE ENVIRONMENT

Due to the ease with which the windfalls were collected, the group members' governments viewed the oil and gas revenues as costless resources that could be redistributed at will among the people. Much of the oil rents was spent on consumer price subsidies on goods and services (fuel products, housing, utilities and public services). Much was also devoted to product subsidies (electricity, irrigation water, fertilisers, transport and communication). A large part of the oil rent was invested in public industrial enterprises that operated under little or no profit constraints and scant international competitiveness. Subsidies in some countries (e.g. Saudi Arabia, Kuwait, other GCC countries and Iran) ran as high as 10–20 percent of GDP in some years.

According to an internal World Bank study, the free-for-all redistribution of the proceeds of non-renewable resources through consumer and producer subsidies involved high costs. It discouraged conservation, encouraged wasteful consumption and inhibited faster growth. The expenditures incurred by the subsidies had to be made up by distortive taxes on trade and capital in some countries in order to finance regular government spending. The marginal cost of the subsidies thus consisted of (a) the efficiency loss associated with the distortive taxes (b) efficiency loss associated with distorted high consumption patterns and waste of subsidised goods and services and (c) adverse environmental impacts in the form of high levels of air pollution from high-sulphur fuel oil, fresh water depletion, deforestation, soil and pasture erosion, and water pollution from agricultural chemical runoffs.[12]

With regard to these environmental and resource-base considerations, group members' experiences also noticeably differed. Reserves of non-renewable resources (e.g. oil, water, fertile soil) fell in either absolute or relative terms in some countries Water was tapped unsustainably at rates that exceeded 100 percent of renewable supplies in Kuwait, Libya, Qatar, Saudi Arabia and the UAE. The drive for food self-sufficiency for rapidly increasing populations – combined with low or zero irrigation water prices, agrochemical subsidies, guaranteed price supports and limited environmental planning – put unbearable pressure on the quantity and quality of surface and groundwater resources in these countries. According to private estimates, between 1960 and 1990, renewable water resources per capita fell by 57 percent in Algeria, 63 percent in Iran, 64 percent in Iraq, 72 percent in Libya and Saudi Arabia, and 94 percent in the UAE. The largest share of water use was in agriculture, ranging from a low of 74 percent in Algeria to a high of 92 percent in Iraq and Saudi Arabia.[13]

COUNTRY-SPECIFIC COMPARISONS

Not all thirteen OPEC members' experiences during the 1974–94 period can be regarded as typical examples of oil-backed economic development. The exogenous, catastrophic and protracted events in Iran and Iraq, for example, badly distorted their economic activities and performance. These two countries would have been best candidates for investigation as they possessed both oil and non-oil resources, large land areas, indigenous labour forces and temperate climates. But the revolution and the war made their experience totally atypical. Smaller oil-rich countries of the Persian Gulf – Kuwait, Qatar and the UAE – were constrained by inclement weather and scant non-oil resources (including labour), and thus could not qualify as typical models. Much the same could be said about Libya. Ecuador and Gabon had good opportunities for development but lacked much oil. This would leave only the following five countries for a comparative examination of the impact of oil wealth on domestic development paths.

Algeria had an unenviable record in almost all major indicators of economic progress – growth, high employment, internal and external stability, and diversification. Setbacks could be traced directly to its flawed economic strategies and management inefficiencies. Under Houari Boumedienne's socialist, statist, interventionist and anti-market ideology, Algeria wasted an enormous chunk of its oil and gas reserves in building up an economic structure doomed to failure. This ideology involved an inexcusable neglect of prosperous pre-independence agriculture, the establishment of mammoth capital-intensive, high-cost, and bureaucratised industries, hostility to foreign private investment and defiance of international advice in the name of national pride and sovereignty. Refusing to seek IMF assistance until April 1994, and engaging in only a cosmetic structural adjustment programme as late as in 1989, without allowing required market-based changes in the exchange rate or the interest structure, the Benjedid administration brought the economy to the brink of

bankruptcy. And when the half-hearted, post-1989 reforms failed to produce immediate results, the thirty-year experience with failed socialism was papered over and
the country's destiny again placed in the hands of an old FLN stalwart in 1993. For
a brief and disastrous period, state and trade monopolies were reinstated, agricultural
privatisation was slowed down and controls were resumed.[14]

In the aftermath of an eight-year bitter and bloody war of independence from
France, a one-party socialism, dominated by a small, interconnected, elite of some
5,000 triumphant veterans (out of a population of 16 million) was responsible for the
country's sclerotic economy, bloated public sector and widespread corruption. By
another estimate, a group of 'not more than 200 people' (including top army generals,
professional politicians, high-level technocrats and their close collaborators) essentially ruled Algeria after independence.[15] As a consequence, Algeria ended up in 1994
with tepid growth, double-digit inflation, and budget deficit, money-losing public
enterprises (including the mammoth national oil company SONATRACH), an
external debt of $30 billion, an inefficient old-guard bureaucracy determined to retain
power and resist reforms, and a protracted cycle of turmoil, violence and terror.

Between January 1992 and May 1994, Algeria had three presidents and four
prime ministers. The government was faced with a catch 22. It was widely believed
that new foreign credits and investments would help revive the lagging economy and
that an economic upturn would undermine the political momentum of the Islamic
fundamentalist opposition (FIS). But it was equally evident that, in the political and
economic crises that gripped the country, there could be little hope for such necessary capital inflows. Foreign economic assistance was contingent upon Algeria's having
a clear programme of economic reform and restructuring. But no such programmes
could be properly executed when the country's policy-making apparatus was virtually
paralysed by political infighting, lack of recognised leadership, terrorist attacks on
security and professional forces, and a crisis of confidence.[16]

Indonesia's economic performance over the 1974–94 period was respectable,
steady and in line with its basic national priorities. This performance, particularly in
the face of the adverse oil market conditions of the 1980s, won plaudits from the
World Bank and other lending institutions. Thanks to its market-oriented reforms
after the mid-1980s, Indonesia's achievements in economic and social indicators of
development – per capita income, alleviation of poverty, industrial diversification,
and reduced dependence on oil – were impressive and somewhat exemplary, albeit
still fragile. In comparison with other OPEC members, particularly Nigeria and
Venezuela, the Indonesian record was clearly outstanding. Compared to the rest of
the membership, Indonesia still ranked as the wisest in the management of its petroleum resources. Unlike with some other OPEC members, oil gave Indonesia a critical
head start in transforming its oil-based economy to one relying increasingly on
manufacturing exports. In 1983, oil and gas still accounted for nearly three quarters
of all Indonesian exports. In 1994, the figure was about 15 percent. In 1974,
Indonesia's per capita income was less than half of Nigeria's. By 1994, it was more
than three times that of the African nation's. A noteworthy feature of Indonesia's
development strategy was its success in responding to both boom and bust with equal
determination. Thus, when oil prices collapsed in the mid-1980s, the Indonesian

economy was able to make a rather smooth transition toward greater reliance on non-oil exports.

The relatively noteworthy overall economic performance in Indonesia – initially the poorest and most-populous OPEC member – can be attributed to several basic factors: the favourable institutions developed before the oil-price rise; the right management strategy and appropriate macroeconomic policies pursued by the Suharto government; the unusual degree of political stability after Sukarno's ouster in 1968; abundance of labour and non-oil resources; timely adoption of modern technology particularly in farm crop production; and particularly the political and financial support received from the West in the context of the Cold War.

A nation of 190 million people, with 350 languages, living on 13,000 islands, Indonesia advanced from being a distinctly poor country to the lower tier of middle-income nations in the period under review, nearing the threshold of $1,000 per capita income. By implementing deregulation and privatisation, the country was able to attract foreign private investment, soaring to nearly $24 billion in 1994 alone. Political stability was also reasonably maintained, mostly through autocratic mano-euvres (and despite the East Timor's frequent insurgencies).

However, in its pursuit of the three-pronged agenda of growth, equity and socio-political stability, Indonesia's achievement record was not even. Equity, in its both narrow and broad sense, was achieved better than the rest of the membership. In the narrow sense of reducing poverty and narrowing income differences, Indonesia was relatively successful, although the income gap that followed a narrowing trend during the 1975–85 period began to widen again under economic liberalisation. By the government's own figures, the lowest income group in 1993 earned an average of about $240 compared with $1,470 for the people in the highest income bracket. In a broader sense of social justice, equity before the law and the rule of merit, however, Indonesia did not fare well. According to informed opinion, there was virtually no functioning legal system for resolving commercial disputes. Nepotism and favouritism were ubiquitous, making the Indonesian system one of 'crony capitalism.' Nor were Indonesia's relatively dramatic economic advances in the 20-year period matched by similar advances in individual freedoms and political pluralism (including the freedom of the press).[17] In his sixth five-year term as president in 1994, Suharto still ruled the country with an iron hand, dismissing widespread charges of pervasive corruption (including his own family's[18]), massive human-rights violations, political repression, wide disparities in wealth brought about through influence peddling, monopolistic practices and abuse of power. Of the 1,000-member People's Consultative Assembly, 60 percent were appointed by the president, and the other 40 percent 'popularly elected' after first passing a security check by the military.

According to foreign observers, corruption was part of the Indonesian culture – if not the culture itself. It was everywhere and in every form. At the lower levels, it was a means of supplementing an inadequate income; at the higher levels, it was a means of facilitating major transactions; at all levels, it helped overcome bureaucratic or business hurdles – a lubricant for the rusty machine.[19] Transparency International, a German watch-dog organisation, in 1995 ranked Indonesia as the number one corrupt country in the world. Overcrowding, unemployment and a closed political system where dissent was not tolerated were among other factors that detracted from Indonesia's strictly economic achievements, and made them highly vulnerable to external shocks.

Nigeria, in contrast with Indonesia, had perhaps the poorest performance record of all group members. Once one of black Africa's brightest prospects – and initially much better off than Indonesia in terms of poverty and underdevelopment – Nigeria ended up ranking among the world's poorest nations with an estimated per capita GDP around $380 – which was still below its own 1974 level and less than half of Indonesia's 1994 figure. A direct outcome of the windfall spending by the state was the impetus given to the political strength of the central government, a further weakening of democratic institutions and the advent of successive military coups. The rapid expansion of state spending precipitated growth of employment in the public sector and in the urban construction industry, causing labour shortages in the rural areas and agricultural stagnation. Unlike Indonesia – another agriculture-based economy before the oil windfalls – Nigeria's efforts to improve productivity in the traditional farm sector did not go far. As a result, while Indonesian agriculture showed respectable vibrancy and growth throughout the period, the reverse was true in Nigeria.

By all accounts, a substantial part of Nigeria's windfall gains during the three oil booms was absorbed by unsavoury rent-seeking activities. Harmful domestic controls and regulations resulted in the reduced efficiency of the non-oil economy, rising levels of corruption and political repression. Altogether, a large and inefficient public sector, a convoluted regulatory regime, unstable exchange and trade policies, the erosion of fiscal and monetary discipline, political instability and, above all, a vacillating macro-economic policy environment made Nigeria's adjustment process chaotic, costly, trouble-ridden and largely ineffective. Because economic policies were perpetually adopted to cope with crises, their intended impact was blunted and they were often put at cross-purposes with one another. The impact of these crisis-triggered policies was often disastrous. Import bans on foodstuffs created a huge black market and encouraged widespread smuggling into the country, hurting local farmers. Subsidies on pesticides and fertilisers, on the other hand, were frequently smuggled from Nigeria into neighboring countries such as Benin and Cameroon. State-controlled distribution of credit, agricultural machinery and other farm inputs was plagued by inefficiency, mismanagement and corruption. Administrative allocation of foreign exchange and domestic credit opened the doors to rent-seeking activities and their anti-competitive consequences.

In 1994, under its sixth military rule since independence, Nigeria faced international isolation and low national self-esteem; it was badly divided ethnically, with its labour unions dissolved, its newspapers censored and its popular presidential election canceled. According to human rights advocates, of all the various dictatorial regimes experienced by Nigeria since its independence, the 1994 regime ranked as absolutely the harshest. Transparency International ranked Nigeria the most corrupt country in the world in its 1996 survey.

Saudi Arabia put the total cost of its comprehensive physical, economic and social development at $1 trillion by 1994. With an incremental capital/output ratio even as high as 5 to 1, such a colossal investment should have been able to produce some $200 billion worth of additional output a year, in real terms. Yet, the Saudi Kingdom's total nominal GDP in 1994 was estimated at about $120 billion. The bewilderingly low returns on domestic investments must, in part, be attributed to the government's development strategy and public policies. Physical infrastructure had a very high priority in Saudi planning and investments. Nowhere else in the world, and in no time in world history, was a country's physical landscape changed almost beyond recognition within only two decades as was the case in Saudi Arabia. Some of this transformation was widely judged as extravagant, if not wasteful. Such show-case projects as the $3.4 billion, second international airport at Riyadh, the $20 billion Jubail city and Jeddah's $10 billion third airport – claimed to be the world's largest – have been glaring examples. In Jeddah, several miles of resplendent gardens were irrigated with desalinated water several times a day in 110-degree heat. Bureaucratic extravagances, frequently reported in the foreign press, included fleets of official chauffeur-driven cars, travel junkets abroad to attend conferences (but actually largely to visit shopping malls), new luxury office buildings with imported furniture, fancy paintings and the latest modern gadgets. Generous subsidies to public goods and services encouraged people to consume wastefully and in an inexcusably profligate fashion. Urban Saudi families owned two or three telephone lines, three, four or even seven automobiles.

Saudi officials repeatedly argued that 'political and social considerations' precluded the government from a reduction in subsidies or increases in fees and charges for the public goods and services offered to Saudi citizens. The same argument was often used as a justification for the absence of income and most other taxes. These socio-political concerns were, of course, part and parcel of the Saudi style of governance. From the onset of the first oil boom, there was a tacit 'social contract' between the people and the government of Saudi Arabia regarding their mutual obligations and responsibilities. The government was to use the oil windfalls to subsidise health, education, utilities, public services and some private endeavours on a generous scale. In the entire 20-year period, the Saudis never paid market prices for their basic necessities. In exchange for this propped-up lifestyle, the rank and file tacitly agreed to forego political involvement and participation in decision-making, letting the royal family run the Kingdom with an autocratic hand.

The risk in this arrangement lay in its potential non-viability. The state was bound to run out of financial means sooner or later due to rising population, higher quality services, rising service costs, bureaucratic waste and inefficiency, and ultimately, lower oil income. At the same time, an increasingly educated, better-fed, healthier and pampered population living in an increasingly modern, urbanised, industrialised and open society would not find happiness in material comfort; it would crave some stake in running the realm. A time would thus arrive when financial belt-tightening, greater austerity measures and the hard work caused by the slump would begin to call into question the traditional authority.

Unsuprisingly, by the mid-1990s, Saudi Arabia faced a combination of social, economic and political crises for the first time since the early 1970s. Rapid population growth, a rising workforce and a soaring number of college graduates, combined with stagnant oil revenues, huge defence spending, massive government domestic debts and limited economic opportunities at home, posed formidable challenges to the Al-Saud ruling family, which was itself preoccupied with a thorny succession problem. With a fourth of Saudi youth virtually idle, per capita GDP in real terms less than one-third of 1980's peak and lavish social expenditure dwindling, the government was running out of easy options. Even a once-opulent state like Saudi Arabia could not escape the consequences of stagnant oil prices, diminished oil revenues in real terms and the excessive burden of a cradle-to-grave welfare state. Public services increasingly deteriorated. Medical services, once a showpiece, declined as hospital staff staged work slowdowns. Schoolrooms became crammed as school construction failed to keep pace with the rising student body. Some 25 percent of young Saudis could not find jobs. New homes sat dark because the state electric utility could not meet demands for connections. Contractors had to wait for months to be paid for government work.

By King Fahd's own admission, the Saudi economy in 1994 was in a crisis. The government was, indeed, faced with several challenges in some key areas. First, Riyadh had to restrain current spending, bring the budget (which had been in the red since 1982) into balance and to pay off its unpaid bills to contractors, grain farmers and others. Second, the government had to finance and service the foreign obligations previously undertaken (including about $30 billion signed with US defence suppliers in the early 1990s and $6 billion new civilian aircraft orders). Third, a way had to be found to deal with a rapidly growing population – pampered by years of cradle-to-grave welfare benefits and accustomed to seeing its rising expectations accommodated by the state – at the time when state resources were stagnant or declining and the costs of public services were on the rise. Fourth, the private sector had to be induced to play a much larger role in providing social services (e.g. education, health care, personal and social services) as financial pressures forced the government to curtail its involvement. Fifth, marketisation and privatisation had to be furthered by a change in the national mindset, from state paternalism to private initiative. And, finally, adjustment had to be made in cost/price distortions, caused, *inter alia*, by a fixed exchange rate (unaltered since 1985), a generous system of fiscal incentives for domestic, agricultural and industrial production, and mandated wage differentials between private and public sectors and between national and expatriate workers.[20]

Grievances by middle-class professionals, anti-government liberals/intellectuals and dissidents living abroad included the preferential treatment of people with ties to the Saudi royal family, the lavish lifestyles the royals lived at the state's expense and outlandish perks and privileges enjoyed by some 5,000 Saudi princes and high-ranking officials. Stories about payoffs in Saudi oil and weapons deals filled the pages of the Western press almost from the start, and particularly after 1979. As the principal means of acquiring a larger income and wealth, inside information and position, family relations, friendship and special contacts made all the difference. Lucrative economic activity was dominated by a small number of very large firms in trade, agriculture and non-hydrocarbon manufacturing. Exclusive agency rights were granted to favoured importers or producers, based often on personal relations or affiliations with state officials.[21] The poignancy of the domestic opposition's criticism of the Saudi defence build-up (as the motherlode of major corrupt practices) came clearly to the fore when, despite enormous expenditures on defence on a per capita basis, Saudi Arabia proved totally incapable of defending itself against Iraq in 1990 and had to be protected by an army of non-Muslims, with ominous subsequent problems.

When the first oil boom dawned in 1974, a widespread and grave concern in Saudi Arabia, was that a mad rush into industrialisation and modernisation would lead to increased Westernisation, native cultural disintegration, erosion of puritanical values, a decline of religious authority and a threat to the established political order. Following the Iraqi invasion of Kuwait and stationing of US ground forces in Saudi Arabia, it was further said that the Kingdom would never be the same afterwards. While these prophecies did not materialise, the Saudi regime, and particularly the Saudi royal family, became increasingly challenged by some ardent advocates of a fundamentalist resurgence – young Saudis educated in the West. Explanations abound. Many Saudis attribute the rise of Islamic fundamentalism to the breakneck pace of competitive development which caused increased anxiety and insecurity, and led people 'to anchor their lives in religion and cultural identity'.[22] Other analysts point to the influence of Islamic Iran's relentless propaganda against the Saudi royal family. Still others consider a penchant for the conservative *status quo* a rational reaction by the educated youth to Western cultural values as a superior alternative to their own – pushing them toward a reassertion of their own identity. Whatever the real root cause, however, the Kingdom could hardly qualify as a model for oil-supported, politico-economic development and progress.

Venezuela, once the highest income per capita nation in Latin America (and just behind the least-developed European countries e.g. Spain), gradually drifted toward one economic crisis after another, accompanied by political and social instability. In 1994, international observers found the country plagued with massive poverty, 60 percent-a-year inflation, 20 percent unemployment, a dramatic increase in crime, widespread corruption and a subsidised economy where the rich got six times more benefits from the government than the poor, and where some 42 percent of respondents to a Caracas poll were prepared to leave the country for good.[23]

In the meantime, Venezuela experienced several major riots, two military coup attempts, an impeached president and a banking crisis unprecedented in its history. Its economic performance towards the end of the period was judged to be South America's worst – with the dubious distinction of having the highest inflation and the deepest recession. There were shortages of food, medicine and electricity, a rise in crime and a near breakdown of the social order in the barrios. In a poll taken early in 1990, 86 percent of respondents believed Venezuela's oil wealth was wasted. The consensus both at home and abroad was that recurrent economic crises in the aftermath of the second oil boom were largely attributable to economic mismanagement. While the Perez administrations in both the 1970s and late 1980s were generally considered 'the most corrupt', those of his rival were characterised as 'the most incompetent'.

Students of the Venezuelan economy believe that the country was held back in its economic achievement by its traditional populism and postwar state socialism. Populist sentiments led the government, flush with the oil windfalls, to increase the size and scope of its activities. In 1994, one out of every seven workers was employed by the state, with a minimum of productive activity. Similar tendencies were also responsible for the president's power to decide private wages and working conditions. State socialism, minimally altered throughout the years despite changes in political administrations, kept most of the wartime controls of prices and wages. And people became accustomed to expecting the government to keep prices and interest rates low. The price system was largely liberalised in President Perez's second term in 1989, tariffs were reduced, interest rates were raised, and deregulation and privatisation became in vogue. The new restructuring also involved the sale of state banks, the airline and telephone companies. However, many goods and services remained regulated under certain political criteria. In President Caldera's second term, a sharp turn was made to the left, setting up price and currency controls, a *de facto* nationalisation of the banking system, slowing down the privatisation drive, reinstituting subsidies and promising further populist actions.

Public expenditure in Venezuela during the 1974–90 period increased twelve times while GDP grew only 1.2 times. Excess liquidity and unproductive investments raised the consumer price index by a factor of 15. Interest rates, however, rose 3.7 times. And the national currency lost nearly 90 percent of its value. Foreign debt reached a record $37 billion from less than $1.5 billion at the outset. The failure in January 1994 of some high-flying banks, which had run wild between 1990 and 1994, resulted in half of the country's banking industry (nine commercial banks) ending up in government hands. The bailout cost, an estimated $6 billion, represented 11 percent of Venezuela's GNP or 75 percent of the 1994 national budget or more than 150 percent of the capitalisation of all industrials in the Caracas Stock Exchange. At the same time, Venezuelans reportedly held $50 billion in foreign banks, compared to only $12.7 billion in domestic banks – despite an interest rate of 105 percent on one-year certificates of deposit in the face of a 46 percent annual inflation.

A deepening economic crisis threatening social unrest led the Venezuelan government in June 1994 to suspend constitutional guarantees against search and arrest without a warrant and to impose restrictions on the free movement of people and goods, private property and business transactions. Controls were also instituted on prices and currency speculations. Transportation and food subsidies were promised to the poor. President Rafael Caldera's revived populist philosophy put a virtual end to the free-market revolution that had begun under his predecessor, Carlos Perez, and brought back the state-directed populism practised in his previous presidency in the 19780s and during the country's oil boom. But the people's nostalgia for the 1970s which brought the Caldera administration back to power could not be satisfied under the oil bust. The oil revenues in 1994 could not finance the state's far-flung obligations: a public sector employment as high as 15 percent of the work force, public subsidies on food, energy, education and travel, and financial assistance to money-losing state enterprises.[24] Within its first two years, the Caldera government resorted to nine economic plans to cope with a sinking economy.

By the end of 1994, the Venezuelan economy was in shambles. Its inflation was the highest in Latin America. The country was painfully emerging from a recession. The middle class was fast disappearing. Mass poverty was on the rise. The bureaucracy was both highly bloated and increasingly inefficient. The country, which in 1974 was the fifth freest market-economy in the world, by 1994 tied with India and Kenya for 63rd, just ahead of Bangladesh. In a sense, oil windfalls also blunted, if not killed off, native creativity, ingenuity, imagination and productiveness, which could have blossomed to deal with the problems of an oil-less or oil-starved economy. As an astute analyst has observed, Venezuela, which once fancied itself as Latin America's Saudi Arabia, ultimately turned out to be its Nigeria.[25] A keen observer of Venezuelan politics finds the country in the mid 1990s beset by economic deterioration, severe state disorganisation, rent-seeking activities which were becoming the central organising principle of political and economic life, and stop-go economic reforms which were accompanying political changes.[26]

AN OVERALL COMPARATIVE RANKING

Can there be a comparative case-study methodology to measure different economies' overall socio-economic performance in quantitative terms? A group of analysts within the United National Development Programme (UNDP) have constructed a comprehensive socio-economic measure of national progress called the human development index (HDI). The HDI is defined as a composite of three basic components: longevity, knowledge and standard of living.[27] For any country, the HDI ranking may differ from GNP per capita by a plus or a minus figure. When the GNP rank is higher than the HDI rank, it shows that the country still has 'considerable potential' for translating its income into improved well-being for its people. When the HDI rank is larger than the GNP value, the difference indicates that the country has made 'more judicious use' of its income to improve its citizens' capabilities.

By these criteria, the highest negative among OPEC members belonged to Kuwait with a minus 47, while the highest positive difference belonged to Ecuador with a plus 5. Except for Nigeria and Venezuela, all other 10 members showed a negative value (Table 17). The four highest negative-rated OPEC members – with a minus figure of 25 or more – showed the highest gap between the real GNP and HDI ranks. Only 15 others among 175 countries surveyed by the UNDP in 1994 exhibited such a negative ranking. What transpires from these figures is that a majority of OPEC members did, albeit in different degrees, fail to translate their high gross domestic products into corresponding improvements in their peoples' welfare.

NOTES ON CHAPTER VI

1 These are the only years in which data for this category are available and comparable for all group members. Figures for 1974 and 1994 are not at hand for most.
2 *Human Development Report 1996* (New York: UNDP, 1996), p 3.
3 See *Human Development Report 1994*, p 73; and *Trade and Development Report 1996* (Geneva: UNCTAD, 1994), p 47.

4   *Claiming the Future* (Washington: The World Bank, 1995), p 28.

5   W.V. Eeghen, *Poverty in the Middle East and North Africa* (Washington: The World Bank, 1995), p 5.

6   For the year-to-year breakdown of this oil dependence, see International Monetary Fund *Government Finance Statistics Yearbook* (Washington: IMF, annual issue).

7   Claiming the Future, pp 24–5.

8   See *Trade and Development Report, 1996*, annex to Chapter II.

9   Estimates of military expenditures by group members published by agencies in Washington, London and Stockholm differ considerably, and in some years are totally irreconcilable. Figures stated here and in Table 14 are closest to the average figures. See Nancy Happe, *Military Expenditure and Arms Trade* (IMF Working Paper WP/94/69 1994).

10 *Human Development Report*, 1994, pp 54–5.

11 For the origin, objectives and operation of these agencies, as well as books, pamphlets and reports on OPEC aid, see *OPEC Aid Institutions: A Profile* (Vienna: The OPEC Fund, 1995).

12 Bjorn Larsen, *Environmental and Natural Resource Management in the Middle East and North Africa* (Washington: The World Bank, 1995).

13 See 'Special Report: Water', *Middle East Economic Digest*, 24 January 1997.

14 See Alan Richards, 'Containing Algeria's Fallout Through Prosperity', *Middle East Quarterly*, June 1995.

15 See *US News and World Report*, 22 August 1994, p 45; and *New York Times*, 6 June 1995.

16 See *Middle East Economic Digest*, 13 May 1994. See also Mohammed Akacem, 'Algeria: In Search of an Economic and Political Future', *Middle East Policy* (No. 2, 1993).

17 See E.A. Gargan, 'Family Ties that Bind Growth', *New York Times*, 9 April 1996, p 1ff.

18 Suharto's six children reportedly controlled more than 260 companies with assets worth several billion dollars. See *Washington Post*, 10 March 1993.

19 'The Extended Family', *The Economist*, 15 August 1987, p 12.

20 See 'Saudi Arabia', *Middle East Economic Digest*, 11 March 1994; and 3 November 1995.

21 See 'Royal Payoffs', *Wall Street Journal*, 4 May 1981.

22 See Judith Miller, 'Economy Gives Saudis Growing Pains', *New York Times*, 27 December 1983.

23 Tod Robertson, 'Venezuelan Economy...', *Washington Post*, 2 October 1995.

24 In 1994, gasoline cost 10¢ per gallon, and its doubling to 20¢ started violent protests. Public subsidies were estimated at $450 million a year. The state-owned industrial complex, CVG, was losing over $1 billion annually.

25 Matt Moffett, 'Populist Disaster', *Wall Street Journal*, 16 August 1995.

26 T.L. Karl (1997), pp 184–5.

27 For the yardsticks measuring these components, see *Human Development Report 1994*.

VII

The Litmus Tests

The cases examined in this study, and OPEC members' 20-year performance records, offer an opportunity to test several rival hypotheses regarding the impact of natural, economic and political phenomena on the process of growth[1] and development. These competing hypotheses may be divided into five main categories. First, the influence of resource-based wealth on the development process and the validity of the Dutch Disease syndrome. Second, the role of non-resource factors such as initial GDP, the annual rate of fixed capital formation, openness of the economy, domestic income inequalities and foreign exchange availability. Third, the significance of non-economic elements such as the nature of governance, the pernicious lure of unearned income and the very nature of 'oil-stateness'. Fourth, the impact of the so-called developmental *vs* patrimonial strategy on the growth outcome. And fifth, the deep-rooted influences of history and culture as the key to growth and wealth. Benefitting from these tests, certain 'performance indicators', based on OPEC members' experiences, will be suggested at the end, and the study's findings will be further summed up.

THE 'CURSE' OF NATURAL WEALTH

Both economic theory and history offer clear evidence that abundant natural assets should and can help a country prosper. Theoretically, resource abundance in the form of exportable commodities earning hard currencies might be expected to benefit an economy through larger, diversified, domestic investment in physical and human capital, acquisition of foreign technology, and a 'big push' toward industrial development. Empirically, too, there is evidence that agriculture made Argentina, Canada and New Zealand rich; Western Europe thrived on plentiful deposits of iron ore and coal; Australia and South Africa benefitted from minerals, and the Soviet Union lasted long after its time because it sat on top of some of the world's largest natural resources.

The disappointing performance of resource-rich economies, including the record of some OPEC members, has however, given rise to the new hypothesis that the abundance of natural resources (mines, agricultural land, water) tends to stifle growth rather than promote it.[2] In a reaffirmation of the Dutch-Disease phenomenon (Chapter II), the new thesis in vogue asks whether there is 'a curse to easy riches'. Historical examples are cited to show how gold-rich Spain was overtaken by resource-poor Holland; why resource-abundant Russia was surpassed by poorer Japan; why Argentina and Brazil were left behind by South Korea, Hong Kong and Taiwan, and OPEC members by the oil-less and resource-poor developing countries of Southeast Asia (before the 1997–8 debacle).[3] The puzzle was: Why should countries with abundant resources perform less well than others even if these riches were to be of no additional help? The gist of this 'riches-may-be-a-curse' thesis is an alleged perverse association between natural resource intensity and economic growth – a phenomenon that stands quite apart from other variables that may explain cross-country differences in economic expansion. The Sachs/Warner analysis suggests that growth rates in a sample of 97 countries between 1971 and 1989 were higher among less endowed countries than among the richer ones.

In the case of OPEC members, however, the findings of this study fail to confirm this hypothesis in its uniform application. Based on the volume of oil deposits as a relative indicator of natural wealth, Saudi Arabia and the UAE, for example, which had the highest ratios of oil exports to GDP in 1974, did in fact grow more sluggishly than Indonesia, Ecuador, Algeria and Iran which were not as richly endowed. But Kuwait and Iran achieved higher growth rates than Nigeria and Venezuela which were less abundantly blessed. The large endowment with oil (and other natural resources in the case of Iran) also did not uniformly retard growth. And there were reasons for poor performance in other group members unrelated to oil reserves. The wealth thesis did not thus hold true in an intra-group comparison.

The possible adverse effects of the oil bounty on OPEC members' performance may be more clearly seen in a comparison of collective and individual growth rates before and after 1974. The annual growth rates under the first oil boom (1974–9) were lower in six members than their 1965–73 rates. But, more significantly, none of the members, between 1980 and 1994, could even come close to the growth rates achieved before 1974. Their annual real growth rates, when not negative, were often a fraction of the pre-1974 rates (Table 10). As indicated before, OPEC members as a whole also lagged behind the resource-poor, newly industrialising nations of Southeast Asia during the 1974–94 period.

A related manifestation of the wealth's 'curse' is depicted by the Dutch-Disease syndrome. The group's experiences between 1974 and 1994, again, only partially validate the significance of this syndrome. Nearly all members underwent real exchange-rate appreciation, rising domestic wages and prices relative to foreign inputs, a strong tendency to invest in non-traded sectors and a preference for import-substitution industries over export-promotion activities. The macro-adjustment models – both the Dutch-Disease and others – correctly identified changes in aggregate domestic demand through price and income effects. But, as was clearly seen in Chapter IV, supply responses did not closely reflect those models' predicted behaviour. The government's overriding role in adopting the national strategy (allocating oil-revenues, distributing credit among various sectors, choosing capital or labour-intensive industries for public investment, instituting domestic wage/price policies and trade protection or promotion) played a crucial part in reducing or magnifying the influence of market forces. In sum, the supply response was influenced by a large number of other factors beside relative wage/prices and exchange rates – the two variables that took centre-stage in the macro models.

The inherent tendency, singled out by Dutch-Disease, for domestic resources to move out of traded activities to non-traded sectors did, in fact, take place in many member countries during the three oil booms. Physical infrastructural expansion and modernisation, private housing and commercial construction and investments in human capital formation – activities that were all in non-traded sectors – did indeed constitute the main recipients of oil windfalls in all countries. However, the tradable sector (apart from oil and gas) – particularly manufacturing and agriculture – did not necessarily decline. In fact, the share of industry in GDP improved in every OPEC member except Nigeria. Agriculture, too, raised its share in Iran, Iraq, Kuwait, Libya, Qatar, Saudi Arabia and the UAE, and it kept its position in Algeria and Venezuela. A number of exogenous factors such as government subsidies and tariff protection, changes in domestic income distribution, rising farm productivity and the intrinsic immobility of the resources involved in small land holdings effectively shielded small-scale, family-operated farms from adverse terms-of-trade developments, rising production costs or potential resource outflows.

These experiences demonstrated that the Dutch-Disease model corresponds best to the case of small industrial countries where free-market forces are dominant and state controls are at a minimum – a situation very different from OPEC's case. In nearly all group members, even those with a large free-market, private sector economy (e.g. Kuwait, Qatar, Saudi Arabia, the UAE), government intervention (particularly in such areas as direct public investment, sector-specific subsidies, price control, credit policy, exchange and trade regulations, and immigration) was able to offset and counterbalance the impact of rising wages, prices and real exchange rates appreciation.[4] In sum, the Dutch-Disease phenomenon could not offer a convincing explanation for the membership's overall poor growth record.

NON-RESOURCE FACTORS

A series of other hypotheses have linked growth to such variables as the size of GDP at the outset, annual investment rate, open-market orientation, domestic income distribution and the supply of foreign exchange. In the OPEC's case, however, none of these criteria offers adequate explanation of differences in performance. If we take initial GDP as a prime variable in determining subsequent growth, the link is not clearly evident. Iran's economy, with nearly $46 billion GDP in 1974, was the largest within the group, followed by Nigeria's ($30 billion), Saudi Arabia's ($28 billion), Venezuela's and Indonesia's (about $26 billion) and Kuwait's ($13 billion) (Table 19). Yet Indonesia achieved the highest average real growth rate, followed by Iran, Kuwait, Nigeria, Venezuela and Saudi Arabia in that order. Libya initially had the same GDP as Algeria ($13 billion), yet its growth rate was about one-fifth of the latter's.

The rate of annual fixed domestic capital formation also was not a clear-cut determinant of growth. The group members' experiences indicated that GDP growth rates were not a direct function of the investment/GDP ratio but, rather, of the choice of investment projects and their completion within projected time horizons and cost estimates (i.e. the incremental capital/output ratio). Thus, Gabon with the highest annual investment allocation had the most dismal growth rate, and Algeria with the second highest ratio lagged behind Indonesia and Iran, which had smaller ratios but showed higher growth rates. Among low-absorbers, too, Kuwait had the lowest average annual investment/GDP ratio yet achieved the highest average yearly growth rate. With roughly the same investment ratio Kuwait also had a much higher growth rate than Nigeria (Table 18). For the entire group, changes in economic growth did not always correspond directly to increase or decrease in investment during either the upswing or downturn. During the boom, the hurried and haphazard increases in investment were, unsuprisingly, met with decreasing returns. During the bust, however, a decrease in the rate of investment was accompanied also by declining capital productivity.

A World Bank study finds part of the reason for this phenomenon in the state dominance over the economy and in its role as a primary agent for receiving oil windfalls and allocating them to investment. With large shares of new investment going into non-tradables (e.g. housing) and quasi non-tradables (e.g. public industries protected from foreign competition), the economies of the group members were believed unprepared to cope with the post-1980 terms-of-trade shocks. The pursuit of misguided policies for rationing investment resources during the depressed oil market of the 1980s resulted in a deficient incentive system for domestic investment, while aggravating inflation and private capital flights.[5] The choice of inappropriate projects resulted in the extremely low level of output obtained from increased investments. The incremental capital-output ratio (i.e. additional output per unit of new investment) in Algeria and Iran, for example, was about one-sixth and one-third, respectively, of Indonesia's. According to World Bank estimates, public enterprise losses in Algeria reached $5 billion in 1990.[6] A reverse correlation was also found between the public sector's share of total investments (e.g. more than 40 percent) and the low productivity of investment and smaller overall growth.

The character of the economic regime is considered another determinant of growth. Free market advocates place great stock on the direct correlation between economic growth and economic freedom, arguing that free countries are likely to grow faster than those which are repressed. Economic freedom is identified with such factors as non-government intervention, open and liberal economic policies, and absence of black-market activity. In the roster of 150 countries compiled by the Heritage Foundation, no OPEC member was classified as 'free'. Indonesia, Kuwait, Saudi Arabia and the United Arab Emirates were 'mostly free'. Algeria, Ecuador, Gabon and Venezuela were 'mostly unfree'. Iran, Iraq and Libya were categorised as 'repressed'. Between 1980 and 1993, while the 'free' countries achieved a per capita GDP growth of nearly 3 percent and the 'mostly free', a rate of nearly 1 percent, the others had a negative per capita growth record.[7]

The association between high growth rates and the openness of markets to foreign competition (i.e. liberal trade and exchange policies) is mildly confirmed by some members' performance, but not all. Algeria, which followed perhaps the most restrictive economic policies in most of the period, had modest growth rates up to 1990. On the other hand, Gabon, which maintained a fairly liberal system, was a very poor example, lagging behind Libya which had extensive restrictions. The star performer, Indonesia, however, owed much of its rapid expansion to its gradually liberalising stance, particularly after accession to the GATT in 1985. Iran's notable growth rates during the 1970s, as compared with its very poor development during the 1980s, could also in part be traced to its different foreign economic postures before and after the 1979 revolution. At the same time, the smaller group members around the Persian Gulf which subscribed to an open, liberal and market-oriented policy, did not fare particularly well except for Kuwait. Nigeria, which pursued a considerably restrictive trade system, achieved a modest growth rate, but Venezuela, which followed a relatively free trade policy and joined the GATT in 1990, trailed the African member's growth performance. Indonesia, as a significant recipient of foreign private investments, prospered rather handsomely, while Nigeria, which was also a large recipient, fared less well. The growth in per capita income also failed to support the free-marketeers' thesis. The 'mostly free' nations of Kuwait, Saudi Arabia and the UAE, for example, trailed behind the 'mostly unfree' countries (e.g. Algeria and Ecuador) as well as the 'repressed' Iran.

Availability of foreign exchange is regarded as another stimulus to growth. However, OPEC members' 20-year experience refutes this widespread postwar conviction that one impediment to economic development is the shortage of foreign exchange. In OPEC's case, this study's findings show that the absence of indigenous technology, the dearth of complementary domestic resources and talent, the difficulty in absorbing borrowed know-how and inadequate national resource mobilisation in the service of genuine, sustainable growth were the critical factors. Nor was initial income inequality essential for rapid growth, as development economics once assumed on the grounds that the rich at the top of the income ladder are more likely to invest than to spend. The insatiable desire of the well-to-do in all group members for high-priced luxury imports and their tendency (particularly in the open economies of the Persian Gulf) to transfer their wealth abroad told a different story.

The impact of internal distribution on growth is also difficult to ascertain in the case of OPEC members. Data on income shares are available for only a few countries. But published figures fail to confirm either the old conventional wisdom or a latter-day thesis. The old orthodoxy which held that large income gaps in the early stages of a country's development were helpful to growth (because the rich were likely to save more and provide more funds for investments) is only partly supported by OPEC's experiences. Venezuela and Nigeria which showed relatively less equitable records had, in façt, two of the lowest growth rates among the group (Tables 9 and 11). By contrast, Ecuador, which had an equally poor income record, achieved a relatively higher growth rate than most. At the same time, the recent 'unconventional' dictum[8] that countries with relatively more equal distribution of income can be expected to achieve higher growth rates (because of better health and education and greater labour efficiency) is also only partially backed by OPEC's performance. Accordingly, Indonesia with less income inequality than all the rest also had the highest rate of expansion within the group. By contrast, Algeria with a better equity record than Ecuador achieved a lower average growth rate during the period. Other countries also showed a similarly mixed picture.

QUALITY OF GOVERNANCE

The nature of a political system and the quality of governance are believed to affect economic growth in four different manners. One thesis posits a positive association between authoritarian discipline and economic performance – the thesis usually defended by undemocratic regimes. Another thesis points to a direct correlation between the spread of democracy and rising living standards – the belief strongly supported by free-market advocates. The third claims a close connection between the 'effectiveness' of bureaucracy (e.g. absence of corruption) and a way out of poverty.

The first political hypothesis suggests that economic reforms, structural adjustment and growth performance can best be achieved by a centralised government system. The gist of this thesis (which before the 1997 economic turmoil in Thailand, Malaysia, South Korea and Indonesia was referred to as 'Asian values') is that a strong central government (read: an authoritarian regime) can devise and deliver much faster economic growth than its rivals due to greater 'efficiency'. Simply put, this thesis holds that authoritarian leaders commonly carry strong national commitments to economic growth, can deal resolutely with opposing vested interests, have the will and power to pursue stabilisation policies and are able to minimise bickering, delays and procrastination on the part of legislators or armchair bureaucrats.[9] Democracies, it is argued, by their very nature allow protracted debates and deliberations, slow down badly needed adjustments, interrupt the continuity of policies through periodic changes in administration and result in delays and inefficiencies. The so-called 'Asian Tigers' further argued that democracy and human rights were Western concepts inapplicable elsewhere and inimical to growth.

OPEC members' records detailed in Chapter V fail to show any clear link between polity and performance. However, the quality and behaviour of leadership in both 'democratic' and 'non-democratic' countries, as well as a host of other socio-political factors (e.g. stability) made a distinct difference. Richard Cooper and his colleagues, in their study of 18 developing countries, (including two OPEC members, Indonesia and Nigeria), also found 'no evident correlation between economic performance and form of government or degree of authoritarianism'.[10] Measured by the average real growth rate as a representative criterion of performance between 1974 and 1994, Indonesia, which had the best record, was a semi-military, authoritarian and centrally-directed economy – giving credence to the 'dictators-do-it-better' thesis. Similarly, Iran and Algeria, both authoritarian regimes, achieved relatively modest growth rates, also supporting the claim. On the other hand, the Arab countries around the Persian Gulf – all traditional, non-democratic and state-dominated, had some of the lowest growth rates. Confusing the issue further, Ecuador and Venezuela – the two members with a greater semblance of democracy – achieved growth rates comparable to those of Saudi Arabia and Nigeria which were in the other camp. At the same time, Gabon, another semi-democratic country, had a worse record than Libya. In Venezuela, both the old military regime and the two subsequent democratically elected governments, employed similar traditional populist-statist policies that failed to produce either efficiency (e.g. economic growth) or equity (e.g. poverty eradication). At the same time, it produced a modest growth rate during the 1980s in the hands of leaders who largely disavowed free enterprise capitalism.

The opposite hypothesis, emphasising the salutory effects of democracy on the growth process, contends that (a) representative governments can stimulate better and more genuine economic development because leaders are made accountable to the electorates through periodic elections (b) extensive consultations and scrutiny usually result in better policy formulation and more useful projects (c) people's participation in the electoral process forces leaders to present and support their agenda in an orderly, peaceful and transparent manner (d) public policies backed by the majority promise greater overall success and (e) cronyism, connections, rent-seeking activities and corruption are minimised, and impartial fiscal control and discipline are enhanced.

Testing this hypothesis in the OPEC's case shows that a link between democracy and economic growth is even murkier. Although no member ever admitted to being 'undemocratic', differences in the type of governance were too obvious to ignore. According to a special survey published in 1993, four of the group members failed to show any close link to democracy. On a scale of zero to one – where zero meant no individual political rights and one meant virtually full rights – Algeria, Indonesia and Iraq received a zero mark, and Iran obtained only 0.17.[11] In another report dealing with political rights and civil liberties, only Ecuador in the OPEC group was judged a 'free' state; Gabon, Kuwait and Venezuela were ranked as 'partly free'; and all nine others were considered 'not free'.[12] Based on this classification, 'non-democratic' Indonesia and 'democratic' Ecuador achieved higher growth rates than 'partly democratic' Venezuela or autocratic Nigeria. By contrast, 'non-democratic' Algeria and the UAE surpassed partly democratic Gabon and Venezuela. And, non-democratic Saudi Arabia beat partly-democratic Iran. Only Iraq and Libya exhibited a clear link between bad governance and poor performance.

Contrary to early post-World War II prognostications, neither economic growth nor rising living standards emanated from (or resulted in) the rule of law or the prevalence of democratic values. The purported linkage between improvements in education and health on the one hand and the spread of democracy and the development of civil society on the other – even if true in the case of Chile, Portugal,

South Korea, Spain and Taiwan – was not evident in the OPEC members' case. On the contrary, material gains in GDP per capita income, high literacy, longer life expectancy, better nutrition, greater health care, improved housing and other social indicators of human development in nearly all member countries were accompanied by the rise of state power. For example, the conservative, traditionally free-enterprise economies like the Persian Gulf sheikhdoms were effectively converted into virtual statist regimes through the enlargement of public sector investment, employment and social welfare involvement. None of OPEC's authoritarian-like countries in 1974 had moved effectively toward a pluralistic democracy by 1994. In fact, political setbacks occurred in Algeria, Indonesia, Iran, Iraq and Nigeria.

The relationship between rising oil wealth and the prospects for democratisation was somewhat undermined because the fiscally autonomous distributive state was able to create its own loyal clients through public spending at the expense of traditional economic groups and burgeoning civil societies. The rise of this entrenched neo-patrimonial state resulted in the subordination of social classes and civil associations to the authoritarian power of the purse, and kept them from crystallising into independent entities capable of countervailing centralised patrimonialism.[13] In short, modernisation from above and in the hands of a patrimonial state tended to inhibit the development of 'rational bureaucratic machinery, stable and responsive political institutions, and new patterns of social relationships' necessary for a modern participatory democracy.[14]

Nor was the presumed link between capitalism and democracy clearly established. Capitalist Gabon was only 'partly free', but capitalist Nigeria was 'not free'. Capitalist/statist Ecuador was 'free', but the other mixed economies (Indonesia, Iran, Qatar, Saudi Arabia and the UAE) were 'not 'free, while two others (Kuwait and Venezuela) were only 'partly free'. Statist Algeria and Libya were the only two that seemed to confirm the link. Instead of becoming more democratic and participatory, Saudi Arabia's traditional, tribal and paternalistic regime, founded more on the ruler's power of persuasion and consensus, gradually turned into an autocratic order hard on its dissidents. The Saudis, by their own admission, refused to adopt 'intangible social and political institutions' imported from elsewhere (read democracy), accepting modernisation but rejecting Westernisation. The more tragically ironic fact was that some of the Saudi dissident groups, created by the same autocratic society and now agitating against it, gave no assurance of embracing Western-style democratic institutions, including civil society, freedom of expression and respect for universally recognised human rights. The so-called Committee for the Defence of Legitimate Rights, a London-based Saudi opposition group formed in 1993, was itself accused of condoning widespread human rights violations.[15]

In 1994, Kuwait's embrace of modern democratic rule was still a fragile one. The parliament (National Assembly) had been shut down by the Emir for a total of 11 years since its establishment in 1963. Political parties were still banned. Women had not been given the right to vote. The National Assembly could be suspended by the ruler at any time – as it was in 1976 and 1986. In Venezuela – a country with a greater semblance of democratic rule than others – there was also a reversal. The near collapse of the banking system in Venezuela in early 1994, due to the imprudent operations of some high-flying banks between 1989 and 1994, led President Rafael Caldera to take not only draconian economic measures on wages, prices and foreign exchange, but also to suspend certain constitutional rights. It was Nigeria's first democratic ruler, the civilian Shehu Shagari in 1979, who allegedly turned the country into a 'police state'.[16]

In defending their non-democratic regimes, government officials in Algeria, Indonesia, Iran, Nigeria, Saudi Arabia and others argued that freedom to speak is not more important than freedom to eat. They invariably cited the examples of South Korea, Singapore and Taiwan to back their 'interim' policies, which supposedly justify muzzling the press and denying Western-type individual rights to their citizens in order to safeguard balanced responsibility, ensure political stability and promote evolutionary progress. A publicity piece by Qatar, boasting about progress made under the oil booms, talks of finding 'in the next 15 to 20 years, a form of democracy which will suit our culture and our Islamic religion'.[17] This clear stand in favour of tradition also rebuts another earlier development orthodoxy – that modernisation and Westernisation go hand-in-hand, and modernity requires a reevaluation of traditional ethos. OPEC's Arab members in the Persian Gulf region became extraordinarily modernised in appearance and gadgetry without undergoing similar cultural transfigurations.

In respect to the third political hypothesis – the link between 'good governance' (e.g. clean government) and good economic performance – the OPEC record is again confusing and inconclusive. Indonesia, with a number one ranking among the most corrupt countries in 1994 according to Transparency International, had the highest average growth rate of all during the period. Nigeria, another country with the same distinction in 1995 and 1996, had a moderate growth rate. Venezuela, as the third OPEC member with a similar designation, however, produced a rather low growth rate. Other OPEC members (e.g. Algeria, Ecuador, Iran, Kuwait and Saudi Arabia) reported in the Western press from time to time as countries plagued with widespread and institutionalised corruption also showed performance records that were not exactly correlated with the seriousness of press allegations.

Nor could the magnitude of corruption be associated with the degree of authoritarianism. In general, democratic institutions – parliament, the press, political parties, civil societies etc. – are expected to limit official corruption, arbitrary unpopular actions and perhaps also unproductive public projects.[18] Autocracies, on the other hand, are said to have a tendency to waste national wealth for private gain. Corruption, cronyism and unsavoury rent-seeking activities are often considered by some the hallmark of dictatorial regimes. However, it is also argued that democracy may increase corruptive practices because it tends to raise campaign costs. In OPEC's case, Indonesia and Nigeria showed a link between dictatorial rule and corruption but 'democratic' Ecuador and Venezuela were also in the same company.[19] The same could be said about the absence of correlation between corruption and religious differences. Both Islamic and Christian members of OPEC were alleged to have witnessed corruptive practices in their leaders. Iran, Saudi Arabia and Kuwait were often reported to have been in the same boat as Ecuador and Venezuela.[20]

Another feature of good governance, namely political stability, again did not always correlate with steady growth and economic betterment. Indonesia had the most stable government during the period and the highest average growth rate. Libya, Gabon and the UAE were also governed by a stable leadership and enjoyed a semblance of political stability, but had a deplorable growth record. Nigeria and Iran had the severest political convulsions but still produced positive growth. Algeria and Kuwait were less stable than Saudi Arabia but achieved higher growth rates. So was and did Venezuela. Furthermore, political stability was as much a consequence as it was a cause of economic prosperity. All during the 1970s and 1980s, relative political stability and social cohesion in Saudi Arabia and other Persian Gulf sheikhdoms could be assured by the lavish expenditure of oil revenues on social welfare. Calls for political freedom and socio-economic reform could also largely be deflected by keeping the citizenry materially well satisfied. In other group members, too, the 'stable' leaders solidified their power and secured their own survival by appeasing the urban population that threatened their rule, bought off opponents, built patronage in the cities, and resorted to the traditionally money-based politics.

In short, the oil wealth did not contribute to internal political stability. There was an historic revolution in Iran; two bloody wars involving Iran, Iraq and Kuwait – three of the OPEC's founding members; several *coups d'état* in Nigeria; a protracted civil strife in Algeria; and an almost unending series of 'bread riots' and other violent disturbances in Indonesia, Iraq, Libya, Saudi Arabia and Venezuela. Rapid urbanisation, housing shortages, weakened established mores, rising popular expectations, lopsided distribution of income and tales of official corruption contributed to the rise of socio-political tensions in all group members without exception.

Finally, OPEC member experiences also fail to authenticate a latter-day thesis that commodity-based states share a 'similar path-dependent history and structuration of choice'. The argument, simply put, is that the origins of states' chief revenues, and especially the character of the revenue-producing sector, determine the nature of the state as predatory or developmental.[21] The findings of this study show that such a deterministic approach lacks both strong non-Marxian theoretical underpinnings and empirically representative examples. While similar boom-generated irrational euphoria, irresistible temptations to profligacy, personal rather than institutional power relations and weak and inept bureaucracies can all be detected in all member countries, differences in the choice of development path and the adoption of policy alternatives among members cast doubt on the existence of any omnipotent 'over deterministic dynamic'. The similarities noticed in the type of allocation, adoption of policies and even overall performance among some OPEC members had much less to do with the degree to which these states depended on a single (mineral) commodity status than with the unique characteristics of petroleum itself and, more significantly, the state ownership of this unique commodity.

DEVELOPMENTAL *VS* PATRIMONIAL STATES

The fourth hypothesis tested here is the type of development strategy followed by the various authorities. Published data by the World Bank show that between 1980 and 1994, the average real GDP growth in East Asia and the Pacific was about 8 percent a year, while the most successful OPEC member, Indonesia, achieved only about 6.5 percent during the same period. The majority of other oil-rich members produced barely half as much growth, and a few grew even more sluggishly. What made the OPEC's case doubly disappointing, if not downright deplorable, was the fact that some members in each group (e.g. Iran and South Korea) began to accelerate their development process in the early 1970s at about the same level of per capita income and industrial know-how. Why then did OPEC members with abundant oil wealth fall badly behind the Southeast Asian countries (without many natural endowments)? The answer offered by this hypothesis is that, while the state had an active role in fostering economic development in both groups, its role in the Asian group was developmental but in OPEC's case it was patrimonial.[22] That is, in Southeast Asia, the government's institutions, facilities and policies were utilised to strengthen the domestic entrepreneurial class but in OPEC's case governmental agencies were directly involved in carrying out development projects.

While this thesis may prove sufficiently relevant and convincing in the case of some OPEC members (e.g. Algeria, Iraq and Libya), it certainly could not quite explain the fate of others where private enterprise and private entrepreneurs were both encouraged and active (e.g. Ecuador, Iran, Indonesia and Venezuela). At the great risk of over-simplification, OPEC members' experiences seem to show two crucial contrasts with those of the Southeast Asian 'tigers': the source of capital for domestic investment and the nature of the development strategy. The successful Pacific Rim countries relied mostly on foreign funds and investments for their rapid industrialisation and pursued an export-oriented strategy in their trade policy, while the OPEC laggards basically used their own oil wealth, on mostly public projects, and generally adopted an import-substitution road to diversification. What distinguishes the two modalities in terms of results is that externally-sourced capital is invariably accompanied by advanced technology, modern management know-how and skill-intensive activities. By contrast, natural wealth-generated funds, in addition to their other growth-impeding flaws, almost totally lack complementary technology, management and skills.

Looked at from a different angle, however, the nature and magnitude of risk taken in the two approaches were considerably different and, as later events showed, not indicative of a clearly superior choice between the two. OPEC members (with the notable exception of Indonesia) used their own wealth and, through their governments, commanded their own development strategies: They were, in a true sense, principal masters of their own destiny. What they could not effectively control (i.e. the only risk they faced) was the price of oil. Even the oil price crash of 1986 was partly the result of their own miscalculations and failed strategy. In short, they could have fared better if they had been wiser.

In the case of the Asian 'tigers', however, national economic fortunes and directions were largely beyond their own control due to their closer integration in the global economy. Their remarkable growth and prosperity were basically underpinned by foreign capital and credit for domestic production and foreign markets for the sale of their products.[23] Overseas funds poured in, chasing high returns on investments guaranteed by stable, high-handed governments and national currencies pegged tightly to the US dollar. Foreign capital financed several booming economies all dependent on exports to virtually the same foreign markets. This dual dependency was highly risky for two main reasons: first, apart from systemic domestic economic flaws (i.e. cronyism, lack of proper banking regulations and the prevalence of sweet-heart deals between government and business) productive industrial capacity expanded in virtually the same industries almost everywhere, and a speculative real estate boom emerged under highly fragile credit conditions. When foreign markets could not absorb all the exports, balance-of-payments began to look weak, and local currencies came under increasing pressures. Second, as soon as the supercharged economies began to slow down under an export-led recession, they triggered a speculative run on local currencies by traders hoping to profit from a probable devaluation. And, when the devaluation finally came, foreign investors began to flee in droves, as the 1997 Asian turmoil vividly revealed. In the final analysis, the choice was not simply between developmental and patrimonial approaches but also between more secure, but less profitable, self-reliance and riskier but more rewarding global integration.

CULTURAL AND HISTORICAL LEGACIES

A latter day hypothesis holds that the causes of nations' divergent development des-tinies – the roots of both wealth and poverty – should be sought in cultural factors. According to this controversial thesis, only those countries with cultures that are sufficiently prone to the adoption of modern technologies and assimilation of special modes of socio-economic organisation can aspire to become rich and stay prosperous. The reason that Europe (and specifically England) developed an edge over other regions was rooted in their openness to new ideas, rejection of dogma and willing-ness to borrow from abroad under a type of government that was friendly to the interests of the business class. In other words, the key to national wealth has been the Industrial Revolution which was initiated in England and followed by the rest of the world. History rather than geography seals the nations' fates. To be secure and sus-tainable, prosperity needs to be undergirded by appropriate cultural preconditions. Intellectual curiosity, industrial inventiveness, business initiative and puritanical par-simony are parts of such culture. Religion plays a crucial role in this hypothesis. Belief systems that are rooted in an authoritarian hierarchy (e.g, Catholicism and Islam) are likely to cramp iniative and stifle widespread experimentation. Others (e.g. Protest-antism) that emphasise the value of 'work, thrift, honesty, patience and tenacity' are conducive to prosperity and riches.[24]

Putting OPEC's 20-year experience to this test, which relates to some 600 years or so of world history, would be both a wrong and sterile task here. Nevertheless, the relevance of cultural and historical traits for the overall and individual performance of group members can hardly be denied. While the influence of neither Islam nor Catholicism (the two religions prevailing in the group) can be generalised due to their extreme diversities within the practitioners (Indonesian Islam is hardly the same faith as in Saudi Arabia's or Iran's), the relative absence or presence of some growth-promoting and growth-inhibiting factors related to culture (and history) would be too obvious to deny. Indonesia, for example, had the benefit of entrepreneurial expertise and experience among its ethnic Chinese minority which controlled most businesses. By contrast, most other members clearly exhibited some negative traits identified by the cultural hypothesis.[25]

PERFORMANCE INDICATORS

Even though none of the foregoing hypotheses was totally supported by the OPEC members' performance record, a combination of factors in the form of a composite index could by correlated with successes or failures. In general, the most successful performance – in terms of high growth rate, fuller employment, reasonable internal and external stability, and greater social welfare – seemed to be associated with the largest combination of the following 10 factors: (a) a relatively small population or a low rate of population growth (2) a relatively high rate of investment in both physical and human capital (3) low government consumption (particularly of sophisticated military gear) (4) minimal wage/price distortions (including low barriers to trade and foreign investment) (5) early and adequate adjustment to the oil boom and bust (6) supporting cultural and entrepreneurial factors (7) a sufficiently large domestic market (8) the availability of non-oil resources (including labour) (9) relatively equitable income distribution or else an effective safety net and (10) an efficient, clean government including an impartial judiciary capable of maintaining the rule of law (Table 24).

By contrast, glaring failures in terms of slow growth, high inflation, widening income gaps, balance-of-payment deficits, unemployment or underemployment and foreign debt were the results of a combination of the following six factors: (a) undue state intervention in the economy (b) volatility of regulations and directives (c) poor choice of national strategies and policies (including encouragement of population growth) (d) unsustainable public welfare services and subsidies (e) selection of costly and complicated capital-intensive public projects and (f) excessive tolerance of harmful rent-seeking activities, corruption and waste (Table 25).

None of the group members enjoyed all the positive elements or escaped the pitfalls altogether. Everyone succumbed to the temptation and greed that easy money-making usually spawns among decision-makers. None could avoid the breakdown of bureaucratic discipline and business ethics that sudden wealth often brings. They all found it exceedingly difficult to spend their windfalls wisely, or to deal with their shortfalls expeditiously. The shortage of profitable projects proved as frustrating and ultimately wasteful as did the abundance of investment outlets in need of capital in oil-less countries. All members were engaged in hasty or ill-conceived expenditure of their boom earnings as though the oil bounty was limitless; and they tried their best to maintain the pace of development during the oil bust through foreign borrowing as though the shortfalls were only temporary.

Nevertheless, a broad-brush application of a composite index made up of the foregoing variables seems able to identify the winners and losers better than any single factor. Thus, Indonesia which accommodated eight out of the ten constructive factors relatively better ended up with the best relative performance. By contrast, Gabon, Iraq, Libya, Nigeria and Venezuela which enjoyed no more than three positive elements performed most poorly. Other members' performance could also be correlated with these positive elements, although not always in precise quantitative terms. In the same vein, countries with the largest number of negative elements again proved to be the less successful ones. As Table 25 shows, Gabon, Iraq, Libya, Nigeria and Venezuela suffered from all the negative factors, where Ecuador and Indonesia were more relatively immune to these adverse influences.

NOTES ON CHAPTER VII

1   The concept of growth and its correct measurement has been a matter of controversy for years. While growth itself is defined as an annual addition to real gross domestic product, and its rise is widely regarded as the answer to the world's poverty problems, some economists have challenged both ideas. See Herman Daly (1996).

2   See J.D. Sachs and A.M. Warner, *Natural Resource Abundance and Economic Growth* (Cambridge: National Bureau of Economic Research, December 1995); see also 'The Natural Resources Myth', *The Economist*, 23 December 1995; and H.W. French, 'The Curse of Riches', *New York Times*, 18 January 1998.

3   'Bonanza development' through minerals in Chile and Peru with subsequent dire circumstances for both are other examples.

4   See, for example, Ahmad Jazayeri (1988), pp 98–101; S.J. Scherr, 'Agriculture in an Export Boom Economy', *World Developments*, April 1989; J.J. Struthers, 'Nigerian Oil and Exchange Rates', *Development and Change*, April 1990.

5   See John Page, *From Boom to Bust – And Back?* (Washington: The World Bank, working paper, May 1995).

6   *Claiming the Future* (Washington: The World Bank, 1995), p 26.

7   See *1997 Index of Economic Freedom* (New York: *Wall Street Journal*, 1997).

8   See 'Do Income Gaps Retard Growth', *New York Times*, 20 August 1995.

9   See Deepak Lal (1983).

10 Richard N. Cooper (1994).

11 See R.J. Barro, 'Democracy: A Recipe for Growth?' *Wall Street Journal*, 1 December 1994.

12 See *Freedom in the World* (Washington: Freedom House, 1996) for individual country rankings.

13 See S. Akhavi in F. Tachan (ed.), *Political Elites and Political Developments in the Middle East* (Cambridge: Schenkman, 1974); and J.P. Entelis in N.A. Sherbiny (ed.), *Arab Oil* (New York: Praeger, 1976).

14 See A. Sid Ahmed (1990).

15 See *Washington Post*, 4 July 1994, 18 August 1994 and 18 December 1994.

16 See Wole Soyinka (1996).

17 See 'Qatar', *Wall Street Journal*, 18 September 1996.

18 Cf *Corruption, Governmental Activities, and Markets* (IMF Working Paper No. WP/94/99 1994).

19 See 'A Global Gauge of Greased Palms' and 'The Worldly Business of Bribes', *New York Times*, 20 August 1995 and 8 July 1996.

20 See 'Saudi Trial May Air Allegations of Leadership Corruption', *Washington Post*, 2 January 1996.
21 T.L. Karl (1997), pp 227, 237.
22 For the definition and functions of a developmental state, see *Trade and development Report*, 1996, Part II.
23 Another factor of crucial significance, long overlooked by admirers of the Asian model, was the fact that the 'tigers' growth was mainly due to large inputs of capital and labour, with relatively limited increases in productivity. See Charles Wolf Jr, 'What Caused Asia's Crash?' *Wall Street Journal*, 4 February 1998.
24 For a full exposition of this thesis see D.S. Landes, *The Wealth and Poverty of Nations* (New York: W.W. Norton, 1998)
25 A recent study by the National Review of Economic Research which deals with factors that affect 'good' government finds history and tradition among major determinants. 'Good' government (identified as being efficient in delivering the goods, protecting individual freedoms and interfering minimally with the private sector) is looked at in 150 countries. Among those at the bottom 10 are Algeria and Indonesia. See *Washington Post*, 1 November 1998, C5.

VIII

Summary of Findings

No outcome in the annals of this century's economic forecasts humiliated (even if it did not humble) renowned soothsayers so much as OPEC's fate. On the eve of the third decade after the historic oil price rise of 1973/4, neither OPEC itself nor any of its members looked anything like resembling the composite picture projected by expert analysts in the early 1970s. Contrary to all predictions, OPEC never acquired the exclusive power to set oil prices or restrict any country's access to oil supplies; it could not even effectively cope with its own cycles of boom and bust. While the global petroleum industry was drastically revamped after 1973/4, the 'new oil order' was hardly remade in OPEC's favour. The victory over the oil majors in wresting control of national oil resources was largely pyrrhic as many members (e.g. Algeria, Iran, Iraq, Kuwait, Qatar and Venezuela) were only too anxious at the end to invite the multinational oil companies back for fresh investment in exploration and extraction. Instead of being able to influence the West's policies and priorities, some OPEC members were not even capable of defending their own territorial integrity without Western military support and weapons sales. Far from becoming bankers to the world, nearly all group members within a short decade joined the world's debtors, and by the end of 1994 accumulated an estimated total external debt of some $390 billion. Four members – Algeria, Ecuador, Gabon and Venezuela, were wards of the International Monetary Fund.

World progress was not set back as predicted. The West's political stability, its free institutions and its internal and external security were hardly affected. Confounding the experts, even the impact of the oil price rise on the economies of the major oil importing countries was, by and large, limited, short-lived, country-specific and not altogether negative. The rise in crude prices – affecting the economies of Europe, Japan and the United States through changes in the terms of trade, rising interest rates and deliberately chosen deflationary domestic polices – resulted in a fairly quick adjustment with an ultimately beneficial effect. In a strange kind of way, the much lamented escalation in the nominal price of crude oil effectively put an end to the orgy of energy profligacy which was neither sustainable nor environmentally justified. Rising oil prices encouraged greater fuel conservation in transport vehicles, increased efficiency in energy/output ratios, helped redesign energy-using machinery and stimulated the development of oil substitutes both conventional (coal, hydro-electric and nuclear) as well as exotic (solar, wind and geothermal). Higher oil prices made the exploitation of oil deposits in high-cost non-OPEC areas profitable, largely at OPEC's expense.

In retrospect, most economic analysts and oil economists underestimated long-term price elasticities of oil demand and supply.[1] Contrary to expectations, rising oil prices cut per capita world demand for oil from about 5 barrels in 1974 to about 4 barrels in 1994. During the same period, per capita world reserves of petroleum increased from about 160 barrels to about 190 barrels. Per capita world consumption of refined petroleum products fell by nearly 20 percent in the 20-year period in response to higher oil prices. At the same time, the world's proven oil reserves rose from 650 billion barrels in 1974 to 1,100 billion barrels in 1994.[2] World population rose from nearly 4 billion to over 5.5 billion during the period, and OPEC's population increased from 294 million to 463 million. OPEC's total output in the mid-1980s fell to barely one third of the volume widely projected to be reached. The organisation's terms of trade *vis-à-vis* the West, instead of steadily improving as predicted, deteriorated in real terms to such an extent that the inflation-adjusted crude price in 1994 was actually below the 1973/4 level. And rapid population growth brought OPEC's per capita oil revenue in 1994 down to only about 22 percent of the 1981 peak.

With one country after another facing mounting economic difficulties after the second oil boom, OPEC's influence and power began to fade. OPEC's ability to influence prices – limited at best to temporarily withholding supply or flooding the market backed by excess capacity – was substantially reduced by the mid-1990s. Not only did OPEC become an increasingly residual oil supplier, it could not even maintain its initial residual share. Almost the entire 2mb/d increase in world demand for oil between 1993 and 1995, for example, was met by non-OPEC producers. And with marginal cost of extraction dropping to $7–8/b by 1994 due to advances in exploration technology, it looked as though any increase in non-OPEC supplies for some time in the near future might come largely at OPEC's expense. The dreaded 'oil weapon' was not used by OPEC members against oil users; rather, it became a policy ploy in the hands of some major consuming countries (particularly the United States) to punish oil exporters, e.g. when oil sanctions were imposed on Iran, Iraq and Libya (and threatened against Nigeria) during the 1980s and 1990s.

Nor did OPEC economies fare as projected. During the short span of twenty years, 1974–94, group members earned more than $3 trillion from exports of oil and gas – more than they had ever received from this source before in their entire energy export history. The extraordinary magnitude of these revenues, and their direct accrual to state treasuries, made their allocation a new bureaucratic task for which government planners and policymakers were ill prepared. What made this task infinitely more difficult was the extreme volatility of annual energy revenues, which played havoc with the countries' half-baked and unsophisticated plans. Due mainly to the ups and downs in prices and volumes of exports (see Chapter III), nearly 60 percent of OPEC's total 20-year proceeds was received in ten years of oil booms – 1974–7, 1979–82 and 1990–1. By contrast, during nearly half of the period when the oil market was depressed (1983–9), petroleum revenues amounted to about one fourth of the total. This troublesome volatility of oil income in the face of committed or planned state expenditures required repeated cyclical adjustments in domestic economies.

The challenges faced by OPEC members during their boom and bust cycles, and the responses shown in dealing with both fortune and misfortune, seem to underscore the universality of certain seemingly irresistible forces (or inevitable traps) that shape the process of development in an oil-reliant economy. The astonishing similarity of OPEC members' experiences in their oil-financed development, despite their vastly dissimilar characteristics, seems to confirm the intriguing role of 'oil wealth' in the growth process. When thirteen disparate nations starting from different vantage points, end up with uncommonly similar internal and external imbalances, dislocations and setbacks, the results cannot all be attributed to bad luck. Can oil patrimony be really tied in with oil profligacy?

Some outstanding manifestations of such oil-related forces seem evidenced by the OPEC experience in several key areas: (1) growing power and influence of the state (2) a propensity toward maximum domestic utilisation of oil windfalls (3) a tendency toward hasty spending (particularly investment) decisions (4) a preference for investment in resource-based industries (5) a lack of policy coordination and supportive measures for return maximisation (6) increased annual expenditure on defence and security, even in non-military-led regimes and (7) the emergence of a 'petro-culture' inimical to growth.

GROWTH OF STATE POWER

The clearest tendency among members was for the state to assume, by necessity or by design, an increasingly dominant role in the economy. With the government as the exclusive owner/exploiter of oil deposits, the state management of the rising oil wealth led to the expansion of bureaucracy and substitution of 'planners' preferences' for those of consumers and the private sector. The dominance of state decisions over the free market's impersonal forces influenced both the character and composition of domestic product as well as the direction and pace of national income development. Even in countries (such as post-1979 Iran) where the state's financial stake in the economy (i.e. the ratio of public revenue to GDP) declined or remained the same, the social engineering and regulatory roles of the government rose significantly. Not only in those group members which were ideologically bent on pursuing a 'socialist transformation' or a 'non-capitalist road' (e.g. Algeria, Iraq and Libya) but also in staunchly free-market economies (e.g. Kuwait, Qatar and Saudi Arabia), the government became the architect of socio-economic transformation. Even in relatively more 'democratic' countries such as Ecuador, Gabon and Venezuela, the state acquired an increasingly stronger hand.

Thanks to the enhanced oil revenues accruing directly to the state treasury, a political leadership that was traditionally separated from and independent of the people in most member countries became also economically and financial independent and autonomous. No longer obliged to 'pluck the geese' (i.e. extract funds from domestic taxpayers) in order to finance public outlays, the state's financial strength and spending capacity now enabled it triumphantly to avoid public scrutiny of its expenses and extravagances. Instead of becoming more democratic over time, the government, in nearly all group members, added to its authoritarian rule, at least in the economic arena.

The oil industry's scant backward and forward linkages with the rest of the economy did little directly to help create a sizeable middle class or a vigorous civil society that could provide a check on government power. Oil revenues enabled the government in nearly all members to have more resources at its disposal and to be more self-reliant, stronger, less responsive to the people and more arbitrary in its decisions than would have been the case without oil. Oil income was used to secure political peace, if not loyalty, ensure public employment as the first option, distribute patronage when effective and coopt the opposition whenever possible. In almost all member countries, petrodollars enabled inefficient, corrupt and wasteful governments to cling to power long after they would have been seriously threatened, if not actually overthrown. Growing state power and functions, in turn, helped to reduce individual responsibilities, increased disincentives for work and saving, contributed to a breakdown of traditional family values and reduced the chances for the development of civil society.

With the state becoming in varying degrees the collector of the oil rent not on behalf of the citizenry but as its sole legitimate claimant, both the institutional character of state functions and the nature of private economic decisions underwent profound changes. The state bureaucracy, living off the oil revenues for the most part, emerged as an agent of allocation and redistribution rather than one of extraction and appropriation. At its extreme form, the state took on a corporatist face, in which special groups, activities and sectors were favoured, promoted or discouraged through its distribution policies.[3] At the same time, the private sector, becoming increasingly dependent on state guidance, credit, protection and subsidies, tended to lose, or failed to develop, the dynamic, innovative, risk-taking and independent spirit required for a healthy nation-building effort.[4]

It was only toward the end of the period that a cautious retreat from the state's dominant position got underway via 'privatisation'. Disappointment with the performance of state-owned enterprises in a majority of OPEC members, and the increasing inability of fund-starved treasuries to subsidise money-losing public entities or to recoup their losses, prompted the governments to seek relief through the sale of state enterprises. International financial institutions and other lending agencies also made the sale of inefficient public businesses a condition of further assistance. But privatisation programmes in almost all countries which adopted them (particularly Iran, Kuwait, Saudi Arabia, Venezuela) had very slow progress due to a number of factors. National security interest in 'strategic' industries (e.g. oil) was routinely cited in order to exclude likely and attractive candidates. Few of the enterprises offered for sale promised to become quickly profitable without a substantial increase in product prices or service fees, or without considerable down-sizing of the workforce – both of which faced strong internal political resistance. In some cases the domestic market had inadequate liquidity to absorb the shares issued by state companies; and attracting foreign private investors was not easy. Finally, the genuine privatisation that would loosen the technocratic hold on power was effectively sabotaged by the bureaucratic vested interests that ran the public enterprises and the labour unions that manned them.

DOMESTIC PRIORITY IN WINDFALL ALLOCATION

A majority of members, particularly in the high-absorbing countries, opted to spend their new wealth at home, instead of placing it abroad for slower and more orderly subsequent drawdowns for domestic investment. Some (e.g. Algeria, Iran and Venezuela) even took advantage of their high credit rating in the 1970s to borrow in the international capital market to add to their windfalls for domestic capacity expansion. It was argued that only a broad vision and high growth targets – even if unreachable at first – could best energise and optimally utilise a nation's resources and talent.

This maximalist approach, rationalised by a number of political, psychological and security arguments, was based on a disingenuous belief in the magical power of money to solve all developmental problems. This confidence, rooted in the early postwar literature on the so-called vicious 'circle of poverty and productivity', gained an extraordinary boost from the euphoria of the 1973/4 oil windfalls. In a state of 'lyrical illusion', the sudden oil affluence was seen as the key to unhindered growth and prosperity. Under this blind faith in the purse's power, national plans and public expenditures in nearly all countries were immediately expanded, sometimes several fold, as in Algeria, Ecuador, Iran, Nigeria, Saudi Arabia and Venezuela. Fiscal outlays under the first oil boom were two to five times larger than the corresponding annual averages during the previous five years, 1968–73. Nine out of thirteen OPEC members devoted between one fourth and one half of their GDPs to domestic investment – a feat never realised previously anywhere among developing economies. Growing public outlays were supplemented by the expansion of credit to the private sector, doubling the previous five-year record. This inward bias in the disposition of oil windfalls was naturally accompanied by a good deal of waste.

Two particular misperceptions on the planners' part were responsible for this strategy. First, it was uncritically assumed that oil prices and export volumes were on a steady rise, and that they could expect to earn larger sums each year from crude exports. Second, planned expenditures were uncritically geared to the projected oil revenues rather than domestic absorptive capacity. Ignored or woefully underestimated were such potential bottlenecks as inadequate domestic infrastructure (port and transport capacity, communication facilities, warehousing space, power supply building materials), a colossal shortage of managerial and skilled workforce, and an inefficient administrative superstructure.

There were several reasons for policymakers in all but a few low-absorbers in the Persian Gulf to spend the oil receipts at home fast rather than saving them abroad. To begin with, there was a determination to show the world (and, by implication, the envious non-oil Third World) that the oil exporters were themselves poor developing countries who badly needed capital for their own domestic development and who, while being fortunate in receiving the oil bonanza, could spend every penny for improving their people's welfare. There was also a latent fear that if the windfalls were not rapidly and justifiably absorbed, there might be a global backlash and some attempt by powerful consuming countries to force an oil price rollback. Furthermore, a quick allocation of the oil money to long-awaited and popular projects (like a steel mill or a petrochemical plant) was considered a smart ploy to win or regain domestic political approval and support. Still further, the use of the windfalls for immediate private consumption was sure to alleviate endemic poverty in most member countries and give back to the people a share of 'their own money'. Finally, various vested interests within the bureaucracy (military, industrial, agricultural and energy groups), competing with one another to get a share of the bounty for their own agenda, left no room for rational allocation and 'saving abroad'.

HASTY COMMITMENTS

The speed with which oil windfalls (and, in some cases, additional borrowed resources) were earmarked for domestic use by all but one or two members preempted a deliberate, careful and rational cost/benefit analysis of competing investment projects. Rather, a host of extraneous, non-economic, and politically popular schemes were uncritically adopted and hastily launched. The desire to spend rapidly necessitated moves in four distinct directions: (a) public investments took the lion's share of the oil windfalls – justified by their ease of implementation, combined with such other considerations as national social welfare, national security or inadequate private sector interest in them; (b) government capital formation favoured large projects that were few in number and absorbed a big chunk of funds quickly (thus minimising allocation battles among prospective claimants); (c) state outlays were, in most cases, urban-oriented due to the larger availability of complementary factors; and (d) the largest share of public investment was earmarked for either physical infrastructure or human capital formation due to both clear necessity and the relative ease of implementation. The latter projects were especially favoured in the low-absorbing countries of the Persian Gulf region where such outlays often received more than two-thirds of development budget allocation.

Inadequate planning and the absence of proper risk calculations frequently resulted in substantial cost overruns and completion delays among industrial projects in the traded sector. Even when comparative advantage existed, in the case of feedstock costs (as in petrochemicals), plant profitability and global competitiveness were not assured because there were offsetting disadvantages in other cost components (e.g. skilled labour, management expertise, marketing and various externalities). In many cases, the size of the industrial plant was much larger than warranted by the size of domestic demand; and there were no guaranteed export markets since many members were competing with each other in the same industry and targeting each other's intended markets for their surplus products. And since the technology used in all modern industrial undertakings was totally borrowed, and there was little or no internal technological inputs, OPEC upstarts were exposed to competition from older and more sophisticated global rivals.

Investments in physical infrastructure, while both necessary and useful were again favoured not because of their calculated productive worth but on account of the convenience of undertaking them with the help of foreign contractors and foreign equipment, their symbolic conveyance of the status of modernity and progress, and their close relationships with rent-seeking activities. In some of the countries (e.g. Persian Gulf members), physical infrastructural projects were not a means of preparing or supporting the productive bases but often an end in themselves. Apart from their minimal benefits for improving internal social cohesion and political security, the true value of highly sophisticated transport and communications networks, in countries with extremely harsh climatic and high-cost conditions, was seldom subjected to careful scrutiny. This was as true for a new capital city in Nigeria, the Trans-Gabon Railway and the Man-Made River in Libya as it was for multiple international airports in the tiny UAE. Not only did such infrastructural investments fail to return anything on their invested capital, not even their maintenance costs could be afforded by the non-oil sector for which they were presumably built. There were legitimate questions whether the full costs of the infrastructure outlays for the Jubail complex in Saudi Arabia or Jebel Ali in the UAE could be amortised by their industrial superstructure.[5] These ultra-modern facilities thus committed the nations to future maintenance expenditures to be paid, *in perpetuity*, out of the oil income.

The development of social and institutional infrastructure (e.g. education, health and housing) was universally emphasised by all group members, albeit in varying degrees. Outlays in these areas were urgently needed due to high illiteracy, poor health conditions and shortages of decent housing – all remnants of pre-oil poverty. Expenditures in these sectors were politically popular, and economically justified as the sole guarantee of viable post-oil development. Implementation was also easy since dependence on foreign materials was minimal (except for the Arab members who were short of both human and physical resources). And, finally, improvements in education, health and housing could be racked up not only as development determinants (and as a means of nation-building) but as an end in themselves, i.e. as part of the social welfare objectives. But even in these seemingly rational and necessary undertakings, substance was often sacrificed for form. The rapid increase in the number of school graduates in the absence of corresponding growth in available employment, for example, resulted in a spectacular rise in the size of the civil service in nearly all member countries and a vast cadre of underemployed bureaucrats.

A similar haste and inadequate calculation were shown when the second oil boom was followed by an oil bust. Plummeting oil revenues in the mid-1980s resulted in the postponement of major planned projects and non-completion of others – again without careful evaluation or cost/benefit analysis. Some worthwhile projects were sacrificed for the benefit of more popular but unprofitable ones. In the process of this deflationary adjustment, private investment was often 'crowded out' by public sector outlays. Consequently, undue haste in the disposition of oil windfalls, coupled with unrealistic assumptions as to their future availability and absorbability, left no room for the oil income allocation to be carried out in a calm, deliberate, well-planned or dispassionate manner.

INDUSTRIALISATION AS DEUS-EX-MACHINA

There was commonly a heavy stress in windfall allocation on modern, capital-intensive and high-cost industries related to oil (petroleum refining and petrochemicals) or to cheap energy (steel, aluminum and metallurgy). This preference for resource-based industrialisation was due to the petroleum sector's scant direct forward and backward linkage with the rest of the economy. Unlike other dominant single industries (e.g. automobiles, construction, housing, tourism etc.) which draw their inputs from a wide variety of other smaller industries, the oil sector was too capital-intensive and too technologically advanced to offer any appreciable backward linkages). Forward-linkage projects such as petrochemicals, steel and aluminum which were fuel-based were thus selected nearly everywhere to fill the void. Having long attributed the advanced countries' politico-economic clout to their military/industrial power and having always identified the development process with industrialisation, the conventional wisdom among OPEC state planners was that the only way to get off the dead centre of poverty and backwardness was to industrialise. What had long kept them from pursuing such an industrialisation strategy, it was argued, was the lack of hard-currency funds. Oil windfalls now effectively and painlessly removed this insurmountable obstacle and gave OPEC decision-makers a free land in selecting 'strategic' projects.

This argument was bolstered by other supporting considerations. Administratively, and from a practical standpoint, large metals, petrochemicals or LNG projects seemed very attractive: they offered the opportunity to absorb windfalls rapidly, with a minimum decision time; did away with long and extensive comparative studies of competing projects; maximised import content and compensated for lack of input at home; involved mostly automated operations requiring few skilled labourers; and called for minimal institutional or other reforms. Nationalistic and prestige considerations were another major impetus for the adoption of resource-based industries (RBI). The development of direct-conversion steel, petrochemicals, non-ferrous metals, heavy machine tools and giant agribusiness enterprises meant a shift of national status from a backward, raw material or handicrafts exporter to a high-tech industrial nation. For all member countries (and particularly for the high absorbers), steel mills and national airlines had been at the top of political agenda for a long time.

211

From a political viewpoint, RBI projects provided not only high, popular acceptability as a rational choice but fitted into the chosen import substitution strategy, reducing dependence on the industrial powers and optimising the opportunity for various rent-seeking, sweetheart deals. On the economic side, RBI promised further major benefits: (a) comparative advantages inherent in an abundant, natural, domestic, resource; (b) production linkages to more labour-intensive downstream activities compensating for large initial capital outlays; (c) access to foreign capital markets at attractive interest rates for financing investments; (d) additional foreign exchange earnings; (e) new public revenues from profits on the sale of products or taxes on their consumption; (f) an opportunity to promote the use of modern technology and to enhance the national skills level; and (g) economic diversification as the paramount objective of windfalls use.[6]

OPEC members' experience in their resource-based industrialisation strategy showed substantial differences with respect to allocation and performance. According to a study that comprised eight oil exporters, including four OPEC economies, performance variations could be traced to three basic phenomena: project mix, type of enterprise and macroeconomic policy.[7] For example, countries that earmarked the lion's share of their RBI capital fund to one or two dominant and inefficient projects fared poorly *vis-à-vis* those who favoured smaller and more competitive ones. Thus, large steel complexes, which underpinned and dominated Nigerian and Venezuelan RBI, resulted in relatively poor overall performance. Similarly, the concentration on Olefins plants and large export refineries in Saudi Arabia was clearly excessive in comparison with appropriate development of LNG in Indonesia. Small chemical projects proved more efficient and more resilient than small metal projects. The type of enterprise was another critical factor. The size and scale of projects, the nature of ownership and operation, and access to foreign technology and marketing facilities formed the essential components of this factor. The mega scale projects which were more complex and technically demanding resulted in cost overruns and delays in completion.[8] Poor performance was also related to its sponsorship. State-owned enterprises (SOE), for example, often proved disastrous regardless of their scale, while joint ventures with foreign partners (particularly with more than 25 percent ownership by multi-national corporations) proved less prone to cost overruns, technical shortcomings and marketing difficulties. In the four OPEC members included in the study, the incremental capital output ratio on RBI investment in the mid-1980s ranged from 5.2 in Indonesia, to 8.5 in Venezuela, 10.7 in Saudi Arabia and 39.2 percent in Nigeria, where a value of 3 or less commonly indicates efficient use of capital.

Domestic industrialisation in most members went, of course, beyond RBI projects. While Nigeria, Saudi Arabia and Venezuela relied too heavily on RBI, Indonesia's reliance was moderate. Algeria, Iran and Libya also launched ambitious steel projects with highly questionable results. There were investments in light consumer industries as well. But the focus of public investment was on heavy industries. This preference for capital-intensive industrialisation – in the name of modernisation and progress – was in theory both simple and popular: Heavy industries, it was believed, were needed to establish a modern industrial base. Latest equipment was both more cost-effective than older and a better means of training skilled workers. Agriculture might be neglected at first but it would be helped once the basic industries got into full swing. State-of-the-art technology was needed to promote and ensure international competitiveness. The rise in growth and income would in time lead to high employment.[9] In practice, however, the strategy resulted in the transplantation of advanced and 'encapsulated' technology and industrial discipline into countries with little or no tradition in such sophisticated undertakings and an extremely meagre stock of skilled workers, technicians, engineers and managers. The limited domestic markets could not absorb the output of the minimally profitable scale of production in such technically large-scale industries, and exports were a feasible alternative only when comparative advantages were clearly present. As a result, equipment was not always properly used, management and highly-skilled functions stayed in expatriates' hands and capacity was perennially underutilised.

A former OPEC Secretary General frankly admitted that 'on pure economic grounds' many projects in his own country and in other member nations 'would not have been embarked on'.[10] The dreams of establishing home-based aerospace and automobile industries in Indonesia or a domestically designed and built 'national' passenger car in Iran's Islamic Republic were admittedly beyond pure economic calculation and opportunity cost consideration. The leaders repeatedly emphasised the 'intangible' (e.g. psychological, social, political) benefits from such endeavours. Reference was often made to national pride and to the inspiration that is felt when nations are ready to join the developed world.

LACK OF POLICY COORDINATION

Domestic macroeconomic policies adopted by nearly all group members – whether by design or by default – often proved inadequate to properly manage aggregate demand and in some cases actually aggravated internal crises. The double-barrelled aim of ensuring economic efficiency (through the establishment of a dominant 'state bourgeoisie') and enhancing distributional equity (through welfare transfers and subsidies), without having sufficient administrative and technical capabilities, confronted the member governments with an impossible mission.

Wages and salaries, for example, were raised or allowed to rise for public servants, the armed forces and urban workers, with a view to spreading the benefits of the oil bonanza among its natural claimants, securing industrial peace and gaining political loyalty. Although in many cases, wage rates did not catch up with inflation, in nearly all cases the increase in real wages was not accompanied by a corresponding rise in productivity. Higher labour costs duly spread to small scale private industries, non-oil exports and import-substituting activities, raising the price level. Fast growing public and private spending, in the absence of appropriate taxation and efficient resource mobilisation, put heavy pressure on domestic productive capacity. Public investments in highly capital-intensive industries with a long gestation period and in costly physical infrastructure contributed to enormous expansions of aggregate demand without creating sufficient aggregate supply. Galloping inflation was invariably the outcome. In a drive to combat inflation when it was significant the authorities (in Algeria, Ecuador, Iran, Libya, Nigeria, Venezuela), somewhat wishfully, tried to relieve the obvious symptoms without daring to deal with the underlying causes. Instead of instituting structural reforms at an early stage, raising taxes, reducing budget deficits, adjusting the exchange rate, increasing domestic and international competitiveness, and freeing interest rates, resort was made to wage and price controls, trade and exchange restrictions, credit rationing and other administrative measures.

Rising prices caused domestic currencies to appreciate in real terms. Yet overvalued exchange rates were, except in rare cases (e.g. Indonesia, Venezuela), allowed to prevail unadjusted over an extended period for a variety of reasons. In almost all members' economic culture, a high domestic currency value was a matter of national pride and devaluation was considered a sign of economic weakness, if not humiliation. Therefore, any other policy ploy that was not so highly visible (even if no more defensible and even more disruptive) was politically preferred to outright devaluation unless all other courses of action were exhausted. Furthermore, the choice of an import-substitution strategy in nearly all members' diversification and industrialisation plans required a large degree of home-industry protection: exchange overvaluation was a subtle means of implementing that strategy. At the same time, non-oil (mostly agricultural and raw material) exports which constituted a rather insignificant part of total national exports in nearly all member countries – and often faced low demand elasticity – would not gain much from exchange depreciation. Still further, in those member countries where a free (black or grey) exchange market operated alongside the official market, the existence of high scarcity rents generated by exchange controls and trade restrictions (e.g. foreign currency allocation or import quotas) had established some strong vested interests that successfully fought against exchange devaluation (or trade liberalisation). Finally, some influential pressure groups in a number of member countries, anxious to transfer their local currency assets abroad, would find it advantageous to keep highly appreciated exchange rates in place and would strongly resist any talk of devaluation.

Still further, nearly all OPEC members kept domestic energy prices at or below cost in order to keep inflation down and to enhance the competitiveness of their energy-based industries in world markets. Pricing of energy products on the basis of production rather than opportunity costs, however, caused serious economic distortions in domestic resource allocation, requiring new countervailing public policies. For much the same reasons, interest rates on saving deposits and investment loans were kept artificially low and totally divorced from the opportunity costs of capital at home and abroad. In short, the fiscal, monetary and industry-specific policies needed to correct emerging distortions often proved difficult to adopt. Despite repeated resolves by all members to reduce national economic dependence on oil exports, most countries opted in favour of import-substitution (rather than export-promotion) projects.

Equity considerations prompted the governments to divert a sizeable chunk of the windfalls to the creation of a western-style welfare state, without the benefits of prior industrialisation or adequate government machinery. Under strong political pressure from a largely poor, frustrated and impatient population, demanding to receive a share of the oil bounty, government leaders found it necessary and expedient to increase direct transfers of oil resources to poorer social strata and to expand state subsidies. Subsidies on food, fuel, housing and other amenities had the threefold objective of helping low-income consumers, reducing costs of inputs in various economic activities and protecting infant industries against foreign competition. Relatively modest at first, these outlays gradually gathered a momentum of their own, and began steadily to rise. In the semi-rentier states (e.g. Kuwait, Qatar, Saudi Arabia and the UAE), welfare benefits encompassed nearly all public goods and services. Transfers and subsidies were instituted with great abandon at first, and easily financed while the oil booms lasted. But once oil revenues stopped growing or changed course in subsequent slack periods, these outlays turned into major burdens on the budget and spawned large fiscal deficits year after year. Curiously enough, some of the major beneficiaries of price and input subsidies were the urban elites who constituted the core of the bureaucracies and almost universally resisted the subsidies removal or reduction.

By allowing the exchange rates to remain overvalued in nominal or real terms, the balance of payments came under increasing pressure, and a resort to external borrowing became unavoidable for most members. But foreign credits, mostly short-term and high-cost, simply postponed the day of reckoning. Payment arrears began to accumulate and debt rescheduling became inescapable. The harsh discipline demanded by the international financial institutions as a condition of their further assistance was at first avoided at all costs, and accepted only when all sources of foreign credit actually dried up. Algeria, Ecuador, Gabon, Nigeria and Venezuela eventually became the wards of the International Monetary Fund during the 1980s and 1990s.

Meanwhile, the phenomenal bloating of inefficient and costly bureaucracies, the mushrooming of money-losing state industrial enterprises, the run-away growth of annual public subsidies and entitlements on top of military and defence outlays produced protracted and intractable budget shortfalls. The reluctance to adjust the exchange rate (and thus to increase state revenues in terms of local currency) aggravated the situation. And financing the budget deficits by borrowing from the domestic banking system or via the 'printing press' steadily increased domestic liquidity and fanned the inflationary fires. Windfalls also encouraged policies favouring easy access to imports and a higher propensity to consume. While making import-substitution difficult, expensive and non-competitive, no real incentive was offered to increase non-oil exports in order to pay for imports in the long run. The mechanisation of agriculture and industry was also concentrated on assembly plants needing foreign raw materials and processed goods for continued full capacity operation.

The desire for a larger national population was another unhelpful policy. This longing was shared by nearly all members and was fulfilled by most. The objective was rationalised by a variety of reasons. In the Persian Gulf countries with small populations at the start, larger birth rates were coveted and encouraged as useful, expedient and not worrisome; it was considered a means of eventually reducing dependence on foreign labour. In Iran after 1979, larger families were actively promoted with the avowed goal of creating new defenders and guardians of the 'Islamic Revolution'. In some left-leaning members (e.g. Algeria), population control was initially regarded as a clever ploy to perpetuate Western dominance over Africa and Asia – a policy that had to be countered by all ex-colonies. In others, the policy was defended as a way of increasing national labour resources in the service of development. The average annual growth rate of population in OPEC members between 1973 and 1985 was 4.1 percent, almost twice the world average. Between 1985 and 1994, the average fell to about 3 percent – still considerably higher than the global rate. In the absence of appropriate educational and labour policies, the near population explosion in almost all countries produced significant open or concealed unemployment. Between 1974 and 1982, when oil revenues were rising faster than population, the extra mouths could be easily fed as per capita income kept on rising. After 1983, when the GNP/population ratios were reversed, a gradual decline in the standard of living was much in evidence. Except for Indonesia, which managed to keep annual population growth below 2 percent and enjoyed a thriving economy, per capita income continued to fall even in countries where growth was positive (e.g. Iran, Qatar and Saudi Arabia).

EXCESSIVE DEFENCE BUILD-UP

The ease with which oil revenues were obtained without belt-tightening or sacrifices in civilian development, offered the oil exporters an unusual and unprecedented opportunity to increase military expenditures at will and, in a majority of cases, much beyond national security needs – a feat never imagined, or feasible, without the oil windfalls. Equating national security with military power, many members, particularly in the Persian Gulf region, expanded their military expenditures both absolutely and as a percentage of the national budget and gross domestic expenditures (Table 15).

More than was common in other developing countries, the military build-up focused on the quantity and modernity of the weapons systems at the expense of adequate training and logistical readiness, proper command and control, and leadership quality. As a result, even the highest military spenders (e.g. Kuwait and Saudi Arabia) were still unable to defend themselves against Iraq in 1990–1. Furthermore, sophisticated defence expenditures, in addition to initially diverting precious resources from more productive investments, entailed further ancillary outlays for infrastructure, military training, equipment maintenance, spare parts and perpetual renovation. In countries that were effectively run by the military – Algeria, Indonesia, Iraq, Libya and Nigeria – the leadership could be expected to favour an arms build-up. Yet, curiously enough, the militarist regimes did not always rank highest in their defence/GDP ratios, as Table 15 shows. High military expenditures were often rationalised not only as a deterrent against aggression by hostile neighbors (e.g. Iran and Iraq against each other, and other Persian Gulf countries against Iran and Iraq) but also as a means of achieving other worthwhile objectives. Among the latter, reference was frequently made to the provision of education and employment for the young village recruits as well as of the technological spin-off for civilian industry.[11]

In reality, the reasons for continued upgrading and expansion of national defences with new, state-of-the-art acquisitions were many and often unrelated to true domestic

216

needs. The non-democratic character of government in most of the member countries and their leaders' insecurity of tenure was a prime reason. Ongoing tensions in the Middle East and North Africa, along with border disputes and ethnic and religious conflicts elsewhere, were another. And, most particularly, the authorities' virtual free hand in deciding how to spend the oil windfalls contributed to the rise of a domestic military/business complex. Under the latter's influence, the possession of sophisticated military hardware was promoted as a sign of national prestige and industrial-class status. Intensive and relentless lobbying on the part of foreign competitive suppliers (particularly American, Russian, British and French arms manufacturers) was also a significant factor. Arms sales provided a much needed boost to domestic employment in the industrial countries' defence industries, particularly after the end of the cold war. Individual defence contractors and companies had their own special interests to promote. And unscrupulous or corrupt officials at the receiving end had their fortunes closely tied to such military procurement. Arms deals are aptly considered the world's most bribe-intensive business.

EMERGENCE OF A NEW PETRO-CULTURE

By far the most common (and most pernicious) outcome of the oil bounty was the rise of a new petroculture in nearly all member economies (but particularly among the Arab oil producers). This new culture – variously termed petromania, quick-money fever or the catch-as-catch-can syndrome – was planted, nurtured and pampered by the oil windfalls. It gradually weakened the traditional work ethic among native populations, reduced incentives for risk-taking, hard work and independent entrepreneurship, lowered natural tolerance for temporary deprivation and austerity, encouraged rent-seeking activities and raised popular expectations beyond reasonable means of satisfying them. The new cultural mindset intensified the desire on the part of the rank-and-file for seeking easy solutions through state intervention. With the oil wealth clearly capable of solving some socio-economic problems at the outset, money was gradually seen as the key to solving every problem. Easy money, in turn, reduced financial discipline within the bureaucracy, and allowed careless budgetary allocation, waste, unsavoury practices and widespread corruption.[12] The results could hardly be more injurious to normal development. The state was elevated to an earthly deity. Cost-free or low-price public goods and services were taken for granted as a matter of birthright. Most manual work had to be done by expatriate workers. Reliance on oil money preempted any serious efforts to mobilise domestic resources through taxation or realistic charges and fees for social amenities. The share of non-oil taxes in GDP fell in nearly all member countries.

The growth of a full-fledged welfare state in some member countries – and the spread of a 'welfare state mentality' elsewhere – gave distributional issues greater prominence over production issues. With the state as the sole recipient and dispenser of the oil windfalls, rent-seeking activities became not only highly profitable financially but also extremely smart socially. Highest returns to entrepreneurial talent did not come from directly productive economic activities but from obtaining a piece of the oil rent: a special foreign exchange allotment, a lucrative government contract, an import quota, a commission on arms purchases or an exemption from repatriation of export proceeds. The more absolutist and patrimonial was the state (where the nation's wealth was not distinctly separate from the ruler's), the wider were the possibilities for such rent-seeking preoccupations. A common tendency in almost all member economies for the services sector to gain prominence over other sectors, and for private entrepreneurs to prefer trade over industrial or agricultural endeavours, was largely rooted in the high and quick profits obtainable through clever 'deals'. The magnitude of the oil rent itself, and the impetus afforded to rent-seeking ventures, also created an inhospitable overall climate for 'real' production.

Worse still, the petroculture undermined the traditional virtues of self-reliance, both at the personal and national levels; it desensitised the people *vis-à-vis* genteel business manipulations, welfare handouts and big-brotherly tutelage. A secure job in the bureaucracy became a symbol of social standing and prestige. The new oil psychology also drove an alarming wedge between privileged, Westernised, progressive elites, who emphasised economic growth as a pre-condition for human development, and cultural traditionalists (or religious fundamentalists in Islamic countries) who regarded the preservation of old moral and spiritual values as the key to human salvation. The tensions between these groups, in addition to aggravating economic difficulties, gradually became a source of continual political turmoil in several member countries (e.g. Algeria, Ecuador, Iran, Kuwait, Nigeria, Qatar and Saudi Arabia). Indonesia joined the group at a later date.

THE END RESULTS

The sudden and dramatic increase in oil revenues and the resulting windfalls placed OPEC governments in the 1970s under intense political pressure to move forward on three strategic fronts all at once: to improve the welfare of the populations; to undertake long sought-after projects of great popular appeal; and to develop the economy in such a way as to reduce future dependence on oil. All this was also to be accomplished through rapid modernisation without losing traditional social and cultural values, i.e. without Westernisation or 'Westoxication'. How was this tall order filled?

Despite ample historical precedents from the failed policies of Spain and Portugal in dealing with their New World's gold and silver fortunes, OPEC authorities succumbed to similar strong temptations, and helped history to repeat itself. Directed from above and force-fed through the bureaucratic process, much development took place but at an inordinately high overall cost and without any assurances for a prosperous future. The hasty expenditure of oil windfalls on a multitude of poorly planned and even more poorly coordinated public projects resulted in a feverish rate of development. Some early pains of growth – production bottlenecks, power blackouts, housing shortage, port congestion and inflation – were stoically tolerated by the rank-and-file in anticipation of better and happier days to come. But when the oil boom turned into a bust, the early euphoria gave way to widespread disappointment, frustration and social unease. Looking back at the two decades of the oil cycles, every accomplishment seems to have carried its own downside.

In the economic arena, the elaborate and expensive physical infrastructure put in place facilitated industrial and agricultural expansion but increased the state's maintenance burden, often beyond budgetary capabilities or tolerance. Food production increased but the declared goal of self-sufficiency often proved elusive. Resource-based industrialisation, even when successful, in the case of petrochemicals, perpetuated the dualistic system of production and an unbalanced wage structure. Industry, apart from oil and gas, remained small, constituting no more than 30 percent of GDP in any of the countries. Including energy, the industrial sector provided less than one-third of total employment in almost all members. Even the relatively small manufacturing sub-sector was frequently underutilised due to excess capacity, small size of domestic market, dependence on foreign inputs and the absence of externalities. There is no evidence of an appreciable increase in overall labour productivity despite massive investments in formal education. Improved health care raised life-expectancy and reduced infant mortality but unemployment and underemployment in the healthy young adult category were aggravated. Per capita income went up in many countries (and even in real terms in some) but domestic income distribution became more skewed in almost all. Services, which grew fastest in most member countries and became a disproportionately large sector, included the so-called informal or underground economy, composed of petty trades and parasitic callings that escaped public scrutiny, supervision or taxation. Indigenous technology, as the sine qua non of genuine, sustainable development, was hardly advanced despite ubiquitous budgetary allocations for 'research'.

At the institutional level, oil windfalls encouraged profligate spending and wasteful outlays, including purchases of ultra-sophisticated military hardware. Fiscal discipline was gradually relaxed and in some cases broke down totally. Domestic taxation was either spared or kept at minimal levels. There was little or no incentive in most group members for the establishment of a broad tax base covering all mainstream economic activities, which resulted in the treasury becoming almost totally unable to replace external oil income with autonomous internal resources when oil prices fell. In the private sector, too, government handouts, easy credit and subsidies interfered with private entrepreneurship, risk-taking and initiative. Interest in long-term productive investment was preempted by short-term, high-return and quick-profit trade and speculations. Despite universal commitments to reduce dependence on oil and promote non-oil exports, the bulk of productive investments in domestic projects was earmarked for the manufacture of import substitutes for which local demand was rising and for which competition could be easily shut off through political intervention and trade protection. No member except Indonesia developed substitute export goods large enough to finance its import needs. Whether or not a similar strategy would have been equally applicable or equally effective in the OPEC case requires further detailed investigation.

In short, a large (albeit fluctuating) portion of the oil windfalls in the early years went for investment in physical, social and human capital, expansion of industrial capacity, improvement of public service facilities, expanded social welfare and

219

increased private consumption in early years. Some relatively small Western industrial and financial assets were also acquired by the governments and private citizens of a few capital surplus countries. But, with rare exceptions, those positive developments were accompanied by serious internal dislocations and heavy external imbalances. As previous chapters showed in detail, economic growth was, at best, anaemic in most member countries. Domestic inflation was not checked in most cases. Unemployment and under-employment were widespread. National savings did not catch up with aggregate investments among the majority of members, and a resource gap developed nearly everywhere. Budget deficits spared almost no member. Other expected concomitants of projected prosperity and welfare – individual freedom, political stability, social cohesion and participatory democracy – remained as mostly pipedreams.

To be sure, the members' miscalculations and mismanagement were aggravated by global oil market volatility, exogenous events unrelated to oil and even some natural disasters. Yet lurking behind every member's setbacks and reversals were certain socio-political factors inherent in unrequited windfall gains. The most familiar of these social factors was the 'easy-come, easy-go' syndrome. Unrequited windfalls, like unexpected gifts or a lottery win, are seldom subjected to the rational spending scrutiny that hard-earned incomes are. Gigantic and sudden inflows of oil revenues following each of the three oil booms gave rise to similar behaviour in all members. The grandiose white elephant projects or military expenditures that were launched in member economies – whether the Man-Made River in Libya, the Trans-Railway in Gabon, the new capital in Nigeria, the largest airport in Saudi Arabia, the biggest mosque in the UAE, nuclear weapons development by Iraq, a 'national car' and a 'national jetline' in Indonesia, a state-of-the-art military stockpile in Kuwait and Saudi Arabia, or a mountain-top resort in Venezuela – would never have seen the light of day if they had had to be paid for by proceeds from exports of domestic manufactures.

Another significant factor responsible for the sub-par performance of group members was the unique susceptibility of the beneficiaries of windfall gains to extraordinary rent-seeking behaviour and corruptive practices.[13] The 'unearned' nature of the gains for which competing rent-seekers were answerable to no one but each other made such activities extremely rewarding. Capturing part of the windfalls through special import quotas, industrial licences, trade franchises, special foreign exchange allocation, or low-cost credits became the object of economic activity in itself, and often more lucrative than any available alternatives. As a result, income and wealth were no longer based on output and productivity but on right connections and insider deals. Resources, time, and talent were devoted to rent-seeking activities instead of socially productive endeavours. And in the battle over the distribution of the oil rent, the balance of power was invariably tipped in favour of urban interest groups, consumers and importers against rural people, producers and exporters. The fight over the allocation of windfalls by powerful groups often resulted in political instability and a disruption of productive processes. The magnitude of rent-seeking activities was, in turn, directly related to the size of per capita oil windfall and the quality of governance.

What did OPEC members gain from their 20-year game, and at what cost? For the reasons detailed in the foregoing sections, the oil windfalls failed to ensure rapid growth in most group members and probably even stifled it in a few. But slow growth, or even stagnation, did not mean that the members' gains in the terms-of-trade were of no benefit to them. Windfalls may not have been good for growth but they served other objectives, and produced other desiderata. Increased private consumption, for example, was a clear beneficial outcome for nearly all. As a percentage of GDP, between 1974 and 1994, private consumption nearly tripled in the UAE, more than doubled in Kuwait and Saudi Arabia, nearly doubled in Gabon, rose by more than 50 percent in Iran and went up in every other group member except for Indonesia (Table 18). In other words, the current generation living in these countries during the two decades enjoyed the highest standard of living in their known history, largely out of the oil rent. With public consumption also up in seven of the thirteen members (mostly around the Persian Gulf), the population also received more and better public services, at little or no cost, in the form of education, health care, housing and other social amenities.

In short, most of the people in most of these countries became better fed, healthier, more educated, more comfortably housed and better prepared to face the challenge of the twenty-first century. Some enterprising citizens in most countries (and in some more than others) managed to accumulate sizeable foreign assets, and ensured their own future economic security regardless of their country's fate. The countries themselves, although still largely in the non-democratic camp, had their major physical and social infrastructures in place for a take-off towards sustained economic and (hopefully) political advancement.

In the four small Arab members of the group on the Persian Gulf, subsistence living by small farming and fishing was changed virtually overnight to the enjoyment of the world's highest living standards. In terms of the transformation of the physical landscape, transport and communication means, foreign travel and access to the products and facilities of the industrial north, there have been no parallels in modern history. The desert wastelands were turned into urban marvels of architectural modernity, with suburban tree stands and desert farms, thanks to the latest water desalination technology and modern irrigation systems. In other group members not affected by exogenous factors (such as war, civil strife or natural disaster), even private per capita consumption was up at least for part of the period.

Could all this have been achieved at less cost than the real resources transferred to OPEC members? The answer is probably yes. A number of works referenced in this study have tried to make the case that a great proportion of the potential gains to the exporters may have been nullified by changes induced in the global economy and the inappropriate policies followed by the recipients due to market uncertainty and volatility under the oil boom and bust. It is also argued that the extra capital formation in the oil exporters was probably not very efficient due to higher incremental capital/output ratios.[14] The argument that, other things being equal, the global economy as a whole (including the OPEC members) might have been better off with as much rise in the real price of oil as occurred between 1974 and 1994 but without the oil boom and bust may have some validity. But concluding that the whole exercise was a negative-sum game and entirely an OPEC creation might be far-fetched. The oil windfalls may not have been optimally productive in the hands of the recipients but, without the resource transfer, resource-owners would have continued to subsidise major resource-users' profligate lifestyles forever. In other words, returns to investment in the major oil-importing countries might have been considerably larger, but world income distribution would have stayed tilted in favour of the industrial countries. Global income distribution following the oil price rise and the resource transfer might not have been ideal. But to the extent that poorer nations benefitted at the expense of richer ones, the cause of universal fairness may have been served. The industrial countries might have produced more wealth out of the transfers but the fruits of that wealth might not have reached OPEC members or other poor developing countries that were the beneficiaries of OPEC generosity.

The downside of the oil windfalls was not so much the low returns to OPEC investments, useless white elephant projects, resource waste and moral corruption – all of which were, of course deplorable in a world still plagued with mass poverty. The real pity is that despite all such unavoidable negatives, most members were, after 20 years of effort, still not on a secure path toward sustained growth and prosperity. Algeria, Ecuador, Gabon and Venezuela were hardly on their way to a meaningful economic recovery. Iraq's economy was in a shambles. Nigeria was in worse shape than at the start. Libya was struggling under stiff UN sanctions. Iran was hardly better in per capita income than in 1974. Kuwait, Saudi Arabia and the UAE had, of course, a more secure future due to their enormous oil reserves lasting well into the next century – as long as competitive oil substitutes were not found. Qatar's small economy could be saved by gas reserves, as long as they lasted. Indonesia was perhaps the member with the least conspicuous economic problems but was inherently threatened by an alarmingly high external debt, a vulnerable currency, import and marketing monopolies controlled by corrupt officials, business empires belonging to the leadership's offspring and cronies, questionable tax exemptions and elites favoured credit allocated by a poorly regulated banking system. Outward prosperity was perpetually challenged by political tension and unease.

The economic structure of all member countries, particularly those with relatively low petroleum reserves, faced the test of long-term viability once oil was exhausted or oil prices were significantly depressed. Most of the new industries supported by low-cost energy supply, relatively cheap imported inputs, cheap credit, state subsidies and tariff protection were likely to encounter input supply difficulties and operating problems if there were an adverse turnaround in oil export revenues. The enormous cost of maintaining the expensive ultra-modern infrastructure could also prove intolerable without oil revenues in the low-absorbing countries, unless they could generate banking, transport, recreation and other services independent of the oil sector.

For all members, the adjustment to the new trend was most difficult. Unlike the earlier 'petrolisation' of the economy – rising incomes, conspicuous consumption, luxury imports, easy credit and unbridled government spending – which was easy and pleasant to get used to, the 'depetrolisation' proved to be excruciatingly difficult. Addiction to food and other imports, rising budget deficits resulting from generous state subsidies, growing unemployment due to unmatched jobs and skills, mounting inflation following excess liquidity, servicing foreign debt piled up from earlier dependence on short-term credits, rising local bad debts and widening income inequalities all became much harder to deal with. The 20-year period of trials and tribulations, however, produced some eye-openers for them and other aspiring new oil exporters. First, members came to the painful realisation that while earlier oil nationalisation made them feel good at the time, it did not do them much good, as it denied them ensured access to markets, direct benefits of modern technology, and readily available capital. Second, eagerness to raise oil prices to their short-term limits encouraged non-OPEC supplies and thus lost them market share. Third, while they initially denied that their organisation was a cartel, and later argued that theirs was not a typical predatory variety, they could not ultimately escape what all cartels do: inter-group cheating, loss of membership and declining power. Finally, all members came to feel that the loss of wealth and prosperity was vastly more intolerable and demoralising than never having enjoyed them at all. The lessons for the newcomers are not hard to draw.[15]

NOTES ON CHAPTER VIII

1   See, for example, Mahdi al-Bazzaz, 'Middle East Oil Revenues', *Middle East Economic Digest*, 15 March 1974.

2   See OPEC, *Annual Statistical Bulletin, 1995* (Vienna: 1995).

3   See K.A. Chaudry, 'The Price of Wealth: Business and State in Labor Remittance and Oil Economies', *International Organisation* (Winter, 1989); and Karl (1997), Chapters 3 and 10.

4   See H. Beblawi and G. Luciani (eds) (1987).

5   See M.S. El Azhari (ed.) (1984), Chapter 7.

6   While petrochemicals are often considered the classical first stage of economic diversification in OPEC members, some question whether these industries represent true diversification away from oil, or just the same industry in other forms. Furthermore, while such RBI may be appropriate for countries with 100 or more years of oil reserves, for others, which may run out of oil in two or three decades, RBI does not apply as they may need to import their raw material inputs. In other words, comparative advantage in exhaustible resources is itself exhaustible.

7   Richard M. Auty, 'The Internal Determinants of Eight Oil-Exporting Countries' Resource-Based Industry Performance', *Journal of Development Studies*, April 1989.

8   K.J. Murphy (1983).

9   See Richards and Waterbury (1990), Chapter 2.

10 See Peter Truell, 'Gulf Nations' White Elephant Projects', *Wall Street Journal*, 18 July 1983. Anecdotal tales of bureaucratic inefficiencies include those of a state-owned airline that kept offices in some US cities where it did not even fly, building hydroelectric plants in areas with low rainfalls, and flaring off natural gas at wellheads while diesel pumps were deteriorating on the pipelines.

11 See *Human Development Report, 1994*, p 50.

12 Allegations of bribery and corrupt practices regarding virtually every OPEC member have been regularly made in the popular press. However, the recent series of annual reports by

the Berlin-based Transparency International, called the Corruption Index, specifically places Nigeria as the world's most corrupt nation, followed closely by such other OPEC members as Indonesia and Venezuela. Other surveys also include Algeria, Iran and Saudi Arabia on their lists. See *New York Times*, 20 August 1995; and *Washington Post*, 8 July and 17, 1996.

13 P. Mauro, 'Corruption and Growth', *Quarterly Journal of Economics* (August 1995).

14 Alan Gelb (1988), p 143.

15 See Jahangir Amuzegar, 'OPEC as Omen', *Foreign Affairs*, November/December 1998.

Supplement

The second half of the 1990s was a remarkably eventful period for both OPEC and its members. The organisation itself, widely written off as moribund if not dead, once again proved the pundits wrong. By demonstrating its undiminished market vigour, it revealed a capability which no oil analyst had foreseen or expected. OPEC's decision in December 1997 to raise its self-imposed output ceiling (unchanged since 1993) by 10 percent to 27.5 mb/d coincided with several adverse developments in the global petroleum market. A combination of mild winter weather, a severe recession in Southeast Asia, and Iraq's resumption of oil supplies pushed the benchmark oil price in the New York Exchange from about $19/b to less than $13/b by mid-March 1998 – the lowest price since November 1988. As a result, OPEC members lost about $55 billion of export revenues in 1998, and ended up with only about half of their 1997 share. Two initial attempts by Saudi Arabia, Venezuela and Mexico aimed at cutting output failed to reverse the downward trend, and the oil price fell to about $10/b in mid-June 1998. Some spot cargoes were sold even more cheaply. A political rapprochement between Saudi Arabia and Iran, and a new entente in early March 1999 between OPEC and non-OPEC members (e.g. Mexico, Norway, Russia and Oman) finally succeeded in turning the tide. By early March 2000, the price of the West Texas Intermediate crude in New York reached a nine-year high of $34.37/b. After a new OPEC decision in March 2000 to raise output by 1.7 mb/d, the benchmark price was brought down to below $30/b. Thus despite many premature announcements of its death, OPEC proved to be alive and consequential at the turn of the new century.

Between 1994 and 2000, several member countries also went through some momentous domestic political changes. In Algeria, the election of President Bouteflika in April 1999, along with the amnesty law of July and the referendum of September 1999, promised an early end to the seven-year-old civil strife, and the beginning of a period of more rapid growth, calmer labour market, and overall macroeconomic stability. In Iran, a pro-reform candidate won an upset victory in the May 1997 presidential election, putting an end to a 17-year conservative rule. Subsequent municipal and parliamentary elections in which reformist candidates gained landslide majorities promised the dawn of a new liberal era. In Indonesia, following a series of disturbing political upheavals and food riots in early 1998, and under popular pressure over an allegedly rigged parliamentary election in 1997, President Suharto was forced to resign – ending his 32-year dictatorial rule. The elections of a new president and a new parliament brought the country face-to-face with a different and uncertain future. In Nigeria, the sudden death of President Abacha in 1998 allowed his successor to hold new national elections and adopt a new constitution, culminating in the democratically free election of former General Obasanjo in May 1999, and the establishment of a new politico-economic reform agenda. Venezuela's congressional and presidential elections in late 1998 resulted in a new administration under President Hugo Chavez, ending a traditional two-party rule long characterised by inefficiency, mismanagement and corruption.

The other members escaped such fundamental political transformations, but broadly suffered from the 1998–9 oil bust. These economic and political developments naturally affected national management of oil wealth, and the conduct of domestic economic policy. The following sections briefly trace domestic economic developments in all but one remaining OPEC members.

ALGERIA

In 1994, when Algeria launched a new IMF-assisted reform package, its currency was highly over-valued and pegged to a basket, interest rates were negative, a generalised system of subsidies costing more than 5 percent of GDP, combined with extensive price controls, were responsible for shortages and black markets, and external trade was subjected to a complicated system of exchange restrictions.

During the 1994–2000 period, Algeria managed – despite considerable domestic political and security turmoil – to reverse its previous macroeconomic imbalances and to shed some of the vestiges of its state-controlled and mismanaged economy. Helped by the IMF, the World Bank and the Western economic community, its growth performance turned positive; inflation was brought down; the external balance was strengthened; price controls were reduced while the trade and exchange systems were liberalised. In short, the transition from a centrally planned and poorly managed economy to a market-oriented, more diversified and more competitive system was achieved against many odds.

Algeria signed a second one-year standby agreement with the IMF in 1994, under which about half a billion dollars was borrowed with a promise of extensive macro-economic adjustments and structural reforms. A year later, more than $1 billion was borrowed under the Extended Fund Facility (IFF) for a period of three years. And, to offset the effects of the oil price decline, another quarter of a billion dollars was obtained under the Compensatory and Contingency Financing Facility (CCFF) in 1999. The Algerian government, in addition to receiving project financing, sought and received assistance from the World Bank in such areas as privatisation methods, banking reform, social safety net policies, housing finance and improved management of public enterprises.

Thanks largely to these efforts, the growth rate, which had remained negative between 1990 and 1994 (averaging -0.9 percent), rose about 3.5 percent a year on average between 1995 and 2000. Inflation, which hovered around 27 percent in the early 1990s, was brought down to an annual average of 10.8 percent. The unemployment rate, however, stuck stubbornly at about 30 percent, owing in large part to massive layoffs following the restructuring of inefficient public enterprises and a steady 3 percent annual growth of the labour force. Under the reform package, the exchange rate regime was shifted from a peg to a managed float; an inter-bank exchange market was introduced; prices and interest rates were gradually decontrolled; generalised subsidies were replaced by a social safety net; and exchange and trade systems were liberalised. External current account convertibility was established in 1997. A new stock exchange was set up to facilitate both domestic privatisation efforts and attract private investment.

On the flip side, heavy reliance on hydrocarbon exports continued despite note-worthy efforts toward diversification. Low oil prices in 1998 and the first half of 1999, and a sharp drop in export receipts, typically resulted in increased budget deficit, a weakened balance-of-payments position and a loss of official reserves. There was a fresh need to reduce development expenditures, maintain high real interest rates, and allow the Algerian dinar to depreciate vis-à-vis the U.S. dollar. The public sector economy also remained uncomfortably large, with some 75 percent of total industrial production remaining in the hands of money-losing state enterprises. External debt stood at more than $28 billion and the debt/service ratio remained high at about 40 percent of exports of goods and non-factor services. Domestic fuel prices were kept artificially low and well below international standards. Interest rates were also set administratively, with little regard to market fundamentals. A high degree of protection was still maintained against imports. And, despite substantial progress in economic stabilisation and internal political order, the Algerian economy in 2000 still faced the twin challenges of slow growth and high unemployment, causing stagnant standards of living. Further, success in correcting existing imbalances required considerably less dependence on oil and gas exports, and a faster imple-mentation of remaining structural reforms. Energy export revenues beyond those budgeted each year had to be saved in a revenue stabilisation fund to be drawn upon in case of falling oil prices. The Algerian authorities at the time were engaged in preliminary discussions on an Association Agreement with the European Union, and eventual membership of the World Trade Organisation. The Algerian government also sought a rating of the country's sovereign risk in order to return to international financial markets. Algeria's Human Development Index ranking during the five-year period improved to 107 from 109 among 174 countries observed by the United Nations Development Programme.

INDONESIA

The economic and financial turmoil that engulfed Southeast Asia in 1997–8 struck its heaviest blow on the Indonesian economy. In addition to putting an end to General Suharto's long dictatorial rule, Indonesia's banking and industrial corporate institutions were shaken to their roots. Amidst an unsettled political situation, severe financial and external disruptions, the sharpest GDP decline in the region, food shortages, and a loss of confidence in the banking system, the IMF, the World Bank and Western creditors came to Indonesia's rescue. As part of a $43 billion global aid package, a standby agreement was signed with the IMF in November 1997 for a total of SDR8.4 billion, of which nearly half was immediately drawn. Nevertheless, the financial situation during the first half of 1998 deteriorated steadily, and the rupiah lost nearly 70 percent of its value, inflation reached 45 percent, and the negative interest rate spread rose to about 35 percent.

Against this background, a new Extended Fund Facility (EFF) was signed with the IMF in August 1998 for a total of SDR5.4 billion. In January 2000, having already drawn some SDR3.8 billion from that fund, the Indonesian authorities asked for the remaining EFF's replacement with a new three-year arrangement. Between 1995 and 2000, Indonesia obtained some $6.8 billion in loan commitments from the World Bank on various development projects, and nearly $6 billion from the Asian Development Bank.

As conditions for the use of IMF resources, Indonesia agreed to exercise tighter fiscal control, refrain from domestic bank financing of the budget deficit, and engage in corporate restructuring, government deregulation, elimination of private monopolies, privatisation of state enterprises, and improved governance. As a result, by January 2000, a semblance of macroeconomic stability was restored: the consumer price level was stabilised; interest rates reverted to pre-crisis levels; trade and current accounts shifted into surplus; the rupiah regained some of its lost value; and the financial sector began to stabilise. An agreement was reached with the Paris Club creditors for the rescheduling of overdue debts.

After 1994, Indonesian GDP showed an annual average of nearly 6 percent in 1995–7, but plummeted more than 14 percent in 1998, and achieved an anaemic rise of 2.6 percent in 1999–2000. The consumer price index rose nearly 8 percent a year in 1995–7, but rose sharply by 65 percent in 1998 before it was brought down to 6 percent a year in the subsequent two years. The government budget showed a slight surplus of nearly 1.3 percent of GDP on average in the first two years, but went into deficit to the tune of 2.4 percent in the last three. Between 1995 and 1997, the current account was in deficit equivalent to 2.5 percent of GDP, but showed a surplus of about 2 percent in the subsequent two years. The overall balance showed a small surplus every year except in fiscal year 1997/8 for about 7 percent of GDP. There was near full employment, with annual joblessness ranging from 2 to 5 percent of the labour force. Gross official foreign assets were equal to about six months of imports. The total external debt in 2000 was equal to 88 percent of GDP, and the debt/service ratio hovered around 30 percent of exports of goods and non-factor services. Non-oil/gas exports accounted for about 72 percent of total exports.

Indonesia's 1997–8 crisis was thus caused mostly by political instability and a lack of confidence on the banking system's solvency rather than any economic fundamentals. Nevertheless, in 2000 a crisis-free restructuring of the banking and corporate sectors still remained to be finalised. Glaring weaknesses in the bureaucracy and the judiciary continued because of weak political leadership and the new democracy's growing pains. Speeding up economic growth, preventing inflation from rekindling, and reducing the colossal debt burden were yet to be achieved. In the best of circumstances, GDP was not expected to reach its pre-crisis level before the year 2003. In short, the Indonesian economy remained fragile, and the new government faced the thankless task of fighting corruption, secrecy, and residual crony capitalism. A government of national unity, hoped for after the election of President Wahid, remained mired in corruption, incompetence and sectarian violence. Indonesia's Human Development Index ranking fell to 109 from 105.

IRAN

In 1994, the Iranian economy was still broadly state-dominated, price-controlled, trade-restricted, exchange-rate-regulated, under-taxed, overly-subsidised, heavily dependent on oil export revenues, and short on incentives for private sector growth. Some 80 percent of the economy was directly or indirectly affected by the government, public enterprises, state-owned financial institutions, and prestatal foundations (bonyads). Prices were controlled and/or influenced by direct or implicit subsidies nearing 20 percent of GDP. Financial rates of return and charges were administratively set. The exchange regime had multiple rates. Trade was subject to various non-tariff restrictions. The private sector suffered from unfair business practices.

For most of the period under review, the Islamic Republic faced fairly comprehensive U.S. economic sanctions, and had only very limited access to Western export credit facilities and international capital markets. Even part of a large loan earlier signed with Japan was withheld under pressure from Washington. Not until 1999 and 2000 were some American sanctions removed, and not until May 2000 did the country obtain two infrastructure loans totalling $232 million from the World Bank. During this period, however, several oil and gas contracts were signed with Western energy companies for nearly $4 billion under the so-called 'buy back' mechanism, and there was another $1 billion of foreign investment in the non-energy sector on a build-operate-transfer basis. Due to a drastic decline in international oil prices in 1998, the Central Bank was unable to service the country's already rescheduled debt and had to seek new accommodations with the creditors – tarnishing the country's pre-1979 70-year impeccable credit record for the second time in the decade.

Under these circumstances, economic growth, reflecting high oil price volatility and low domestic investment, averaged about 3.5 percent a year in real terms. Unemployment remained high – officially claimed to be in the low teens, but privately estimated to run at more than 20 percent of the labour force, as GDP growth typically lagged behind the 4 percent average annual expansion of the workforce. Inflation averaged 26 percent a year, coming down from the initial high of 49 percent in 1995 to 20 percent in 1999–2000. The true figure was probably higher, since the official consumer price index covered a number of price-controlled items. The external position followed the vagaries of international oil prices, with the current account remaining positive, except during the 1998–9 oil bust. By keeping annual imports down at half their 1992 level, at the expense of unused industrial capacity, the $23 billion accumulated external debt at the beginning of the period was brought down to less than $11 billion towards the end. The public sector's debt to the banking sector, however, more than doubled during the period. The value of the Iranian rial was allowed to depreciate vis-à-vis all foreign currencies. The free market rate declined from less than IR4500 to the dollar to around IR8500 in early 2000.

Meanwhile, notable steps were taken toward long-promised structural reform. The multiple exchange-rate regime was gradually liberalised so that by the year 2000 there were only two formal rates for all transactions, with the one applicable to non-governmental payments being determined each day by market forces. Promises were also made toward early unification. A number of impediments to trade and exchange were also removed. All customs duty exemptions, discounts and agency-specific preferences were officially outlawed. Nationalised banks were allowed greater flexibility and new private banks and non-bank credit institutions were authorised for on- and off-shore operations. A number of other institutional reforms (including extensive privatisation of state enterprises, demonopolisation of certain economic activities, replacement of quantitative trade restrictions by tariffs) were legislated under the Third Five-Year Development Law of 2000. The government obtained a sovereign risk rating from Moody's and planned to enter the international capital market later in 2000. The Third Plan (2000–5) was aimed at an 'orderly' transition to a market economy as a means of raising growth and creating job opportunities. During the period under review, the Iranian government received a number of technical assistance missions from the IMF and the World Bank in the areas of taxation, monetary and exchange regulation, statistics and sector analyses.

Despite great hope generated by the creeping democratisation trend since the election of President Khatami in 1997, the Islamic Republic of Iran continued to suffer from a number of structural deficiencies in various aspects of the economy. A large, dominant and inefficient government, mismanaged public enterprises, extensive growth-impeding regulations, judicial inadequacies, maldistribution of wealth and income, and such other woes as nepotism, cronyism and corruption con-tinued to lie at the root of slow growth, high inflation, double-digit unemployment, fiscal deficits and external payments imbalances. Such social scourges as the rising crime rate, drug addiction, prostitution and suicide were easily traceable to economic causes. Iran's Human Development Index ranking declined to 97 from 86.

KUWAIT

Having partly recovered from the destructive effects of the Iraqi invasion by 1994, Kuwait moved ahead to reduce lingering financial imbalances, replenish lost foreign assets, and strengthen its banking system, plagued with bad debt. Real GDP, which had performed strongly in 1994 due to reconstruction activity, achieved a lacklustre sequel to 2000, with annual growth averaging less than 1 percent, due in part to a sharp decline in 1998 oil export prices. Non-oil GDP (including construction, trade, real estate and financial services) showed a stronger performance, with average real annual growth of 2.4 percent. With its population growing at 3.4 percent annually, however, real per capita GDP followed a steady decline.

During the period under review, Kuwait's economy continued to be dominated by state-based crude oil extraction and refining and petroleum-based petrochemicals. Oil accounted for some 40 percent of GDP, 94 percent of exports and 67 percent of budget revenue. With its exchange rate effectively pegged to the U.S. dollar, inflation during the period was kept at 1.8 percent a year on average. Current account and overall balances were generally positive, and international reserve assets (excluding assets in the Fund for Future Generations) rose to the equivalent of four months of imports, roughly equal to total external debt. Some $6 billion a year of net factor income from abroad raised gross national product by about 20 percent.

Owing to this significant financial cushion, the Kuwait economy coped much better than other members with the 1998–9 oil price drop. Like other members of the GCC, Kuwait relied heavily on a large expatriate workforce, with 83 percent of its 1.2 million labour force consisting of foreign workers manning 95 percent of private sector employment needs. By contrast, about 93 percent of Kuwaiti nationals in the labour force were employed in the public sector. With some 10–20 thousand new job-seekers each year, and the treasury bent on ending perennial budget deficits (exclusive of income from foreign investment), Kuwait faced increased unemployment among its nationals as public sector hiring reached its budgetary and productivity limits. Given internal labour market segmentation, characterised by marked differences between pay scales in the public and private sectors, the dilemma confronting the authorities was whether to replace low-paid expatriate workers with high-paid nationals (and reduce global competitiveness) or continue to maintain an open-door policy (and face the socio-political consequences of higher unemployment among natives). Also threatening the fiscal balance was Kuwait's annual military expenditures, which at 9.2 percentage of GDP was the second highest in OPEC and fourth largest in the world.

Beginning in 1994, Kuwait adopted a comprehensive privatisation programme for the sale of government equity shares in non-energy companies as well as the transfer of management of public utilities, ports and health services to the private sector. During the period under review, the Kuwait Investment Authority sold equity shares in 28 companies for a total of nearly $2.5 billion. Nevertheless, in the year 2000 the Kuwaiti private sector accounted for less than 45 percent of GDP, about three quarters of total employment, and 70 percent of non-oil exports. While the private sector was dominant in construction, real estate, manufacturing, retail trade and financial services, it was still heavily dependent on public sector expenditure, making it vulnerable to oil price fluctuations. To spur economic growth, the government submitted a reform package for approval by the National Assembly designed further to privatise telecommunications, transport, aviation and foreign investment, as well as to reform labour law. Kuwait was also opening up oil activities to foreign oil companies under services contract agreements.

Despite its marked ability to withstand the oil shock, Kuwait still suffered from structural imbalances caused by the twin dominance of oil and state in the economy. Oil dependence made economic planning hazardous, and the state role limited private sector activity. In non-economic areas, however, the Kuwait Human Development Index ranking improved from 51 to 36.

LIBYA

Libya's economy continued to be dominated by the hydrocarbon sector. More than 90 percent of export earnings, more than 50 percent of budget revenues, and at least a quarter of GDP were derived from oil and gas. Up until April 1999, Libya was under comprehensive U.N. economic sanctions. The United States kept its own special sanctions throughout the period under review. The economy was largely controlled by the state, and heavily subsidised, with the private sector being encouraged from time to time, but tightly restricted between these times. In mid-1996, there was a reversal of efforts towards wider private sector participation in the economy under-taken after 1987, and many private businesses were closed. Inconsistent economic policies and periodic anti-corruption drives undermined business confidence and kept the private sector stagnant and inhibited. Trade and exchange systems were tightly controlled, and many transactions took place through the grey market. A draft bill prepared in 1996 to encourage foreign investment in the non-energy sectors remains to be enacted. In the oil sector, however, the government was more 'progres-sive' than many OPEC members, continuing to offer attractive exploitation contracts to foreign oil companies on a production-sharing basis. Crude oil output remained fairly constant during the period, coming from oil fields with high production costs.

Libya's economic growth between 1994 and 1999 was held back by the compre-hensive U.N. sanctions, estimated to have cost the country upward of $25 billion (mostly due to shortages of imported parts and equipment). Average real annual growth during the period is estimated at 0.7 percent. Per capita income shrunk by more than a third (but was still the highest in North Africa). In the absence of official data, inflation has been estimated at 27 percent a year between 1995 and 1999 (assuming heavy state subsidies on food in the major urban areas), and 13 per-cent at official retail prices. Wages lagged behind price rises every year. The budget, perennially in deficit, was proclaimed 'balanced' after 1996 by covering shortfalls with draw-downs on foreign reserves. Annual normal expenditures, however, were kept nearly constant during the period (despite considerable yearly inflation), and marginally changed in line with ups and downs in oil export receipts. The budget was also constantly drained by the earmarking of more than 20 percent of annual expenditures for the Great Man-made River Project, whose Phases III and IV still remain to be constructed.

External balances followed fluctuations in oil prices, with the current and overall accounts showing a surplus in every year except 1998, and largely affected by expatriate workers' remittances and repatriation by foreign oil and gas companies. The Libyan dinar was devalued by 16 percent in 1998 to LD0.45 to the dollar, with a free market rate of between three and seven times the official rate prevailing afterwards. By the end of the decade, Libya had more than $6.6 billion of reserves and a total external debt of about $3.3 billion. Figures on unemployment were not officially available, but foreign private estimates put the figure at around 30 percent while the country was still served by between one and two million foreign workers. A policy of reducing expatriate workers was put into effect in 1995, but the economy remained in need of both skilled foreign labour and manual workers for diverse reasons.

Despite moves towards economic diversification (the last one in the short-lived 1994–6 economic development plan) and considerable outlays in education, Libya, at the turn of the century, was still a perilously dependent economy, unable to feed itself with its paltry farm sector (accounting for less than 6 percent of GDP), and its total annual non-energy exports able to finance only 28 days of necessary yearly imports. With the removal of the U.N. sanctions, however, the economy's future was somewhat improved. Curiously enough, Libya's Human Development Index ranking improved to 72 from 79.

NIGERIA

Since 1994, Nigeria has witnessed the most fundamental domestic political transformation among OPEC members. A military government under General Abacha widely reputed for its fiscal profligacy, wasteful public expenditures, aimless monetary policy and rampant corruption was first overtaken by a reform-minded military officer in 1998, and later by a democratically elected government in 1999. Up until the changeover, Nigeria suffered from a multiple exchange rate system, currency overvaluation (with the heaviest subsidy on the official rate), over-regulation, loss-making public enterprises, oil price subsidies, budget deficits, heavy trade restriction, overall economic mismanagement and an absence of the rule of law. The economy was further severely handicapped by the high costs of doing business arising from institutional impediments.

The new Obasanjo government, determined to build public confidence through a reversal of past excesses, has moved to establish fiscal discipline, cancel dubious concessions, halt wasteful capital spending, rid the government of corrupt military and civilian high officials, and retrieve some of the funds misappropriated during the previous regime. A serious privatisation programme was announced to get rid of the state holdings in banks, public utilities, industrial enterprises, airways, oil refining and marketing.

In the six-year period under review, real GDP growth averaged 2.6 percent a year, barely exceeding population growth of 2.4 percent. Annual inflation averaged more than 20 percent. With petroleum accounting for 90 percent of export earnings, and over 70 percent of government revenue, the external current account and the fiscal balance fluctuated in tandem with the price of oil. The external balance was marginally positive on average, but the budget showed a deficit of 1.2 percent of GDP. The growth of the labour force was nearly double the rate of population growth as baby-boomers entered the job market. Military expenditure as a percentage of GDP, at 0.7 percent, was the lowest among the major OPEC members.

In the year 2000, however, Nigeria was still no better off than in 1970. Per capita income was less than $300. Income was unevenly distributed, with the top 20 percent of Nigerians accounting for more than 50 percent of national income, the bottom 20 percent for only about 5 percent. Seventy percent of the population (80 million) lived below the poverty line. The government's domestic debt neared 40 percent of GDP, some of it arrears to local contractors. External debt was heavy, at about $28 billion – equivalent to 94 percent of GDP, most of it also in arrears. Nigeria still followed a highly restrictive trade and exchange regime. Social tensions among ethnic and religious groups in the North and Southwest remains high.

Medium-term economic prospects still remained weak despite a favourable outlook for oil prices due to uncertainties regarding the government's ability to establish needed reforms, among them achieving economic growth in the 7–8 percent range (or three times the latest five-year record); observing fiscal prudence; disengaging from loss-making public enterprises; adjusting the exchange rate to market conditions; improving the efficiency of existing institutions (particularly the judiciary system); and increasing the competitiveness of non-oil export industries.

In comparison with Indonesia, which was in the same per capita income category as Nigeria in 1970, differences at the end of the century were indeed astounding. By 2000, Nigeria had only 22 percent of Indonesia's per capita income, 45 percent of its farm productivity, and 10 percent of its manufacturing value-added. It had also fallen behind its Asian partner in per capita private consumption, infant mortality reduction and the national literacy rate.

In 2000, Nigeria signed a new standby agreement with the IMF after operating under a Staff Monitored Programme for a while as a prelude to the use of Fund resources. Economic policy reforms, more than a sustained high oil price, remained at the root of the country's economic recovery. The country's Human Development Index ranking dropped from 139 to 151.

QATAR

While the Qatari economy continued to be dominated by hydrocarbons, and the non-energy economy was also heavily dependent on the energy sector, oil's share of GDP by the end of the decade was about 40 percent, and oil revenue as a share of total revenue was 70 percent. During the 1995–2000 period, Qatar was able to triple its oil reserves to more than 12 billion barrels from 3.7 billion barrels, and was a persistent quota violator within OPEC. The country's massive gas reserves (equivalent to 100 billion barrels of oil) was increasingly exploited through the export of LNG (an estimated 9.6 million tons in 2000) thanks to $10 billion of investment. LNG and petrochemical exports accounted for some 40 percent of total exports in 2000.

As recently as 1997, the IMF expressed concern over Qatar's recurring budget deficits and their effects on exchange rate stability. Acting on IMF recommendations, Qatar established the official Doha Securities Market, and issued two domestic bonds to convert its domestic debt into a tradable asset. In May 1999, Qatar issued its first ten-year bond on the international markets: some $1 billion at 9.5 percent interest. In June 2000, the government issued a new 30-year sovereign bond – a Middle East first – for $2.5 billion to finance its external debt. Subsidies were removed from food and fuel and reduced on utilities. Nominal fees and charges were also levied for education and health care. The Ministry of Water and Power was replaced by a corporation as a move toward further privatisation and the removal of all subsidies. Qatar's public Telecommunication Corporation was partly privatised in

1998, but the programme for the privatisation of other public monopolies has not moved much forward.

Economic growth during the period was about 4.2 percent a year in real terms on average, accompanied by a more than 6 percent a year increase in nominal per capita income – giving Qataris the highest per capita in the GCC – about $23,500. The consumer price index, except for 1996, was generally subdued, averaging about 3.7 percent a year, thanks to tight monetary policy and the government's lid on spending. The budget deficit was brought down from 10 percent of GDP in 1995 to only one percent in 2000 by a strict austerity programme in both current and capital expenditure. On 600,000 b/d of oil production, Qatar needed an oil price of $18/b to be in surplus, and able to service its accumulated internal debts after years of uninterrupted budget deficits (due largely to high subsidies on utilities and free social services). Thanks to a surge in oil prices, Qatar achieved a ten-year record high trade surplus in 1999, and the first current account surplus in more than five years. The overall balance in the five-year period, amounting to $2.25 billion, raised foreign exchange reserves. Qatar's total government and state-guaranteed debt stood at $12 billion (or 84 percent of GDP) at the end of the period (with a debt/service ratio of 18), mostly due to heavy gas investment in the off-shore North Field in the Persian Gulf. The central bank managed to keep the Qatari riyal pegged at QR3.64 to the dollar. Foreign reserves (excluding funds in Qatar Investment Office) amounted to $1.9 billion at the end of the decade.

Qatar's economy at the dawn of the new century exhibited considerable fragility due to oil price uncertainties, a heavy debt burden and the necessity of several more years of high revenues from its off-shore gas and petrochemical investment. Qatar's Human Development Index ranking improved to 42 from 56.

SAUDI ARABIA

Since 1994, Saudi Arabia's economy has exhibited, on the one hand, continued vulnerability to sudden oil shocks and, on the other, a serious commitment to liberating reforms and the building up the non-oil sector. Constrained by OPEC's production ceiling and relatively low gross domestic investment, GDP between 1995 and 1997 had a lacklustre growth of 1.5 percent a year on average, while the consumer price index was subdued at an annual average of 1.8 percent. The overall fiscal balance showed an average yearly deficit equivalent to 4 percent of GDP, resulting in an increasing domestic debt. A sharp drop in oil prices in 1998 and the early part of 1999 – ironically caused by the Saudi's own insistence in the September 1997 OPEC meeting in Jaharte on an increase in OPEC's self-imposed production ceiling – caused a major setback to the slowly expanding economy. Saudi oil's average price in 1998 dropped from $19.2/b to $11.8/b, causing the Saudi treasury's oil export receipts to decline by 38 percent (or about 16 percent of GDP), to $33.4 billion from $53.2 billion. As a result, the fiscal deficit widened to 10 percent of GDP, the external account showed a large deficit, and real GDP growth fell.

The Saudi authorities moved swiftly to tighten the budget by cutting (mostly development) expenditures and raising revenues, in part, by drawing down foreign investment income and improving collection of non-oil revenues. The strong banking system remained largely unaffected by the turmoil, thanks to the use of ample foreign assets. The central bank, in turn, increased commercial bank access to its facilities and intervened on the spot and forward exchange markets to counter speculative attacks against the Saudi riyal. Nevertheless, the oil shock continued to affect private sector confidence (as reflected by a drop in stock market prices by 28 percent in 1998), although less severely than in previous oil busts.

During the period under review, the central bank remained firmly committed to maintenance of the Saudi riyal's foreign exchange rate against the U.S. dollar, and resisted exchange rate devaluation despite the terms-of-trade reversals in 1998–9. Sizable foreign assets, a credible pegged exchange rate and a strong, well-capitalised, tightly supervised and profitable banking system allowed Saudi Arabia to follow a liberal trade and exchange policy. Yet, while the vulnerability to oil shocks of the economy overall (and the private sector in particular) was lower than in the 1980s and early 1990s, the need for further genuine diversification and fiscal consolidation remained keen.

To these ends, in the second half of the 1990s the Saudi government embarked on a wide range of structural reforms including preliminary steps toward eventual privatisation of port services, telecommunications and electricity. Laws were enacted on income and excise taxation, investment and incorporation, customs administration, wage rationalisation and subsidy reduction. Domestic non-oil revenues were raised through a 50 percent increase in gasoline prices (to near international levels, excluding taxes) and further increases in fees and charges. Expenditures (including military outlays) were also streamlined and scaled back. Yet the overall fiscal deficit, if not remedied, promised to remain high, at about 6 percent of GDP, for some years and posed the greatest challenge to the Saudi economy.

Despite the notable structural changes, however, the Saudi economy in the year 2000 was still an energy-based one in which petroleum accounted for more than a third of GDP, 90 percent of total exports, and 70 percent of budget revenues. Agriculture accounted for less than 10 percent of GDP. While the $30 billion or so in the external debt (22 percent of GDP) was fairly moderate by international standards (with the debt/service ratio comfortably less than 10 percent of exports), and the central bank's foreign reserves, at nearly $50 billion, could finance 12 months of imports, the government's domestic debt, at about 115 percent of GDP, was hard to sustain. Saudi Arabia's annual military expenditure as a percentage of GDP, at 12.8 percent, was the highest within the group and the third highest in the world.

With an estimated 60 percent of the population under 18 years of age, job creation also posed a formidable challenge for Saudi planners. The largely traditional and restricted nature of the Saudi business culture had to be drastically altered to allow its desired membership of the World Trade Organisation. The kingdom's liberal external trade and payments system and the strength of its currency, however, were supportive of that end. Saudi Arabia's Human Development Index ranking was lowered to 75 from 67.

UNITED ARAB EMIRATES

During the second half of the 1990s, petroleum continued to serve as the mainstay of the UAE economy. The UAE's federal government continued its open trade and payments system and joined the World Trade Organisation in 1996. The exchange regime allowed the free transfer of funds on current and capital accounts. The UAE dirham was effectively fixed at Dh3.6 to the dollar. With other members of the GCC, there were no restrictions on the movement of goods, services, capital and labour. A close link between domestic and U.S. interest rates was maintained. Revenues from oil and gas exports accounted for 70 percent of government current, development and defence expenditure. Fiscal deficits were financed by returns from the large foreign currency reserves accumulated from past savings, and set aside for the purpose of coping with oil price fluctuations. The oil sector accounted for 50 percent of export earnings and 30 percent of GDP. Manufacturing, trade, tourism and entrepot services together accounted for 64 percent of GDP. As in some other GCC countries, UAE nationals comprised only 15–20 percent of the total population, and 10–15 percent of the labour force. Also as in most other GCC countries, the bulk of non-government employees were expatriates paid considerably less than nationals. Subsidies and transfers in the federal budget ranged from 3 to 5 percent of GDP.

The UAE still operated as a neo-patrimonial state, where details of the budget, the size and composition of foreign assets, the disposition of oil and gas revenues and other economic data lacked necessary transparency and escaped public scrutiny. While continued progress was made in promoting the private sector, the driving force behind such promotion, as in other GCC states, was the government itself, because such private sector activities as construction, transport and communication, tourism and banking were all dependent on oil and gas revenues. Also, since each emirate followed its own development strategy, duplication of projects and overall inefficiencies ensued.

During the 1994–2000 period, the UAE's real GDP rose by 4.3 percent per year on average. Inflation averaged 2.5 percent. The budget, however, showed consistent yearly deficits, averaging about 11.4 percent of GDP, financed primarily by drawdown of the Abu Dhabi Investment Authority's foreign resources. Except in 1998, the current account was in surplus to the tune of 3.4 percent of GDP. The overall balance was in the black each year. The central bank's reserves, at $12 billion, were comfortably equal to 4.2 months of imports. Due to a high rate of population growth, however, real per capita income toward the end of the decade had declined to 79 percent of the 1990 level.

Compared with other OPEC members during the period, the UAE showed an impressive economic performance, thanks to the adoption of an outward-oriented economic strategy, a stable currency and wise financial policies. Extensive and elaborate social and physical infrastructures financed by oil revenues, and a prudent commitment to diversification – combined with the world's highest per capita share of oil deposits – gave the country a solid base for future growth. A divestiture programme in the public sector was pursued, and private sector participation in shipbuilding, banking, insurance and utilities were undertaken through joint ventures. However, restrictions on foreign private investment in the emirates and labour market rigidities were major impediments to faster development. The UAE also had arguably one of the world's highest per capita military expenditures as a percentage of GDP. The UAE's Human Development Index ranking improved to 45 from 62.

VENEZUELA

Venezuela started the second half of the 1990s with an economy mired in a high rate of inflation, a large budget deficit, high unemployment, low labour productivity, weak global competitiveness in its resource-based industries, widening income distribution and a large real appreciation in the value of the bolivar. The government signed a new standby agreement with the IMF in July 1996, and adopted a Staff Monitored Programme in 1998. The 1998–9 oil shock affected the economy more than some other OPEC members as domestic political uncertainties regarding congressional and presidential elections weakened private sector confidence. Both the exchange and stock markets came under speculative attacks, and international reserves declined. The average bank lending rate reached 75 percent during 1998 before coming down to 50 percent by year end. The bolivar, under the crawling exchange rate regime, reached nearly 600 to the U.S. dollar by March 1999, from about 470 in 1996 and 150 in 1994. Inflation reached nearly 100 percent in 1996 before decelerating to less than 30 percent in early 1999, while unemployment hovered around 15 percent. During the period under review, Venezuela availed itself of the resources of the World Bank and the Inter-American Development Bank to the tune of $3.5 billion, of which about $1 billion was disbursed. At the same time, the country received technical assistance from the IMF in the areas of taxation, accounting and internal control, banking legislation and monetary management.

Throughout the period under review, the Venezuelan government sought a comprehensive stabilisation and structural adjustment programme to reduce fiscal and monetary imbalances, mitigate exchange revenue fluctuations, raise stagnant overall productivity, attract foreign private investment, and improve access to international capital markets. To these ends, the administration which came into power early in 1999 moved towards a wholesale restructuring of the long-ailing economy. An Enabling Law was passed in April 1999, authorising President Chavez to modify by decree previous laws on specified economic issues, including the public sector's role in the economy, the budget deficit, the tax system, the customs law and sectoral issues.

Already on the books was the Organic Law for the Social Security System enacted in 1997, under which the existing all-encompassing system was divided into separately autonomous financial systems covering pensions, health, unemployment, housing and recreation. A Macro-economic Stabilisation Fund was enacted in 1998 to smooth out the treasury's receipts from fluctuating oil export revenues caused by oil price changes. A new regulatory framework for public utilities, provisions for the equal treatment and protection of domestic and foreign investors, and the privatisation of public industrial enterprises were also positive steps undertaken during the period. On the flip-side, the tightening of trade restrictions and the use of an exchange rate crawl to underwrite the bolivar's real effective appreciation in order to combat inflation served as significant obstacles to economic recovery and high employment.

During the 1994–2000 period, Venezuela's annual real economic growth averaged an anaemic 0.5 percent with considerable ups and downs. Consumer prices rose at nearly 48 percent a year, with the highest rates in 1995–6. The central government's budget deficit averaged about 1.8 percent of GDP, while the overall public sector deficit (including non-financial public enterprises) reached 2.8 percent. The current, capital and overall balances were positive for the whole period, but fluctuated from year to year. Net public external debt declined marginally in the six-year period in dollar amount, and in relation to GDP as total external assets increased. The continued over-valuation of the bolivar, however, was a significant contributor to the slow growth of the economy. Repeated devaluations failed to achieve desired increases in global competitiveness, partly due to a lack of policy coordination (particularly allowing substantial wage increases). The unemployment rate rose about one percentage point to reach over 11 percent by 2000. As a result, Venezuela's real per capita income in 2000 was below the of 1975 figure.

Venezuela had arguably the worst record in domestic income distribution, as the bottom 70 percent of the population in 1998 received just over 10 percent of the national income, while the top 10 percent received 70 percent. The share of population falling below the poverty line exceeded 60 percent. Social indicators in education and health also worsened during the period. Venezuela's Human Development Index ranking declined to 65 from 46.

SUMMARY

OPEC members' economic performance in the 1994–2000 period exhibited significant variations, reflecting not only underlying politico-economic differences and domestic political changes, but also the zeal for the adoption of specific reform measures. In all countries, undue dependence on crude oil exports and energy-related fiscal revenues, state dominance over the economy, limited effective diversification, a lack of meaningful competition in the private sector, structural weaknesses in the budget, labour-market rigidities, and a glaring disconnection between the education curriculum and necessary national market skills still perpetuated a high vulnerability to external shocks. However, in countries with more open trade and exchange regimes (e.g. the GCC), the overall situation improved during the period. Countries which took diligent measures to broaden their revenue base and restrain current expenditure, pursue a serious privatisation programme, strengthen financial supervision and reduce labour-market rigidities were also more successful than the others.

More specifically, Algeria was arguably more successful than most in moving toward sustained growth and stability. Despite a still highly fragile socio-political order, the economy seemed to be on the right track. In 2000, the Algerian government's medium-term economic agenda placed its emphasis on speeding up banking system reform, privatisation of public enterprises (e.g. telecommunications, energy and public utilities), business deregulation, tax and customs reform, and judicial revision (including clearer definition of property rights and land ownership).

A noteworthy move toward the establishment of civilian rule, democratic political participation, a free press and reduced corruption and cronyism in Indonesia served as a necessary (albeit not sufficient) step towards economic salvation. Nevertheless, sectarian violence, separatist movements and the new democracy's inadequate political maturity still threatened Indonesia's cohesion and cooperation, and kept its economic direction uncertain.

In Iran, not only was the economy still mired in anaemic growth, and plagued with high inflation and unemployment, the Khatami administration's early advances in socio-economic liberalisation (which were promised to spearhead an economic revival) underwent serious setbacks at the hands of the powerful conservative forces within the leadership. The new Third Development Plan also had a painful start.

Kuwait was no less dependent on foreign powers for security protection against its enemies, on foreign teachers for domestic education, or on foreign workers for its private sector operation in 2000 than it was in 1974. Further democratisation and secularisation were still thwarted by its conservative national assembly. Social scars from the 1990 invasion had not yet totally healed.

The other Arab members walked a tightrope between traditional values and the Internet culture. The Persian Gulf members spent much of their national incomes on sophisticated military gear from major powers, and welcomed Western oil companies to share in their energy resources, partly in the hope of ensuring their political survival and security. The presence of foreign navies in the area, however, was a source of socio-political discontent among domestic nationalists and an impediment to greater openness to the West.

In Venezuela, the end of the old two-party system, which was generally identified with corruption and mismanagement, had yet to produce beneficial results. The new Chavez administration, bent on remaking the political regime, was initially greeted with a decline in investment, a sharp drop in economic growth, capital flights and brain drain. With the new political restructuring, mandated by a national referendum, a new constitution and fresh national elections, a serious and credible economic strategy was promised under the banner of 'peaceful revolution'. The outcome remained to be seen.

In sum, the majority of the challenges faced by the members still run deep, and defied quick, easy or popular solution.

Supplemental Table Selected Data 1994–2000

	Growth[1]	Inflation[2]	Unemployment[3]	Government Finance[4]	External Debt[5]	Reserves[5]	Population Growth Rate
Algeria	3.5	10.8	30.0	1.0	26.8	9.6	2.8
Indonesia	1.3	18.6	3.0	-0.8	149.1	26.5	1.8
Iran	3.1	26.1	16.0	-2.8	10.4	9.0	1.9
Kuwait	1.0	1.8	...	-9.4[6]	4.4	4.4	3.4
Libya	0.7	26.6	30.0	-2.8	3.3	6.6	2.8
Nigeria	2.6	20.0	...	-3.8	28.3	5.6	2.4
Qatar	4.2	3.7	...	-5.7	12.2[7]	1.9	0.6
Saudi Arabia	1.5	2.0	...	-5.6[6]	28.0[8]	47.0	3.5
UAE	4.3	2.5	...	-11.4[6]	0.0	12.0	6.9
Venezuela	0.5	48.0	11.0	-1.8	28.5	14.0	2.7

Notes

1 Average annual real GDP.

2 Average annual change in consumer price index.

3 Average yearly joblessness in total labor force.

4 Average annual change in overall fiscal balance as percentage of GDP.

5 In billions of U.S. dollars in January 2000.

6 Excluding foreign investment income.

7 Public and private sector.

8 Private sector.

Source: IMF, *International Financial Statistics*; author's estimates.

Tables

Table 1 Physical Infrastructure

	Area (1000 sq. km)	Paved roads (1000km)			Railway tracks (km)			Main telephone lines (1000)				Power generating capacity (1000kw)			Electricity production (1,000,000kwh)	
		1970	1980	1990[1]	1970	1980	1990[1]	1975	1980	1990	1994	1970	1980	1990[1]	1970	1994
Algeria	2382	33	39	44	3933	3907	4653	172	311	794	1123	750	2006	4657	1979	19,883
Ecuador	281	3	4	6	990	965	965	176	227	490	660	304	1118	1657	949	8256
Gabon	268	0.1	0.5	0.6	0	224	683	..	10	21	30	40	175	279	97	933
Indonesia	1904	21	56	116	6640	6637	6964	219	376	1070	2528	907	2786	11,480	6981	53,414
Iran	1648	10	34	59	4412	4567	4996	814	1025	2255	4402	2197	5300	17,554	6758	79,128
Iraq	438	5	14	26	2528	1589	2372	..	275	675	666	680	1200	9000	2750	27,060
Kuwait	18	..	..	3	0	0	0	103	157	331	406	..	..	6790	..	22,798
Libya	1760	..	..	..	..	..	..	..	..	269	265	..	..	..	..	17,800
Nigeria	924	15	30	31	3504	3523	3557	..	163	260	325	805	2230	4040	9946	15,530
Qatar	11	..	..	..	..	..	..	..	..	170	107	..	..	..	..	..
Saudi Arabia	2150	9	22	..	577	747	1380	141	407	1234	1708	316	5904	18,510	1060	91019
UAE	84	..	..	..	..	..	..	..	..	733	607	..	..	..	..	18,870
Venezuela	912	18	23	26	295	280	445	578	860	1495	2311	3172	8471	18,647	12,707	73,116

Notes

1 Last year for which data are comparable for the majority of members.

Source: The World Bank, *World Development Report 1994*; *World Development Indicators 1997*; United Nations Development Program (UNDP), *Human Development Report 1997*.

Table 2   Population and Labour Force

| | Population (millions) | | Labour force[1] (millions) | | Composition of labour force (%) | | | | | | Unemployment (%) |
| | | | | | Agriculture | | Industry[2] | | Services | | |
	1974	1994	1974	1994	1974	1994	1974	1994	1974	1994	1994
Algeria	16	27	3.5 (22)	8.5 (30)	57	18	17	33	26	49	27
Ecuador	6.5	11	1.8 (28)	3.8 (34)	46	33	19	19	35	48	12[4]
Gabon	0.9	1.3	0.5 (55)	0.6 (46)	73	51	10	16	17	33	...
Indonesia	132	190	58 (43)	85 (45)	70	51	9	14	21	35	3
Iran	32	58[3]	10 (31)	17 (29)	49	30	26	26	25	44	11
Iraq	11	20	3 (27)	5 (25)	50	14	20	19	30	67	...
Kuwait	0.9	1.5	0.3 (33)	0.8 (21)	2	1	34	26	64	73	...
Libya	2	5	0.5 (25)	1.2 (24)	41	20	21	30	36	50	...
Nigeria	73	108[3]	29 (40)	35 (32)	72	48	10	7	18	45	3
Qatar	0.2	0.5	0.1 (50)	0.3 (60)	17	3	24	28	59	69	...
Saudi Arabia	7	18	2 (28)	5.5 (30)	68	48	11	14	21	38	7
UAE	0.5	1.8	0.3 (60)	1 (55)	21	5	32	38	47	57	...
Venezuela	12	21	4.5 (37)	8 (38)	30	13	24	25	46	62	9
OPEC	294	463	113 (38)	172 (37)							

Notes

1 Figures in parentheses denote percentage of total population.

2 Includes mining, manufacturing, construction and utilities.

3 Figures in other sources are considerably different.

4 Figures for urban areas was 7.1 percent.

Source: Calculated from data published by the World Bank, UNDP and national authorities.

Table 3   Basic Oil Data

	First commercial production	Joining OPEC	Peak output year (mb/d)	Proven oil reserves (million barrels)		Reserves' life span (1994)
				1974	1994	
Algeria	1958	1969	1.2 (1978)	7700	9980[4]	36
Ecuador	1972	1973[1]	0.309 (1991)	5700	2015	16
Gabon	1957	1973–5	0.312 (1993)	652	2350	24.5
Indonesia	1885	1962	1.68 (1977)	15,000	4980	10.5
Iran	1913	1960	6.67 (1976)	66,000	94,300[4]	72
Iraq	1927	1960	2.78 (1989)	35,000	100,000	130[3]
Kuwait	1946	1960	3.28 (1972)	81,450	96,500	132
Libya	1961	1962	3.32 (1970)	24,480	22,800	45
Nigeria	1958	1971	2.3 (1979)	20,900	20,990	31.5
Qatar	1949	1961	0.508 (1979)	6000	3500[4]	27
Saudi Arabia	1948	1960	9.90 (1980)	173,150	261,375[4]	89
UAE	1962	1967[2]	2.38 (1991)	33,920	98,100	124
Venezuela	1925	1960	3.70 (1970)	18,570	64,880	75

Notes

1 Ecuador left OPEC in December 1992.

2 Abu Dhabi, UAE, 1974.

3 At the 1990 rate of production.

4 Estimates by *BP Statistical Review of World Energy 1996* differ slightly from OPEC's own.

Source: OPEC, *Annual Statistical Bulletin* 1996.

Table 4 Structure of Production
(percentage of GDP)

	Oil		Agriculture		Industry[1]		Services	
	1974	1994	1974	1994	1974	1994	1974	1994
Algeria	28	22	12	12	18	22	42	44
Ecuador[3]	13	12	22	12	16	24	49	52
Gabon[3]	46	32	12	8	12	17	30	43
Indonesia[3]	12	12	38	18	19	30	31	40
Iran	43	18	11	21	12	18	34	43
Iraq	49	18[2]	8	16	15	17	28	49
Kuwait	72	42	0.3	0.5	6	11	22	47
Libya	57	21	2	8	3	11	38	60
Nigeria[3]	23	14	28	38	19	10	30	38
Qatar	91	31	0.5	1	1	19	8	49
Saudi Arabia	77	32	1	7	5	20	17	41
UAE	78	34	0.8	2	5	20	16	44
Venezuela[3]	40	22	5	5	23	23	32	50

Notes

1 Including manufacturing, construction, and utilities.

2 1989.

3 Sectoral data differ noticeably in different sources. The 1994 figures in this table are the average for 1993–5.

Source: Calculated from data published by national authorities, the World Bank, the International Monetary Fund and UNDP.

Table 5   Structure of Merchandise Exports

	Total exports (US$m)			Oil exports (as percentage of total exports)		Non-oil exports (as percentage of imports)		Oil exports (as percentage of GDP)	
	1973	1974	1994	1973	1994	1973	1994	1974	1994
Algeria	1887	4687	8880	81	71	16	27	32	15
Ecuador	532	1124	3820	47	40	71	47	18	9
Gabon	298	906	2713	71	89	45	23	50	32
Indonesia	3217	7434	40,053	50	15	59	106	20	3
Iran	6207	21,541	19,432	90	81	17	36	46	21
Iraq	1941	6601	485	95	87	11	13	53	2
Kuwait	3816	10,962	11614	92	90	28	17	81	43
Libya	4005	8259	7790	98	92	4	15	62	33
Nigeria	3463	9216	11347	84	97	30	5	29	26
Qatar	617	2015	3213	97	82	9	30	99	36
Saudi Arabia	8986	35,556	42,614[1]	99	89	1	19	95	32
UAE	2156	7380	24040	86	48[1]	35	58[1]	88	32
Venezuela	4680	11,153	16,089	92	71	12	40[2]	40	19

Notes

1 Including re-exports.

2 Average for 1993–5.

Source: IMF, *International Financial Statistics Yearbook* 1996; OPEC, *Annual Statistical Bulletin* 1990, 1994 and 1996; author's estimates.

Table 6　Health Profile

| | Life expectancy at birth (years) | | Infant mortality (per 1000) | | Population access to: (percentage) | | | | | | Population per: | | | | Public expenditure on health | |
| | | | | | Safe water | | Sanitation | | Health care | | Physicians | | Nurses | | | |
	1974/5	1994	1974	1994	1974/5	1993/4	1980	1994/5	1980	1993/4	1974	1994	1974	1994	% GNP 1960	% GDP 1990
Algeria	53	68	139[1]	51	77	79	...	77	...	98	4400[1]	1064	2500	330	1.2	5.4
Ecuador	60	69	70	47	36	71	43	64	...	88	2840	960	2880	600	0.4	...
Gabon	46	54	138[1]	91	...	68	...	76	...	90	5250[1]	2500	570[1]	1471	0.5	0.7
Indonesia	48	63	118[1]	53	11	62	23	55	...	80	18160	7143	8630	1754	0.3	...
Iran	51	68	120	40	51	84	60	82	50	80	2570	2536	1910	1150	0.8	1.5
Iraq	53	57	104	146	66	44	...	38	...	93	2370	1667	3310	1370	1.0	...
Kuwait	67	75	44	17	89	100	100	100	100	100	1140	690	290	220	...	...
Libya	53	64	122[1]	64	87	97	...	98	...	...	1140	962	340	328	1.3	...
Nigeria	41	55[1]	163	82	...	40	...	63	40	66	25440	5882	6230	1639	0.3	1.2
Qatar	60	71	68[1]	19	97	91	...	100	...	100	1947	667	570	186	...	...
Saudi Arabia	45	70	119[1]	27	64	95	76	86	85	97	6660	710	5510	460	0.6	3.1
UAE	61	74	87	17	...	95	75	95	96	95	1100[1]	600	580	225	...	...
Venezuela	65	72	46	22	79	79	52	55	...	...	870	640	470	330	2.6	2.0
All Developing Countries	52	62	...	64	55	70	...	39	...	80	8252	5767	2873	4715	0.9	2.0

Note

1 1970.

Source: National authorities; The World Bank: *World Development Report 1978, 1996; World Development Indicators, 1997*; UNDP: *Human Development Report 1997*; author's estimates where figures differ in original sources.

Table 7   Education Profile

	Adult literacy (percent of total population)		School enrolment (percent of school age group)						Book titles published (per 100,000 people)	Daily newspapers (copies per 100 people)	GNP share spent on education (percentage)		Scientists and technicians (per 1000 people)
			Primary		Secondary		Tertiary						
	1974	1994	1970	1993	1970	1993	1980	1993	1992–4	1994	1980	1994	1992
Algeria	35	59	76	99	11	60	6	12	1	5	7.8	5.6	...
Ecuador	69	89	97	123	22	55	37	20	7	7	5.6	3.0	0.2
Gabon	33[1]	63	85	134	8	...	...	3	...	2	2.7	3.2	0.2
Indonesia	62	83	80	115	16	38	4	10	3	3	1.7	1.3	...
Iran	50	69	7[2]	109	27	57	4	12	16	2	7.5	5.9	0.1
Iraq	34	56	93[2]	...	35[2]	...	9[2]	...	...	3	3.0	...	...
Kuwait	55	78	89	65	63	51	11	14	11	40	2.4	5.6	1
Libya	37[1]	75	145[2]	...	45[2]	...	72	...	...	1	3.4	...	0.6
Nigeria	25[1]	56	37	76	4	20	2	...	1	2	6.4	1.3	0.1
Qatar	60[1]	79	110	90	52	83	...	...	69	15	2.6	3.4	0.6
Saudi Arabia	15	62	45	78	12	46	7	14	...	5	4.1	6.4	...
UAE	55	78	93	110	22	69	2	10	14	16	1.3	2	...
Venezuela	82	91	94	99	33	34	21	30	17	21	4.4	5.1	0.2
All Developing Countries	43	61	...	...	...	...	...	...	...	4	3.8	3.6	0.8

Notes

1 Year other than specified in the column heading.

2 1975.

Source: The World Bank, *World Development Report* (various issues); UNDP, *Human Development Report* 1994, 1995, 1996, 1997.

Table 8   Social Welfare Indicators

	Per capita income (US$)				Daily calorie consumption (per capita)	Energy consumption[2] (per capita) (kg of oil equivalent)		Telephone lines[3] (per 1000 people)	Radios (per 1000 people)	TVs (per 1000 people)
	1974 (GNP)		1994 (GDP)							
	WB1	OPEC[1]	WB1	OPEC[1]	1992	1971	1994	1994	1994	1994
Algeria	730	820	1530	1510	2897	255	906	41	236	70
Ecuador	480	480	1478	...	2587	202	565	59	327	130
Gabon	1960	2375	3034	3865	2250	805	652	30	147	50
Indonesia	170	200	917	928	2755	71	366	13	148	150
Iran	1250	1410	1019	1182	2861	714	1565	66	237	120
Iraq	1110	1155	...	902	2122	...	...	34	218	70
Kuwait	10,030	13,835	15,180	15,430	2535	7264	8622	226	445	410
Libya	4440	5630	...	4500	3310	...	...	51	226	100
Nigeria	280	410	326	383	2125	39	162	3	196	40
Qatar	7240	12,500	12,820	12,498[1]	...	...	...	215	428	430
Saudi Arabia	2830	4055	6586	6886	2751	1061	4566	96	294	250
UAE	11,060	18,265	14,752	19,473	...	4151	10,531	276	312	290
Venezuela	1960	2250	2748	2833	2622	2072	2186	108	443	180
All Developing Countries	...	...	970	...	2546	255	570	33	178	140

Notes

1 Differences between the two estimates are largely (but not exclusively) due to different estimations of the size of population.

2 Figures for the same items differ in different sources, and in the same source in different annual reports.

3 Discrepancies with Table 1 are due to different estimates in different sources.

Source: The World Bank, *Atlas* 1976, *World Development Report* 1996, 1997; OPEC, *Annual Statistical Bulletin* 1996; UNDP, *Human Development Report* 1997; author's estimate.

Table 9   Population and Income Trends
(average annual percentage change)

	Population			Real GNP			Real per capita income		
	1960–73[1]	1973–85	1985–94	1960–73	1973–85	1985–94	1960–73	1973–85	1985–94
Algeria	3.2	3.1	2.5	4.9	5.8	0.1	1.7	2.6	-2.4
Ecuador	3.4	2.9	2.3	5.3	4.0	3.3	1.9	1.1	1.0
Gabon	1.2	1.5	1.8	5.9	0.9	-0.5	4.7	-0.6	-2.3
Indonesia	2.0	2.3	1.6	4.4	6.3	7.6	2.4	4.0	6.0
Iran	3.2	3.5	3.3	9.6	2.9	2.3	6.4	-0.6	-1.0
Iraq[2]	3.2	3.6	2.9	6.1	2.5	-12.7	2.9	-1.1	-15.6
Kuwait[2]	9.4	5.7	-0.5	7.3	6.0	-1.8	-2.1	0.3	-1.3
Libya[2]	3.7	4.0	3.6	14.2	1.3	-0.6	10.5	-2.6	-4.2
Nigeria	2.5	2.8	2.9	6.1	0.3	4.1	3.6	-2.5	1.2
Qatar[2]	8.5	6.8	4.5	13.6	-2.3	2.3	5.1	-8.5	-2.2
Saudi Arabia[2]	1.7	4.8	3.6	10.4	6.6	2.5	8.7	1.7	-1.2
UAE[2]	9.9	4.3	3.3	29.2	3.3	0.8	19.3	-1.0	-2.5
Venezuela	3.4	3.3	2.5	5.4	1.3	3.1	2.0	-1.9	0.6

Notes

1 Published data for the same period in the World Bank's *Atlas* 1975 and *World Development Report* 1985 differ considerably.

2 Estimates adjusted for discrepancies in original sources.

Source: The World Bank, *Atlas* 1975, 1987 and 1996; national authorities; author's estimates.

Table 10   Trends in Real Gross Domestic Product[1]
(annual average percentage change)

	1965–73	1974–79	1980–90	1990–94	1974–94	
					GDP	GDP per capita
Algeria	7.0	8.5	1.6	-0.4	2.8	0.0
Ecuador	6.4	5.7	2.0	3.5	3.3	0.6
Gabon	7.9	1.5	0.5	-2.1	1.0	-0.6
Indonesia	8.1	7.0	6.1	7.6	6.7	4.7
Iran	10.4	8.5	0.4	6.2	3.8	0.4
Iraq[3]	4.4	6.3	-0.2[2]	-20.0[2]	-3.5	-6.8
Kuwait[3]	5.1	7.0	0.9	7.2	4.0	1.4
Libya[1]	7.7	5.9	-2.3[2]	3.0	1.0	-2.8
Nigeria	9.7	6.0	1.6	2.4	2.9	0.0
Qatar[3]	...	4.0	-0.4[2]	-0.8[2]	0.6	-5.0
Saudi Arabia[3]	11.2	9.2	1.4	4.3	4.0	2.1
UAE[3]	...	5.4	-1.1	4.3[1]	1.8	-2.0
Venezuela	5.1	5.0	1.1	3.2	2.6	-0.3

Notes

1 Figures may not exactly match those of Table 9 due to overlapping differences between GNP and GDP data, and other discrepancies in the original sources.

2 1990–92.

3 Estimates are adjusted for discrepancies in the original sources.

Source: The World Bank, *World Development Report* 1983 and 1996; *World Development Indicators* 1997; International Monetary Fund, *World Economic Outlook*, April 1998; national authorities; author's estimates.

Table 11 Domestic Income Distribution
(percentage share of income or consumption)

	Survey year	GINI index	1 Lowest 10%	2 Lowest 20%	3 Second quintile	4 Third quintile	5 Fourth quintile	6 Highest 20%	7 Highest 10%	7:1	Poverty[1] Year	National line (%)	Internat. line (%)
Algeria	1988	38.7	2.8	6.9	11.0	15.1	20.9	46.1	31.5	11.2	1988	...	1.6
Ecuador	1994	46.6	2.3	5.4	8.9	13.2	19.9	52.6	37.6	16.3	1994	35.0	30.4
Gabon	...	...	...	...	...	...	...	...	...		...	...	...
Indonesia	1993	31.7	3.9	8.7	12.3	16.3	22.1	40.7	25.6	6.6	1990–	15.1	14.5
	(1976)	...	...	(6.6)	(7.8)	(12.6)	(16.9)	(49.4)	(34.0)	...	93		
Iran	1994	...	1.0	3.0	...	...	...	50.0	30.0	30.0	1994	20.1[2]	7.0
Iraq	...	...	...	...	...	...	...	...	...		...	...	...
Kuwait	...	...	...	...	...	...	...	...	...		...	...	...
Libya	...	...	...	...	...	...	...	...	...		...	...	...
Nigeria	1992/3	45.0	1.3	4.0	8.9	14.4	23.4	49.3	31.3	24.0	1992–	34.1	28.9
											93		
Qatar	...	...	...	...	...	...	...	...	...		...	...	...
Saudi Arabia	...	...	...	...	...	...	...	...	...		...	...	...
UAE	...	...	...	...	...	...	...	...	...		...	...	...
Venezuela	1990	53.8	1.4	3.6	7.1	11.7	19.3	58.4	42.7	30.5	1990–	31.3	11.8
	(1970)	...	(1.3)	(3.0)	(7.3)	(12.9)	(22.8)	(54.0)	(35.7)	...	91	...	...

Memorandum Items

Brazil		63.8											
Belarus		21.6	4.9										
Slovak Rep.				11.9		15.8		67.5	51.3				
Panama			0.5						18.2				
Norway					18.9		25.6						

Notes

1 National poverty line differs in different countries. International poverty line is set for all countries at $1 a person per day in 1985 international prices, adusted to local currency in terms of purchasing power. Figures show percentage of population below the poverty line.

2 26.3 percent rural and 14 percent urban.

Source: The World Bank, *World Development Report* 1997, *World Development Indicators* 1997; national authorities.

Table 12   Price Trends
(annual percentage change)

	Consumer price index				GDP deflator		
	1965–73	1980–90	1973–85	1990–5	1980–90	1985–94	1990–5
Algeria	3.8	9.1	12.2	27.1	8.0	22.3	25.8
Ecuador	6.2	35.8	17.8	40.0	36.4	47.5	37.2
Gabon	3.5[1]	5.1	13.1	5.8	1.9	3.3	13.0
Indonesia	63.0	8.3	17.4	8.8	8.5	8.9	7.6
Iran	5.5	18.1	16.3[1]	27.7	14.6	23.4	32.3
Iraq	3.2	...	10.4[1]	...	10.3	170[3]	...
Kuwait	4.6	2.9	9.2	2.3	-2.4	-0.5[1]	-2.0
Libya	9.4	...	10.8	...	...	21.0[2]	...
Nigeria	10.3	21.5	13.0	49.1	16.6	29.6	47.1
Qatar	...	...	5.4[1]	2.0	...	3.1[1]	...
Saudi Arabia	5.1	-0.8	14.1	1.7	-3.7	2.8	1.0
UAE	...	...	8.7	...	0.7	3.4[1]	...
Venezuela	3.3	20.9	11.7	43.8	19.3	36.4	38.4

Source: The World Bank, *World Development Report* (various issues); IMF, *International Financial Statistics Yearbook* (various issues); author's estimates from national publications. Discrepancies reflect overlaps and absence of reliable data.

Notes
1 Author's estimates.
2 1990.
3 For 1990–4, 270 percent on average.

Table 13   Fiscal Balance

	Government revenue (as % of GDP)		Government expenditure[1] (as % of GDP)		Number of years		
	1974	1994	1974	1994	Budget deficit	Domestic borrowing	Foreign borrowing
Algeria	42.7	...	39.0	...	21	21	...
Ecuador	12.2	15.5	12.3	15.2	14	18	7
Gabon	23.9	...	28.1	...	13	10	11
Indonesia	17.1	18.1	18.3	18.7	16	14	19
Iran	46.1	25.0	24.9	25.1	19	21	...
Iraq	...	...	...	...	14	...	...
Kuwait	71.5	...	29.9	56.0	4	...	...
Libya	50.0	20.0	55.0	27.0	14	14	
Nigeria	19.2	14.4	14.3	22.1	15	13	11
Qatar	69.6	...	24.0	...	14[1]	...	...
Saudi Arabia	...	...	...	...	15	12	1
UAE	2.6	11.8	2.4	11.6	12	...	...
Venezuela	39.5	18.9	35.7	23.1	13	13	13

Notes

1 Including net lending and equity participation.

Source: The International Monetary Fund, *Government Financial Statistics Yearbook* (various issues); *International Financial Statistics Yearbook* 1997; author's estimates.

Table 14   External Balance (1974–94)

	Trade account	Current account (yrs in deficit)	Overall account	External debt (millions of US$) 1974	1994
Algeria	5	11	10	3.3	29.9
Ecuador	2	19	16	0.3	14.3
Gabon	0	8	14	0.5[1]	3.9
Indonesia	0	18	8	6.3	96.5[2]
Iran	7	11	11	4.1	22.7
Iraq	3	11	11	4.0	92.0[3]
Kuwait	2	2	6	...	9.5
Libya	1	10	11	...	4.8
Nigeria	1	10	16	1.2	33.5
Qatar	0	8	3	...	2.5[4]
Saudi Arabia	0	13	11	...	29.0[5]
UAE	0	0	3	...	12.0[6]
Venezuela	2	8	10	1.5	36.8
OPEC	0	11	...	21.2	387.4

Notes
1 1980.
2 Excluding unreported offshore bond borrowings.
3 Estimated but not identified.
4 Including net lending and equity participation.
5 Including public enterprises.
6 Excluding undisclosed foreign assets held by Abu Dhabi Investment Authority.

Source: National authorities; International Monetary Fund, *Balance-of-Payments Yearbook* 1996; World Bank, *World Debt Tables* 1979, 1984–5 and 1996; OPEC, *Annual Statistical Bulletin* 1996; estimates.

Table 15   Defence Outlays and Military Manpower

	Military expenditure (US$m)			Military expenditure as % of GNP					
	1974–79	1980–85	1986–94	1970	1975	1980	1985	1990	1994
Algeria	3805	7570	8650	2.0	2.5	2.9	2.5	2.0	3.3
Ecuador	1015	1315	3430	2.0	2.4	2.2	2.8	3.2	3.5
Gabon	200	430	1130	1.5	0.8	3.1	2.8	3.4	2.9
Indonesia	6640	10,960	15,870	3.3	3.8	3.1	2.4	1.5	1.4
Iran	70,070	47,045	53,170	7.8	17.6	6.6	7.7	6.0	2.4
Iraq	26,575	87,350	86,040[2]	12.9	17.4	22.5	41.2	61.3	14.6
Kuwait	4330	6275	60,790	2.9	5.4	3.0	5.7	53.1	11.1
Libya	10,080	20,210	18,430	2.8	7.0	8.5	20.5[1]	14.3[1]	4.3
Nigeria	11,720	8360	2405	5.9	5.5	2.7	1.5	0.9	0.8
Qatar	1390	3110	3320	19.9[1]	3.5	9.3	...	13.2[1]	3.9
Saudi Arabia	51,375	127,860	193,190	11.9	17.4	14.4	22.7	20.6	14.2
UAE	2685	11,575	19,840	1.4[1]	0.4	6.2	6.7	7.3	5.7[1]
Venezuela	2990	4085	9970	1.8	1.9	1.3	2.4	2.1	1.3

	Military expenditure as % of govt exp.						Armed forces (per 1000 people)					
	1970	1975	1980	1985	1990	1994	1970	1975	1980	1985	1990	1995
Algeria	8.1	7.8	10.0	6.3	7.7	7.5	5.8	5.0	5.4	7.7	5.0	4.5
Ecuador	8.8	19.5	15.0	16.9	20.1	21.8	2.7	2.9	4.3	4.9	5.4	5.3
Gabon	6.1	1.8	8.0	6.6	13.7	9.6[1]	2.0	5.0	6.2	6.9	8.4	5.3
Indonesia	24.5	17.9	12.7	10.3	7.8	7.1	2.9	1.9	1.6	1.6	1.5	1.4
Iran	26.1	39.1	19.7	34.1	30.1	7.4	8.5	11.6	7.8	7.3	7.7	8.4
Iraq	37.6	29.8	26.9	...	...	...	10.0	13.8	32.6	50.2	75.4	21.4
Kuwait	7.6	17.8	11.0	13.6	134.0	22.2	14.3	25.0	8.8	9.3	3.3	8.9
Libya	9.9	16.7	23.4	40.0[1]	28.0	...	9.0	10.0	17.5	24.6	19.7	15.8
Nigeria	36.6	15.7	9.3	9.4	3.6	5.0	9.0	8.8	1.7	1.8	1.1	0.8
Qatar	9.9	5.8	20.1	...	29.2[1]	9.2[1]	20.0	25.0	26.0	20.3	24.3	19.3
Saudi Arabia	27.5	34.1	26.7	27.0	60.6	41.5[1]	10.5	13.0	7.8	6.1	9.2	9.1
UAE	43.0	10.5	41.4	43.5	65.8	50.1[1]	33.3[1]	35.0	44.0	28.0	29.3	21.5
Venezuela	9.0	5.7	4.9	10.8	9.0	5.6	4.2	4.3	3.7	4.0	4.0	3.6

Notes

1 Year nearest to the one specified in the column heading.

2 1986–91.

Source: The US Arms Control and Disarmament Agency, *World Monetary Expenditures and Arms Transfer* 1982, 1987, 1996; The International Institute for Strategic Studies, *The Military Balance* 1995/6.

Table 16   Official Development Assistance
(average annual, US$m)[1]

	1974–79	1980–85	1986–94	Total
Algeria	83	81	29	1162
	(0.32)	(0.15)	(0.05)	...
Ecuador	...	...	...	...
Gabon	...	...	...	...
Indonesia	...	...	...	...
Iran	345	-95	0.75	1160
	(0.56)	(-0.9)	(-)	...
Iraq	236	177	0.75	2250
	(1.0)	(0.5)	(-)	...
Kuwait	983	1,042	460	15,310
	(5.9)	(3.7)	(1.8)	...
Libya	147	166	69.5	2355
	(0.97)	(0.55)	(0.31)	...
Nigeria	40	73	22	836
	(0.08)	(0.08)	(0.08)	...
Qatar	217	117	2.5	1810
	(7.97)	(1.75)	(0.05)	...
Saudi Arabia	3565	4022	2284	62,510
	(6.4)	(3.3)	(2.3)	...
UAE	997	473	234	2920
	(7.8)	(1.6)	(0.75)	...
Venezuela	74	123	29	1370
	(0.2)	(0.18)	(0.06)	...

Note

1 Figures in parentheses denote percentage of donor's GNP.

Source: The World Bank, *World Development Report* 1996; UNCTAD, *Handbook for International Trade* 1994; OPEC
Fund for Internatinal Development, *Three Decades of OPEC Ai*d, 1987; OPEC, *Aid Institutions – A Profile* 1995;
OECD Secretariat

Table 17   Human Development Ranking

	GDP group ranking[1] 1993–4	Human development index[2] 1994	Real GDP per capita ranking[3] (RGDPC) 1994	RGDPC minus HDI[4] 1994
Algeria	LMI	82	65	-17
Ecuador	LMI	72	77	5
Gabon	UMI	120	95	-25
Indonesia	LI	99	92	-7
Iran	LMI	70	61	-9
Iraq	LI	126	102	-24
Kuwait	HI	53	6	-47
Libya	UMI	64	56	-8
Nigeria	LI	141	142	-1
Qatar	HI	55	22	-33
Saudi Arabia	UMI	73	41	-32
UAE	HI	44	27	-17
Venezuela	UMI	47	48	1
		Memorandum Items[5]		
Canada[5]	HI	1	8	7
Sierra Leone[6]	LI	175	171	-4

Notes

1 LI = low income; LMI = lower middle income; UMI = upper middle income; HI = high income.

2 The "human development index" (HDI) is based on three components: life expectancy at birth; educational attainment; income.

3 Real GDP per capita is calculated on the prchasing power parity (PPP) basis, and differs from figures in Table 8.

4 A negative value shows that the HDI rank is worse than the real GDP per capita rank.

5 Highest rank among 175 nations.

6 Lowest rank among 175 nations.

Source: UNDP, *Human Development Report, 1997.*

Table 18   Structure of Demand
(average annual percentage of GDP)

	Consumption						Gross Domestic Investment			
	Private			Public						
	1974–9	1980–5	1986–94	1974–9	1980–5	1986–94	1974–9	1980–5	1986–94	1974–94 Average
Algeria	48.1	45.8	51.6	12.8	14.8	17.3	40.5	33.3	29.8	33.9
Ecuador	62.2	63.1	69.2	13.7	13.0	10.7	22.5	21.2	20.9	21.4
Gabon	26.4	28.2	45.4	13.0	16.3	19.2	52.3	34.2	29.3	37.4
Indonesia	62.7	58.1	58.3	10.5	10.5	10.3	23.5	27.3	30.4	27.5
Iran	41.5	60.0	63.5	21.3	17.3	12.5	25.0	22.8	16.8	20.8
Iraq	46.2	43.6	53.0[1]	18.8	30.8	32.7	25.5	33.5	17.0[1]	...
Kuwait	26.3	42.6	55.2	11.3	18.0	56.2	15.1	19.6	20.6	18.7
Libya	26.8	31.0[2]	...	25.5	28.0	...	25.5	26.6[2]	...	...
Nigeria	61.3	68.2	67.0	10.5	11.6	5.2	25.2	16.4	14.8	18.2
Qatar[3]	15.0	22.0	43.0	10.0	14.0	18.0	17.0	16.5	20.0	18.0
Saudi Arabia	20.4	36.1	45.1	19.6	28.3	33.0	23.5	24.7	17.8	21.4
UAE	15.3	22.2	43.3	10.2	14.6	18.8	31.2	27.2	24.5	27.2
Venezuela	50.5	61.0	67.5	13.5	13.2	9.2	35.8	21.1	19.6	24.6

Notes

1 1986–9.

2 1980–2.

3 Estimates.

Source: Compiled from data published by national authorities, the World Bank, the IMF, and the United Nations.

Table 19   Gross Domestic Product
(current market prices in US$m)

	1973	1974	1975	1976	1977	1978	1979	1980	1981	1982	1983
Algeria	8101	13,299	15,597	17,797	21,030	26,434	33,274	42,541	44,424	45,120	48,902
Ecuador	2348[1]	3484	4211	5154	6476	7433	8960	11,152	13,214	12,817	11,753
Gabon	658[1]	1544	2158	3009	2809	2389	3030	4281	3860	3618	3464
Indonesia	15,682[1]	25,802	30,465	37,275	45,817	52,452	51,208	72,486	91,841	94,214	87,093
I.R. Iran	24,372[1]	45,753	51,523	66,589	84,267	73,486	84,704	91,639	100,658	123,625	149,726
Iraq	5388	12,420	16,151	19,169	20,460	24,465	44,572	53,586	37,289	42,282	42,129
Kuwait	4255[1]	13,004	12,016	13,124	14,141	15,505	24,842	28,688	25,255	21,586	20,870
SP Libyan AJ	6426[1]	13,116	12,768	16,575	19,466	19,213	26,502	35,585	31,552	29,880	28,816
Nigeria	12,768[1]	29,845	35,384	44,003	50,805	56,813	71,593	93,040	82,680	76,810	78,986
Qatar	654	2000	2513	3284	3618	4052	5634	7829	8661	7597	6485
Saudi Arabia	7806[1]	27,977	38,833	58,091	63,942	73,430	114,799	156,489	155,144	121,150	10,787
UAE	2853	7855	9972	12,936	16,181	15,630	20,941	29,619	32,988	30,618	28,030
Venezuela	16,332[1]	26,184	27,561	31,496	36,272	39,394	48,386	59,218	66,440	67,861	67,598

	1984	1985	1986	1987	1988	1989	1990	1991	1992	1993	1994
Algeria	52,983	58,087	53,231	64,609	56,736	54,930	62,031	45,723	47,853	49,713	41,978
Ecuador	11,834	14,782	10,357	9766	9587	9389	10,033	11,800	12,700	14,300	16,600
Gabon	3331	3508	3591	3474	3830	4214	5489	5434	5913	5354	5250
Indonesia	87,638	87,297	82,249	75,889	84,268	94,460	106,014	116,682	128,123	157,963	176,968
I.R. Iran	124,874	166,584	198,248	133,076	122,232	119,597	117,481	119,124	107,335	73,766	71,108
Iraq	47,819	49,819	48,457	57,586	64,444	67,640	74,945	10,115	16,806	17,973	17,973
Kuwait	21,707	21,446	17,905	22,368	20,690	24,314	18,399	11,011	20,136	24,197	24,995
SP Libyan AJ	25,583	24,330	21,864	25,768	22,774	22,869	26,953	29,423	25,566	22,600	22,050
Nigeria	83,229	81,080	54,251	27,174	32,401	30,557	32,449	32,848	33,692	31,819	41,602
Qatar	6870	6153	5017	5104	5717	6488	7360	6884	7646	7157	7374
Saudi Arabia	99,718	86,678	73,215	73,551	76,142	82,996	104,670	118,035	123,204	118,515	120,168
UAE	27,703	27,022	21,574	23,699	23,672	27,267	33,641	33,914	35,412	35,522	36,220
Venezuela	60,177	59,865	62,972	46,856	60,380	49,169	48,763	53,615	60,754	60,435	60,581

Notes

1 Gross National Product

Source: OPEC *Annual Statistical Bulletin* 1991, 1994, 1996; author's estimate.

Table 20   Values of Total Exports
(US$m)

	1973	1974	1975	1976	1977	1978	1979	1980	1981	1982	1983
Algeria	1887	4687	4700	5259	5944	6326	95,511	13,871	14,396	13,176	12,583
Ecuador	532	1124	974	1258	1436	1558	2104	2481	2451	2327	2348
Gabon	298	906	983	932	1343	1107	1848	2173	2000	2161	1888
Indonesia	3217	7434	7102	8547	10,853	11,643	15,590	23,950	25,165	22,328	21,146
I.R. Iran	6207	21,541	20,181	23,541	24,260	22,476	19,807	12,328	11,821	20,452	21,802
Iraq	1941	660[1]	8297	9273	9650	11,064	21,572	26,349	10,140	10,033	8161
Kuwait	3816	10,962	9184	9846	9754	10,427	18,404	19,663	16,044	10,864	11,504
SP Libyan AJ	4005	8259	6834	9554	11,411	9895	16,076	21,910	15,571	13,948	12,341
Nigeria	3463	9216	8003	10,117	12,367	10,445	16,733	25,934	17,837	12,176	10,363
Qatar	617	2015	1805	2211	2072	2391	3789	5672	5691	4343	3345
Saudi Arabia	8986	35,556	29,683	38,287	43,464	40,659	63,428	109,073	119,858	79,081	45,859
UAE	2156	7380	7218	8891	10,062	9832	14,801	21,967	21,792	18,224	15,391
Venezuela	4680	11,153	8800	9299	9551	9187	14,317	19,221	20,980	16,590	13,937

	1984	1985	1986	1987	1988	1989	1990	1991	1992	1993	1994
Algeria	12,795	12,841	7832	8225	7810	9570	12,930	12,570	11,130	10,230	8880
Ecuador	2620	2905	2172	1928	2193	2354	2714	2852	3008	3062	3717
Gabon	2011	1951	1269	1286	1196	1598	2464	2559	2082	1921	1872
Indonesia	21,888	18,587	14,805	17,136	19,219	22,160	25,675	29,142	33,967	36,823	40,053
I.R. Iran	17,087	14,125	7171	11,916	10,709	13,081	19,305	18,661	19,868	18,080	19,432
Iraq	9317	10,409	7465	9705	9610	12284	10,314	487	609	472	485
Kuwait	11,623	10,487	7383	8264	7661	11476	7042	1088	6660	10,248	11,614
SP Libyan AJ	11,136	12,314	8215	8694	6673	8034	13,225	11,235	9942	8590	7790
Nigeria	11,849	13,111	5083	7560	6875	7871	13,671	12,264	11,886	11,651	11,347
Qatar	4513	3419	1849	1985	2210	2738	3890	3210	3841	3245	3213
Saudi Arabia	37,542	27,481	20,086	23,199	24,376	28,383	44,416	47,697	50,287	42,395	42,614
UAE	15,990	14,764	10,129	12,765	12,250	17,596	23,544	24,436	24,756	23,550	24,040
Venezuela	15,997	14,438	8660	10,577	10,244	13,286	17,497	15,155	14,185	14,686	16,089

Note

1 Author's estimate.

Source: OPEC *Annual Statistical Bulletin* 1991, 1994, 1996; author's estimate.

Table 21   Value of Petroleum Exports[1]
(US$m)

	1973	1974	1975	1976	1977	1978	1979	1980	1981	1982	1983
Algeria	1522	4267	4295	4791	5562	5856	8746	12,971	13,066	11149	9655
Ecuador	250	613	516	565	494	604	1161	1551	1709	1304	1578
Gabon	212	773	918	924	900	921	1283	1745	1786	1802	1652
Indonesia	1599	5211	5311	6004	7298	7439	8871	15,595	18,164	15,493	13,558
I.R. Iran	5617	20,904	19,634	22,923	23,599	21,684	19,186	11,693	11,491	20,168	21150
Iraq	1842	6534	8227	9201	9560	10,913	21,382	26,096	10,039	9933	7816
Kuwait	3520	10,568	8594	9090	8918	9557	17,294	18,935	14,229	9066	10,069
SP Libyan AJ	3927	8149	6671	9553	11,394	9885	15,941	21,906	14,930	13,692	12,341
Nigeria	2905	8640	7461	9444	11,561	9452	15,624	24,931	17,291	11,883	9941
Qatar	600	1979	1757	2137	2055	2305	3693	5372	5496	4214	2993
Saudi Arabia	8956	35,476	29,473	38,157	43,308	40,332	62,855	108,175	118,998	78,119	44,830
UAE	1867	6948	6762	8383	9258	8661	12,862	19,390	18,761	15,956	13,016
Venezuela	4328	10,548	8324	8763	9110	8740	13,633	17,562	18,609	15,633	13,857
OPEC Total[2]	37,145	120,610	107,943	129,935	143,017	136,340	202,531	285,921	264,568	208,534	162,688

	1984	1985	1986	1987	1988	1989	1990	1991	1992	1993	1994
Algeria	9778	9668	5161	6555	5725	6815	9588	8464	7885	6902	6335
Ecuador	1779	1926	983	724	976	1147	1258	1059	1251	1149	1390
Gabon	1684	1629	723	896	779	1200	1967	1740	1712	1506	1668
Indonesia	12,477	9083	5501	6157	5042	6062	7404	6714	6619	5693	6005
I.R. Iran	16,726	13,710	6255	10,755	9673	12,031	17,906	15,767	16,802	14,251	14,801
Iraq	8863	10,097	6905	9416	9312	11,876	9594	351	482	425	421
Kuwait	10,996	9451	6378	7523	6840	10,432	6385	874	6224	9708	10,482
SP Libyan AJ	10,970	12,132	8202	6293	6070	6573	10,715	10,211	9326	7689	7170
Nigeria	11,534	12,568	4770	7024	6267	7470	13,265	11,792	11,642	10,859	11,040
Qatar	4386	3068	1720	1829	1709	1955	3273	2828	2870	2811	2623
Saudi Arabia	36,285	25,937	18,061	20,427	20,205	24,095	40,130	43,701	44,754	38,621	38,139
UAE	12,037	10,896	6865	7900	7627	10,215	14,846	14,356	14,251	12,118	11,683
Venezuela	14,824	12,956	7178	9054	8158	10,001	13,953	12,302	11,208	11,030	11,473
OPEC Total[2]	152,339	131,273	78,701	94,553	88,384	109,872	150,283	130,159	135,026	122,762	123,230

Notes

1 Includes petroleum products, condensates and NGLs.

2 Totals may not add up due to independent rounding.

Source: OPEC *Annual Statistical Bulletin* 1991, 1994, 1996; author's estimate.

Table 22   Current Account Balances
(US$m)

	1973	1974	1975	1976	1977	1978	1979	1980	1981	1982	1983
Algeria	-447	176	-1658	-882	-2325	-3540	-1631	249	90	-183	-85
Ecuador	7	38	-220	-7	-343	-703	-630	-670	-1012	-1226	-4
Gabon	-43	219	73	52	99	74	248	384	403	309	98
Indonesia	-475	597	-1109	-908	-50	-1414	980	3011	-566	-5324	-6338
I.R. Iran	155	12,267	4708	7660	2816	104	11,968	-2438	-3446	5733	358
Iraq	801	2618	2705	2495	2988	4087	11,117	14,500	-10,600	-11,900	-2000
Kuwait	2583	9066	5929	6929	4561	6130	14,032	15,302	13,699	4963	5311
SP Libyan AJ	67	2700	392	2845	2159	738	3771	8214	-3963	-1560	-1643
Nigeria	-9	4898	42	-357	-1016	-3757	1671	5178	-6474	-7282	-4332
Qatar	346	1720	1190	997	585	923	1288	2647	2384	1124	410
Saudi Arabia	2520	23,026	14,385	14,360	11,991	-2212	10,206	41,503	39627	7575	-16,852
UAE	643	4150	3361	3419	1886	1692	5255	10,069	9207	7001	5257
Venezuela	877	5760	2171	254	-3179	-5735	350	4728	4000	-4246	4427

	1984	1985	1986	1987	1988	1989	1990	1991	1992	1993	1994
Algeria	74	1015	-2230	141	-2040	-1081	1420	2367	1020	800	-1800
Ecuador	-148	149	-613	-1131	-505	-472	-136	-473	-215	-682	-807
Gabon	113	-162	-1057	-449	-615	-192	214	183	-264	-284	-300
Indonesia	-1856	-1923	-3911	-2098	-1397	-1108	-2989	-4260	-2780	-2106	-2792
I.R. Iran	-414	-476	-5155	-2090	-1869	-191	327	-9448	-6504	-4215	4950
Iraq	-1730	-3540	-3040	2300	-140	2850	500	-1000	-1500	-250	-229
Kuwait	6428	4798	5616	4561	4602	9136	3886	-26,478	-450	1938	2489
SP Libyan AJ	-1456	1906	-166	-1128	-1826	-1026	2201	560	1402	-1371	-215
Nigeria	123	2604	211	-73	-296	1090	4988	1203	2268	-780	-2128
Qatar	830	549	-189	-132	-261	-4	306	-100	-9	-659	-502
Saudi Arabia	-18,400	-12,932	-11,795	-9773	-7340	-9538	-4152	-27,546	-17,740	-17,268	-10,480
UAE	7463	6946	2370	4086	2533	4549	5230	1634	3043	183	360
Venezuela	4651	3327	-2245	-1390	-5809	2161	8279	1736	-3749	-1993	2541

Source: OPEC *Annual Statistical Bulletin* 1991, 1994, 1996; and author's estimate. OPEC data for some years differ (sometimes considerably) from IMF's *International Financial Statistics*.

Table 23   Annual Average Exchange Rate
(Units of national currency per US$)

	1974	1975	1976	1977	1978	1979	1980	1981	1982	1983
Algeria (AD)	4.181	3.949	4.164	4.147	3.966	3.853	3.837	4.316	4.592	4.789
Ecuador (SU)	25.0	25.0	25.0	25.0	25.0	25.0	25.0	25.0	30.03	44.12
Gabon (CFAG)	240.7	214.31	238.95	245.68	225.66	212.72	211.28	271.73	328.6	381.06
Indonesia (RP)	415.0	415.0	415.0	415.0	442.05	623.06	626.99	631.76	661.42	909.26
I.R. Iran (IR)	67.625	67.639	70.222	70.62	70.475	70.47	70.61	78.33	83.6	86.36
Iraq (ID)	0.295	0.295	0.295	0.295	0.295	0.295	0.295	0.295	0.299	0.311
Kuwait (KD)	0.293	0.290	0.292	0.287	0.275	0.276	0.27	0.279	0.288	0.291
SP Libyan AJ (LD)	0.296	0.296	0.296	0.296	0.296	0.296	0.296	0.296	0.296	0.296
Nigeria (N)	0.63	0.616	0.627	0.645	0.635	0.604	0.547	0.618	0.674	0.724
Qatar (QRL)	3.947	3.931	3.963	3.959	3.877	3.773	3.657	3.64	3.64	3.64
Saudi Arabia (SRL)	3.55	3.518	3.53	3.525	3.4	3.361	3.327	3.383	3.428	3.455
UAE (VADH)	3.959	3.961	3.953	3.903	3.871	3.816	3.707	3.671	3.671	3.671
Venezuela (B)	4.285	4.285	4.29	4.292	4.292	4.292	4.292	4.292	4.292	4.297

	1984	1985	1986	1987	1988	1989	1990	1991	1992	1993	1994
Algeria (AD)	4.983	5.028	4.702	4.850	5.915	7.609	8.958	18.473	21.836	23.345	35.059
Ecuador (SU)	62.54	69.56	122.78	170.46	301.61	526.35	767.75	1046.25	1534.0	1919.1	2196.7
Gabon (CFAG)	436.96	449.26	346.3	300.54	297.85	319.01	272.26	282.11	264.69	283.16	555.2
Indonesia (RP)	1025.94	1100.58	1282.56	1643.85	1685.70	1770.06	1842.81	1950.32	2029.92	2087.1	2160.75
IR Iran (IR)	90.03	91.05	78.76	144.91	177.97	226.0	304.347	408.587	600.0	1267.77	1748.75
Iraq (ID)	0.311	0.311	0.311	0.311	0.311	0.311	0.311	0.311	0.311	0.311	0.311
Kuwait (KD)	0.296	0.301	0.291	0.279	0.279	0.294	0.288	0.284	0.293	0.302	0.297
SP Libyan AJ (LD)	0.296	0.296	0.296	0.273	0.286	0.3	0.283	0.281	0.285	0.305	0.321
Nigeria (N)	0.767	0.894	1.755	4.016	4.536	7.365	8.038	9.910	17.298	22.065	21.996
Qatar (QRL)	3.64	3.64	3.64	3.64	3.64	3.64	3.64	3.64	3.64	3.64	3.64
Saudi Arabia (SRL)	3.524	3.622	3.703	3.745	3.745	3.745	3.745	3.745	3.745	3.745	3.745
UAE (VADH)	3.671	3.671	3.671	3.671	3.671	3.671	3.671	3.671	3.671	3.671	3.671
Venezuela (B)	7.017	7.5	8.083	14.5	14.5	34.681	46.9	56.816	68.376	90.826	148.503

Source: OPEC *Annual Statistical Bulletin* 1991, 1994 and 1996.

Table 24 Positive Elements in Economic Development[1]

	1	2	3	4	5	6	7	8	9	10
Algeria		+					+	+	+	
Ecuador			+	+	+			+		
Gabon	+	+		+						
Indonesia	+	+	+		+	+	+	+	+	
Iran		+					+	+		
Iraq		+						+		
Kuwait	+	+		+					+	
Libya										
Nigeria			+				+	+		
Qatar		+		+	+				+	
Saudi Arabia		+		+					+	
UAE		+		+					+	
Venezuela		+			+			+		

Notes

1 Headings:

  (1) low population growth;

  (2) high rate of investment;

  (3) low government consumption;

  (4) minimum wage/price distortions;

  (5) prompt adjustments to external shocks;

  (6) supportive cultural factors;

  (7) large domestic market;

  (8) ample non-oil resources;

  (9) equitable income distribution;

  (10) quality of governance.

Source: Tables 9, 18, 20; Chapters IV and V.

Table 25   Negative Elements in Economic Development[1]

	A	B	C	D	E	F
Algeria	+	+	+	+	+	+
Ecuador			+	+		+
Gabon	+	+	+	+	+	+
Indonesia			+		+	+
Iran	+	+	+	+		+
Iraq	+	+	+	+	+	+
Kuwait	+			+	+	
Libya	+	+	+	+	+	+
Nigeria	+	+	+	+	+	+
Qatar	+		+	+	+	+
Saudi Arabia	+		+	+	+	+
UAE	+			+	+	
Venezuela	+	+	+	+	+	+

Notes

1 Headings:
   (A) undue state intervention;
   (B) volatility of regulations;
   (C) poor choice of strategy;
   (D) unsustainable welfare programs;
   (E) multitude of unprofitable projects;
   (F) corruption and waste.

Source: Chapters IV and V.

Table 26   Level of Economic Freedom[1]

	I	II	III	IV	V	VI	VII	VIII	IX	X	Overall[2]
Algeria	5	3.5	3	3	3	3	3	3	3	3	3.25
Ecuador	3	2.5	1	5	2	3	3	3	4	4	3.05
Gabon	5	4.5	3	1	2	2	3	3	3	3	2.95
Indonesia	2	3.5	1	2	2	3	3	3	4	5	2.85
Iran	5	5	5	4	5	5	4	5	4	5	4.7
Iraq	5	5	5	5	5	5	5	5	4	5	4.9
Kuwait	2	1	4	2	4	3	3	1	2	2	2.4
Libya	5	5	5	2	5	5	5	5	5	5	4.7
Nigeria	5	3	2	4	2	4	2	3	4	3	3.2
Qatar	...	...	...	...	...	...	...	...	...	...	...
Saudi Arabia	4	4	4	1	4	3	3	1	2	2	2.8
UAE	2	1	3	1	4	3	3	1	2	1	2.1
Venezuela	4	4	3	5	3	3	3	3	3	5	3.6

Notes

1 Higher values mean less freedom.

Headings:

(I)   Trade policy;

(II)   Taxation policy;

(III)   Government intervention;

(IV)   Monetary policy;

(V)   Capital flow;

(VI)   Banking policy;

(VII) Wage/price control;

(VIII) Property rights;

(IX)   Regulation policy;

(X)   Black market.

2 Average for all categories.

Source: *Wall Street Journal, 1997 Index of Economic Freedom.*

Select Bibliography

This brief bibliography contains only major *books* on the subjects. All consulted journal articles and newspaper reports appear only in notes at the end of each chapter.

Abucar, M.H., *The Post-Colonial Society: The Algerian Struggle for Economic, Social, and Political Change, 1965–1990* (New York, Peter Lang, 1996)

Achime, N.H., *Investment Policy Analysis and the Nigerian Economic System* (Ogbete Enegu, Zwei Consort Publications, 1996)

Adejumabi, Said (ed.), *The Political Economy of Nigeria under Military Rule* (Harare, SAPES Books, 1995)

Adelman, M.A., *The Genie Out of the Bottle: World Oil Since 1970* (Cambridge, MIT Press, 1995)

Ahmad Khan, Sara, *Nigeria: The Political Economy of Oil* (Oxford, Oxford University Press, 1994)

Ahmad, A.S., *Development and Resource-Based Industry* (Vienna, OPEC Fund, 1990)

Aicardi de Saint-Paul, Marc, *Gabon: The Development of a Nation* (London, Routledge, 1989)

Al-Chalabi, Fadhil, *OPEC and the International Oil Industry* (Oxford, Oxford University Press, 1980)

Idem, Opec at the Crossroads (New York, Pergamon Press, 1989)

Al-Eyd, Kadhim A., *Oil Revenues and Accelerated Growth: Absorptive Capacity in Iraq* (New York, Praeger, 1979)

Al-Khalil, Samir, *Republic of Fear* (Berkeley, University of California Press, 1988)

Al-Sabah, Y.S.F., *The Oil Economy of Kuwait* (London, Kegan Paul, 1980)

Al-Yahya, *Kuwait: Fall and Rebirth* (London, Kegan Paul, 1993)

Algeria. A Country Study (Washington, Library of Congress, 1994)

Allan, J.A., *Libya: The Experience of Oil* (Boulder, Westview Press, 1981)

Allen, R.L., *Venezuelan Economic Development* (Greenwich, Jai Press, 1977)

Idem, OPEC Oil (Cambridge, Oelgeschlager, 1979)

Alnasrawi, A., *OPEC at the Crossroads* (Oxford, Pergamon Press, 1989)

Amuzegar, Jahangir, *Iran's Economy Under the Islamic Republic* (London, I.B. Tauris, 1997)

Idem, Oil Exporters' Economic Development in an Interdependent World (Washington, International Monetary Fund, 1983)

Idem, Iran: An Economic Profile (Washington, The Middle East Institute, 1977)

Angaye, G.S., *Socio-Economic Development in Nigeria* (Port Harcourt, Pam Uniwue Publishers, 1995)

Arya, P.L., *Structure, Policies and Growth Prospects of Nigeria* (Lewiston, Mellen Press, 1993)

Asiodu, P.C., *Essays on Nigerian Political Economy* (Lagos, Sankore Publishers, 1993)

Askari, Hossein, *Saudi Arabia's Economy* (Greenwich, Jai Press, 1990)

Axelgard, F.W., *Iraq in Transition* (Boulder, Westview Press, 1986)

Barnes, Philip, *Indonesia: The Political Economy of Energy* (Oxford, Oxford University Press, 1995)

Beblawi, H., and G. Luciani (eds), *The Rentier State* (London, Croom Helm, 1987)

Bennoune, Mahfoud, *The Making of Contemporary Algeria* (Cambridge, Cambridge University Press, 1988)

Beschorner, Natasha, *Libya in the 1990s* (London, Economist Intelligence Unit, 1991)

Betancourt, Romulo, *Venezuala, Oil, and Politics* (Boston, Houghton Mifflin, 1979)

Bevan, David, *Nigeria: Policy Responses to Shocks, 1970–1990* (San Francisco, ICS Press, 1992)

Bhattacharya, Amar, *Indonesia: Development, Transformation, and Public Policy* (Washington, World Bank, 1993)

Bienen, Henry, *Political Conflict and Economic Change in Nigeria* (London, Frank Cass, 1985)

Booth, Anne (ed.), *The Oil Boom and After: Indonesian Ecomomic Policy and Performance in the Soeharto Era* (Singapore, Oxford University Press, 1992)

Bresnan, John, *Managing Indonesia: The Modern Political Economy* (New York, Columbia University Press, 1993)

CARDRI, *Saddam's Iraq* (London, Zed Books, 1989)

Chalmers, Ian, *The Politics of Economic Development in Indonesia* (London, Routledge, 1997)

Clawson, Patrick, *How Has Saddam Hussein Survived?* (Washington, National Defense University, 1993)

Cooper, R.N., *Boom, Crisis and Adjustment* (Washington, World Bank, 1994)

Crane, R.D., *Planning the Future of Saudi Arabia* (New York, Praeger, 1978)

Crystal, Jill, *Oil and Politics in the Gulf* (New York, Cambridge University Press, 1990)

Idem, Kuwait: The Transformation of an Oil State (Boulder, Westview Press, 1992)

Daly, Herman, *Beyond Growth: Economics of Sustainable Development* (Boston, The Beacon Press, 1996)

Deadend to Nigerian Development (Dakar, Codesria, 1993)

Desfosses, Helen, *Socialism in the Third World* (New York, Praeger, 1975)

De Mowbray, Patricia, *Gabon to 1995* (London, Economist Intelligence Unit, 1991)

Dirkse, P.P. (ed.), *Development and Social Welfare: Indonesia's Experiences under the New Order* (Leiden, KITLV Press, 1993)

Ecuador. A Country Study (Washington, Department of the Army, 1993)

El-Azhari, M.S. *The Impact of Oil Revnues on Arab Gulf Development* (London, Croom Helm, 1984)

El Mallakh, Ragaei, *Qatar: Energy and Development* (London, Croom Helm, 1985)

Idem, The Economic Development of the United Arab Emirates (London, Croom Helm, 1981)

Idem, The Absorptive Capacity of Kuwait (Lexington, Lexington Books, 1981)

Enright, M.J., *Venezuela, The Challenge of Competitiveness* (New York, St Martin's Press, 1996)

Entelis, J.P., *Algeria: The Revolution Institutionalized* (London, Westview Press, 1986)

Idem, State and Society in Algeria (Boulder, Westview Press, 1992)

Evans, John, *OPEC, Its Members and the World Energy Market* (London, Longman, 1986)

Farsy, Fouad, *Saudi Arabia: A Case Study in Development* (London:KPI, 1986)

Fathaly, Omar, *Political Devlopment and Social Change in Libya* (Lexington, Lexington Books, 1980)

Gelb, Alan, *Oil Windfalls* (New York, Oxford University Press, 1988)

Ghadar, Fariborz, *The Evolution of OPEC Strategy* (Lexington, Lexington Books, 1977)

Ghanem, S.M.A., *Industrialisation in the United Arab Emirates* (Brookfield, Avebury, 1992)

Goodman, L.W., *Lessons of the Venezuelan Experience* (Washington, Woodrow Wilson Center Press, 1995)

Gurney, Judith, *Libya. The Political Economy of Oil* (Oxford, Oxford University Press, 1996)

Hadiz, V.R., *Workers and the State in New Order Indonesia* (New York, Routledge, 1997)

Hellinger, Daniel, *Venezuela: Tarnished Democracy* (Boulder, Westview Press, 1991)

Henderson, S., *Instant Empire: Saddam's Ambition for Iraq* (San Francisco, Mercury House, 1991)

Hill, Hal, *The Indonesian Economy Since 1966* (Cambridge, Cambridge University Press, 1981)

Idem, (ed.), *Indonesia's New Order* (St Leonards, Allen and Unwin, 1994)

Hirshman, A.O., *The Strategy of Economic Development* (New Haven, Yale University Press, 1958)

Idem, Essays in Trespassing (New York, Cambridge University Press, 1981)

Hobohm, Sarwar, *Indonesia: Industrial Growth and Diversification* (London, Economist Intelligence Unit, 1993)

Idem, Indonesia to 1993 (London, Economist Intelligence Unit, 1989)

Hollinger, W.C., *Economic Policy Under President Soeharto* (Washington, The United States – Indonesia Society, 1996)

Husain, Ishrat and Rashid Faruqee (eds), *Adjustment in Africa* (Washington, World Bank, 1994)

Ibe, A.C., *Problems and Policies of Development for Nigeria* (Agulu, Levrene Publishers, 1994)

Indonesia: A Strategy for a Sustained Reduction in Poverty (Washington, World Bank, 1990)

Indonesia: A Country Study (Washington, Department of the Army, 1993)

Indonesia: Environment and Development Challenges for the Future (Washington, 1994)

Iraq. A Country Study (Washington, Department of the Army, 1988)

Islamic Republic of Iran: Industrialisation Revitalization (London, Economist Intelligence Unit, 1995)

Ismael, J.S., *Kuwait: Dependency and Class in a Rentier State* (Gainesville, University Press of Florida, 1993)

Jacquemin, A.P., *et al* (eds), *Market, Corporate Behavior, and the State* (The Hague, Nijhoff, 1976)

James, Duncan, *Nigeria to 2000* (London, Economist Intelligence Unit, 1993)

Jazayeri, Ahmad, *Economic Adjustment in Oil-Based Economies* (Aldershot, Avebury, 1988)

Johany, A.D., *The Saudi Arabian Economy* (Baltimore, Johns Hopkins University Press, 1986)

Kanovsky, Eliyahu, *The Economy of Saudi Arabia* (Washington, Washington Institute for Near East Policy, 1994)

Karl, T.L., *The Paradox of Plenty: Oil Booms and Petro-states* (Berkeley, University of California Press, 1997)

Karshenas, Massoud, *Oil, State and Industrialization in Iran* (Cambridge, Cambridge University Press, 1990)

Kelidar, Abbas (ed.), *The Integration of Modern Iraq* (London, Croom Helm, 1979)

Khader, Bichara (ed.), *The Economic Development of Libya* (London, Croom Helm, 1987)

Khalifah, Anid-Amar, *The United Arab Emirates Socio-economic Development and its Foreign Manpower Requirements* (Geneva, Institut Universitaire de Hautes Etudes Internationales, 1988)

Khouja, M.W., *The Economy of Kuwait* (London, Macmillan, 1979)

Kublah, A.A., *OPEC: Past and Present* (Vienna, Petroeconomic Research Center, 1974)

Kubursi, Afif, *Oil, Industrialization and Development in the Arab Gulf States* (London, Croom Helm, 1984)

Lal, Deepak, *The Poverty of Development Economics* (London, Institute of Economic Affairs, 1983)

Landes, D.S., *The Wealth and Poverty of Nations* (New York, W.W. Norton, 1998)

Libya. A Country Study (Washington, Department of the Army, 1993)

Looney, R.E., *Economic Origins of the Iranian Revolution* (New York, Pergamon Press, 1982)

Idem, Economic Development in Saudi Arabia (Greenwich, Jai Press, 1990)

Marks, Jon, *Algeria: Toward Market Socialism* (London, *Middle East Economic Digest*, 1990)

Martz, J.D. (ed.), *Venezuala: The Democratic Experience* (New York, Praeger, 1986)

MERI Report, *United Arab Emirates* (London, Croom Helm, 1985)

MERI Report, *Saudi Arabia* (London, Croom Helm, 1985)

Mikdashi, Zuhair, *The Community of Oil Exporting Countries* (London, Allen and Unwin, 1972)

Mills, E.S., *Growth and Equity in the Indonesian Economy* (Washington, The United States – Indonesia Society, 1995)

Mofid, Kamran, *Development Planning in Iran* (Cambrigeshire, MENAS Press, 1987)

Idem, The Economic Consequences of the Gulf War (London, Routledge, 1990)

Montagu, Caroline, *The Private Sector of Saudi Arabia* (London, Comet, 1994)

Moser, G. *et al, Nigeria: Experience with Structural Adjustment 1986–94* (Washington: International Monetary Fund, 1997)

Murphy, K.J., *Macroprojects in the Third World* (Boulder, Westview Press, 1983)

Myint, H.A., *The Economics of Developing Countries* (New York, Praeger, 1964)

Nafi, Z.A., *Economic and Social Development in Qatar* (London, F. Pinter, 1983)

Naim, Moises, *Paper Tigers and Minotaurs: The Politics of Venezuela's Economic Reforms* (Washington, Carnegie Endowment for International Peace, 1993)

Niblock, Tim (ed.), *State, Society and Economy in Saudi Arabia* (London, Croom Helm, 1982)

Idem, Iraq (London, Croom Helm, 1982)

Nigeria. A Country Study (Washington, Library of Congress, 1991)

Nnadozie, E.U., *Oil and Socioeconomic Crisis in Nigeria* (Lewiston, Mellen University Press, 1995)

Obadan, M.I., *Whither Structural Adjustment in Nigeria?* (Ibadan, National Centre for Economic Management and Administration, 1993)

Okigbo, P.N.C., *National Development Planning in Nigeria 1900–92* (London, Currey, 1989)

Okowa, W.J., *Oil, Systematic Corruption, Abdulistic Capitalism and Nigerian Development Policy* (Port Harcourt, Pam Unique Publishers, 1994)

Olayiwola, P.O., *Petroleum and Structural Change in a Developing Country: The Case of Nigeria* (New York, Praeger, 1987)

Onokerhoraye, A.G., *The Impact of the Structural Adjustment Programme on Grassroots Development in Nigeria* (Benin City, Nigeria, 1995)

Onosode, G.O., *Three Decades of Development Crises in Nigeria* (Lagos, Malthouse, 1993)

Osada, Hiroshi, *Oil Prices and the Indonesian Economy* (Nagoya, Japan, Nagoya University, 1988)

Pangestu, Mari, *Economic Reform, Deregulation, and Privatization: The Indonesian Experience* (Jakarta, Centre for Strategic and International Studies, 1996)

Peck, M.C., *The United Arab Emirates* (Boulder, Westview Press, 1986)

Persian Gulf States: Country Studies (Washington, Library of Congress, 1994)

Philip, George, *The Political Economy of International Oil* (Edinburgh, Edinburgh University Press, 1994)

Planning and Development in Modern Libya (Cambrigeshire, MENAS Press, 1985)

Presley, J.R., *A Guide to the Saudi Arabian Economy* (London, Macmillan, 1989)

Quandt, William, *Saudi Arabia's Oil Policy* (Washington, The Brookings Institute, 1982)

Rahnema, S. and S. Behdad (eds.), *Iran After the Revolution* (London, I.B. Tauris, 1996)

Randall, Laura, *The Political Economy of Venezuelan Oil* (New York, Praeger, 1987)

Razavi, Hossein, *The Political Environment of Economic Planning in Iran. 1971–1983* (Boulder, Westview Press, 1984)

Richards, Alan and John Waterbury, *A Political Economy of the Middle East* (Boulder, Westview Press, 1990)

Robins, Philip, *The Future of the Gulf: Politics and Oil in the 1990s* (Brookfield, Dartmouth University Press, 1989)

Robinson, Richard, *Indonesia: The Rise of Capital* (North Sydney, Allen and Unwin, 1986)

Robinson, T.J.C., *Economic Theories of Exhaustible Resources* (London, Routledge, 1989)

Roland-Holst, D.W., *Stabilization and Structural Adjustment in Indonesia* (Paris, OECD, 1992)

Rouadjia, Ahmed, *Le Grandeur et Decadence de l'Etat Algerien* (Paris, Karthala, 1994)

Rouhani, Fouad, *A History of OPEC* (New York, Praeger, 1971)

Reudy, John, *Modern Algeria* (Indianapolis, University of Indiana Press, 1992)

Salazar-Carillo, Jorge, *Oil in the Economic Development of Venezuela* (New York, Praeger, 1976)

Schats, S., *Nigerian Capitalism* (Berkeley, University of California Press, 1977)

Schodt, D.W., *Ecuador: An Andean Enigma* (Boulder, Westview Press, 1987)

Schwarz, Adam, *A Nation in Waiting: Indonesia in the 1990s* (Boulder, Westview Press, 1994)

Shihata, I.F.I., *The Other Face of OPEC* (London, Longman, 1982)

Skeet, Ian, *OPEC: Twenty Five Years of Prices and Politics* (Cambridge, Cambridge University Press, 1988)

Sluglett, M.F. and P. Sluglett, *Iraq Since 1958* (London, I.B. Tauris, 1990)

Soyinka, Wole, *The Open Sore of a Continent* (New York, Oxford University Press, 1996)

Tachan, F., *Political Elites and Political Development in the Middle East* (Cambridge, Schenkman, 1974)

Terzian, Philip, *OPEC: The Inside Story* (London, Zed Books, 1985)

Venezuela: A Country Study (Washington, Department of the Army, 1991)

Vine, Peter and P. Caseuy, *United Arab Emirates* (London, Immel, 1992)

Waterbury, John, *et al*, *Fragile Coalitions: The Politics of Economic Adjustment* (New Brunswick, Transaction Books, 1989)

Woo, W.T., *Macroeconomic Policies, Crises, and Long-term Growth in Indonesia. 1965–90* (Washington, World Bank, 1994)

Yates, D.A., *The Rentier State in Africa: Oil Rent Dependency and Neocolonialism in the Republic of Gabon* (Trenton, Africa World Press, 1996)

Yergin, Daniel, *The Prize* (New York, Simon and Schuster, 1991)

Yesufu, T.M., *The Nigerian Economy: Growth Without Development* (Benin City, University of Benin, 1996)

Zartman, I.W., *The Political Economy of Nigeria* (New York, Praeger, 1983)

Index

This abbreviated index includes only selected relevant entries in the text. Names of authors in footnotes are not included. Nor are the names of persons, places or companies that appear in the text but are not subject of discussion or analysis.